IT'S STILL DEBATABLE!

USING SOCIOSCIENTIFIC ISSUES TO DEVELOP SCIENTIFIC LITERACY

K–5

IT'S STILL DEBATABLE!

USING SOCIOSCIENTIFIC ISSUES TO DEVELOP SCIENTIFIC LITERACY

K–5

Sami Kahn

National Science Teaching Association

Arlington, Virginia

National Science Teaching Association

Claire Reinburg, Director
Rachel Ledbetter, Managing Editor
Andrea Silen, Associate Editor
Jennifer Thompson, Associate Editor
Donna Yudkin, Book Acquisitions Manager

ART AND DESIGN
Will Thomas Jr., Director
Joe Butera, Senior Graphic Designer, interior design
Himabindu Bichali, Graphic Designer, cover design

PRINTING AND PRODUCTION
Catherine Lorrain, Director

NATIONAL SCIENCE TEACHING ASSOCIATION
David L. Evans, Executive Director

1840 Wilson Blvd., Arlington, VA 22201
www.nsta.org/store
For customer service inquiries, please call 800-277-5300.

Copyright © 2019 by the National Science Teaching Association.
All rights reserved. Printed in the United States of America.
22 21 20 19   4 3 2 1

NSTA is committed to publishing material that promotes the best in inquiry-based science education. However, conditions of actual use may vary, and the safety procedures and practices described in this book are intended to serve only as a guide. Additional precautionary measures may be required. NSTA and the authors do not warrant or represent that the procedures and practices in this book meet any safety code or standard of federal, state, or local regulations. NSTA and the authors disclaim any liability for personal injury or damage to property arising out of or relating to the use of this book, including any of the recommendations, instructions, or materials contained therein.

PERMISSIONS
Book purchasers may photocopy, print, or e-mail up to five copies of an NSTA book chapter for personal use only; this does not include display or promotional use. Elementary, middle, and high school teachers may reproduce forms, sample documents, and single NSTA book chapters needed for classroom or noncommercial, professional-development use only. E-book buyers may download files to multiple personal devices but are prohibited from posting the files to third-party servers or websites, or from passing files to non-buyers. For additional permission to photocopy or use material electronically from this NSTA Press book, please contact the Copyright Clearance Center (CCC) (*www.copyright.com*; 978-750-8400). Please access *www.nsta.org/permissions* for further information about NSTA's rights and permissions policies.

Library of Congress Cataloging-in-Publication Data
Names: Kahn, Sami, author.
Title: It's still debatable! : using socioscientific issues to develop scientific literacy, K-5 / Sami Kahn.
Other titles: It is still debatable!
Description: Arlington, VA : National Science Teaching Association, [2019]
Identifiers: LCCN 2019021981 (print) | LCCN 2019022402 (ebook) | ISBN 9781681406299 (print)
Subjects: LCSH: Science--Social aspects--Study and teaching (Early childhood)--United States. | Science--Social aspects--Study and teaching (Elementary)--United States. | Technology--Social aspects--Study and teaching (Early childhood)--United States. | Technology--Social aspects--Study and teaching (Elementary)--United States. | Curriculum planning--United States.
Classification: LCC Q175.5 .K2564 2019 (print) | LCC Q175.5 (ebook) | DDC 372.35--dc23
LC record available at *https://lccn.loc.gov/2019021981*
LC ebook record available at *https://lccn.loc.gov/2019022402*

Dedication

This book is dedicated to teachers—current and future—who wake up each morning and choose to change the world ... one student at a time.

CONTENTS

About the Author
xi

Acknowledgments
xiii

Prelude: "A Cup of Inspiration"
xv

Unit I

Introduction: *It's Debatable!* for the Next Generation
1

Unit II

Framework for This Book
7

Unit III

Strategies for Promoting Inquiry, Argument, and Inclusion
15

Unit IV

A Guide for Reading and Implementing the Lesson Plans
39

Unit V
Lesson Plans
55

Grades K–2

Lesson 1
Leave It to Beavers: *Should We Relocate the Beaver Dam?*
56

Lesson 2
Swingy Thingy: *What Makes a Great Playground?*
75

Lesson 3
Take a (Farm) Stand: *Can Plants Help Us Fight Hunger?*
95

Lesson 4
Monkey Business: *Do We Need Zoos?*
126

Lesson 5
Soaky Doaky: *What's the Best Way to Clean Up Spills?*
154

Lesson 6
Bee-ing There for Bees: *Are Bees Disappearing?*
186

Grades K–3

Lesson 7
Weather or Not: *Should We Rebuild in Twisterville?*
215

Grades 3–5

Lesson 8
Eggstreme Sports: *Is Football Too Dangerous for Kids?*
249

Lesson 9
Marsh Madness: *What's Your Plan for Bullfrog Pond?*
282

Lesson 10
Finders Keepers?: *Who Owns the Dinosaur Bones?*
320

Lesson 11
Blast From the Past: *Do We Still Need a Space Program?*
347

Lesson 12
"Mined" Your Own Business: *Was the California Gold Rush Good for the United States?*
391

Lesson 13
Fueling Around: *Which Energy Sources Are Best?*
413

Lesson 14
Watch Your Step: *Should Distracted Walking Be Illegal?*
456

Unit VI
Developing Your Own SSI Lessons
481

Unit VII
For Teacher Educators: Including *It's Still Debatable!* in Your Preservice and Inservice Elementary Science Courses
487

Unit VIII
Finale: Embracing the Controversy in Your Classroom
497

Appendix: *NGSS* Lesson Plan Alignment Matrices

501

Index

519

About the Author

Dr. Sami Kahn is a 30-plus-year veteran science educator with extensive experience in classroom teaching, professional development, and curriculum development. She is proud to share that she has taught science to students in almost every grade, from kindergarten through college. Dr. Kahn currently serves as Executive Director of the Council on Science and Technology at Princeton University where she works to promote scientific literacy for all through STEM education research, course development, and outreach. An award-winning teacher and scholar, she uses her background in science education and law to inform her research and teaching on inclusive science practices, socioscientific issues (SSI), argumentation, and social justice. Dr. Kahn has authored numerous journal articles, including several in *Science and Children*, and has coauthored three books on enhancing scientific inquiry experiences for children and adults, including the NSTA Press book *It's Debatable! Using Socioscientific Issues to Develop Scientific Literacy, K–12* (2014). Her service to the field includes leadership positions with the National Science Teaching Association and the Association for Science Teacher Education. Dr. Kahn holds an MS in ecology and evolutionary biology from Rutgers University, a JD in law from Rutgers School of Law, and a PhD in curriculum and instruction with a specialization in science education from the University of South Florida, where she served as a presidential doctoral fellow. Before coming to Princeton, Dr. Kahn held positions at Ohio University, Collegiate School in New York City, and Rutgers University.

Acknowledgments

One point that is definitely *not* debatable is that it takes the contributions of many highly talented individuals to develop a successful book. I am grateful to have had the assistance of such a group in the development of *It's Still Debatable!*

First, I thank the outstanding team at NSTA Press. Under the inspired leadership of Claire Reinburg, the exceptional editorial staff gave me immeasurable guidance and support. Special thanks to Andrea Silen for her expertise and patience!

Next, I would like to acknowledge the talents of some outstanding students, teachers, colleagues, and friends who have contributed ideas, expertise, and technical support to the book: Cathe Blower, Kelly Bornmann, Crystal Cole, Cassie Comer, Sarah Cross, Sabrina Douglas, Kyleigh Falcone, Julie Barnhart Francis, Madalyn Green, Sara Hartman, Liz Keogan, Kaitlin Krugman, Molly Mason-Hurst, Julia Rabold, Alycia Stigall, and Lindsay Zeisler.

I am also tremendously grateful to the following schools for promoting the development of lessons and in some cases supporting field testing: Amesville Elementary School, Amesville, Ohio; The Plains Elementary School, The Plains, Ohio; the Collegiate School, New York City; Sacramento Country Day School, Sacramento, California; Union Furnace Elementary School, Union Furnace, Ohio; and Ohio University, Athens, Ohio.

And finally, I thank my wonderful husband, Sanford Starr, and my beautiful daughter, Rachel Helene Kahn, for their patience, support, and unending love. I am truly blessed.

Prelude
"A Cup of Inspiration"

I had just settled in at my desk, cup of coffee in hand, when one of my fourth graders who had arrived early at school approached and asked what seemed like a very simple question: "Ms. Kahn, is coffee good for you?"

Now, I have to confess that the first thing that came into my mind was "Of course! It's what keeps me sane!" But the science teacher inside me quickly kicked in, and I replied, "Well, coffee contains substances called antioxidants that help cells do their jobs well. It definitely has some health benefits."

"So, should kids drink coffee?" he queried.

And there it was—a "should" question. On its surface, it seemed innocent enough, but "should" questions aren't always easy, and they often aren't answerable by science alone. In this case, I knew that science could *inform* my answer but not necessarily determine it. Should I simply respond with a *safe* answer like "It's up to children's parents whether they should or shouldn't drink coffee"? I decided to keep the conversation going.

"Well, the other thing about coffee is that it contains caffeine, which is a type of stimulant. It can make people nervous and make it harder for them to fall asleep at night. So a lot of people are against giving coffee to children."

My student nodded rather somberly, but then, showing a bit of glint in his eye, he asked, "But what about *decaf*? Why don't my parents let me drink decaf?"

I wondered … could my eager young student be trying to get me to contradict his parents? But might he also be genuinely interested in the science behind the bean? Although I felt the pull of quicksand drawing me in ever deeper, I was also intrigued.

"Even decaf has caffeine," I answered. "Maybe that's why your parents don't want you to have it. Have you ever asked them why?"

He paused for a moment, looking rather circumspect before replying: "Nope, I didn't ask because I know when it's a *maybe* no versus a *definite* no. This was a *definite* no!"

Prelude

Sensing his disappointment and wanting to take advantage of a teachable moment, I said, "Why don't we both learn a bit more about this coffee subject? I'm curious!"

We proceeded to scour some online articles about the pros and cons of coffee, how coffee actually gets decaffeinated (something I never really thought about), and how caffeine affects children and adults. As we searched, I found myself naturally talking to him about the difference between sources like medical journals or university websites (we noticed the .edu and .org extensions) and Wikipedia. We even found information on coffee companies' websites, sources that, my student astutely noted, "might be trying to sell us on it." Our takeaway from our brief perusal was that there is fairly solid evidence that coffee can lower risks for several different diseases, including type 2 diabetes, Parkinson's disease, and some cancers. But it can also harm tooth enamel, leach calcium from bones, and cause anxiety and insomnia. And adding sugar, cream, and flavors (and whipped cream!) contributes empty calories and fat.

My student and I had a great time learning together by weighing the copious and somewhat confusing information about this everyday product. I was quite surprised by how much science content we discussed in a short time, and we touched on issues such as determining the trustworthiness of sources, how and why different studies might yield different results, and the tentativeness of scientific knowledge. This brief interlude added to my deep belief in the use of debatable socioscientific issues (SSI) in science teaching, especially for elementary-age students, as they are developing the habits of mind that will last a lifetime. They are at an age when their thoughts, feelings, and beliefs about science are being solidified.

Soon, I heard the voices of his classmates arriving, so I turned to my student and said, "We've seen some arguments for and against coffee. What do you think about it?"

He confidently replied, "I'm going to wait until I'm older to try it, just in case it's not good for me. I have a feeling there'll be LOTS more research by then!"

That night, my student shared our findings with his parents, who had, in fact, nixed coffee for the reasons we had identified but hadn't articulated those reasons to their son. He now felt empowered because he could engage in an informed discussion on this subject and could even understand and appreciate his parents' decision, whether he ultimately agreed with it or not. It was on that day that I decided an elementary-level sequel to *It's Debatable!* was needed. And in case you were wondering, it was also on that day that I decided to continue drinking my morning cup of coffee!

—Sami Kahn

Unit I
Introduction: *It's Debatable!* for the Next Generation

Do we need zoos?

Which alternative energies are best?

Is football too dangerous for kids?

I have heard it said that today's students are tomorrow's decision makers. While that is certainly a true statement, it is also true that today's students are *today's* decision makers! Every day, children are deluged with information that has an impact on their lives and choices, from which foods are healthiest to whether a particular product or technology is best. Children receive input from a variety of sources and must learn to distinguish fact from fiction. There is so much information available today that it is sometimes easy to forget that information isn't knowledge, and knowledge isn't wisdom. As elementary teachers, we try to instill in children the ability to weigh information and make thoughtful choices, but it certainly isn't easy. Quality science education demands that our students apply scientific understanding to their everyday lives, yet this is a tall task, particularly since science teaching is often done in isolated contexts, making it difficult to recognize and apply science in everyday situations. Think about it. Do you know any adults

Introduction: *It's Debatable!* for the Next Generation

who took years of science classes but have difficulty applying any of the content to decisions about health, technology, and the environment?

The authors of *A Framework for K–12 Science Education* clearly had this challenge in mind when they noted the importance of preparing students to "engage in public discussions on science-related issues, to be critical consumers of scientific information related to their everyday lives, and to continue to learn about science throughout their lives" (NRC 2012, pp. 1–2). The *Next Generation Science Standards* (*NGSS*) echo the importance of ensuring informed decision making through science in the following opening statement:

> *There is no doubt that science and, therefore, science education is central to the lives of all Americans. Never before has our world been so complex and science knowledge so critical to making sense of it all. When comprehending current events, choosing and using technology, or making informed decisions about one's healthcare, science understanding is key.* (NGSS Lead States 2013, p. 1)

This vision of scientific literacy expects students to make sense of science and apply it to real-world decision making. It also requires teachers to address an extensive array of standards and position them in meaningful contexts for our learners. As daunting a task as that is, there's even more to the story. Elementary teachers know that the real-world scientific questions that students confront can't always be answered easily by science. The messiness of these questions may stem from the fact that scientific understanding is often incomplete and ever changing, a situation that can lead students (and the public) to be confused about the information they receive or, worse, to doubt the integrity of science itself. Children also wonder about moral or "should" questions that can be informed by science but also require consideration of the consequences of their decisions on others. While science teachers could easily be put off by these difficult questions, a growing body of research suggests that these are precisely the types of questions that fascinate and engage students, including those who otherwise might not be interested in science. The use of debatable, science-related societal questions, or socioscientific issues (SSI) (Zeidler 2014), can serve as a powerful teaching framework that addresses science content, the application of that content, and the type of informed citizenship that is envisioned within the *NGSS*.

How does this work? Let's consider the question "Is football too dangerous for kids?" Can you spot the science? Clearly, examining this question would require understanding concepts of forces and motion (What happens when objects collide? How do helmets work?). It would also require an understanding of anatomy

Introduction: *It's Debatable!* for the Next Generation

(Why would we protect our heads? How does the brain work?). Engineering also enters the picture, as we could ask students to examine and test different helmets, and even design their own (Which helmet protects best? What materials work best for this purpose and why?). And yet, even after all this scientific investigation, this issue still leaves open questions like *Who* should decide whether football is too dangerous? Should the rules be different for children of different ages? What are the costs and benefits of youth sports? Approaching the topic of forces and motion in this way creates a rich context for learning by engaging students in an issue that feels real to them, motivates them to investigate the underlying content through science and engineering practices, and allows them to emulate the type of real-world, evidence-based decision-making practices that they will use throughout their lives.

One of the most compelling reasons for SSI implementation in contemporary classrooms is that it clearly supports the conceptual shifts that drove the development of the *NGSS*. The authors of the *NGSS* sought to eliminate superficial, decontextualized science learning and instead support high-quality, real-world, evidence-based investigation and deliberation. These priorities are quite different from those of earlier science education standards and are articulated in Table 1.1 (p. 4), along with how SSI curriculum supports them.

Of course, elementary teachers don't just teach science—they teach children! This means that social and communication skills, literacy and numeracy, and character development though personal and civic responsibility are also key goals. The approach outlined in this book emphasizes each of these critical facets of elementary pedagogy in a seamless manner. Through thoughtful, collaborative problem solving, students begin to think about ways they can support their classroom, school, town, and global community through scientific understanding, civic engagement, conscience, and caring. Typical SSI lessons engage students through hands-on investigations and research in conjunction with discourse through debates, role-playing, in-person or online discussion platforms, and written communications. By interacting with others in this way, students challenge their existing beliefs, collect and examine evidence through multiple research experiences, and develop arguments from evidence.

It should be noted that although moral and ethical issues are broached, teachers do not instruct students in *what* to believe, but rather *how* to integrate information from a variety of sources, evaluate the quality of those sources, and develop perspective-taking skills. Unlike other science-and-society approaches, SSI focuses on empowering students to become agents of change in their schools and communities by integrating diverse viewpoints, appreciating the impact of their actions (or inactions) on others, and finding common ground toward the development

TABLE 1.1.
How SSI Addresses Conceptual Shifts Identified in the *NGSS*

Conceptual Shift in *NGSS*	How SSI Addresses Conceptual Shift
"K–12 science education should reflect the interconnected nature of science as it is practiced and experienced in the real world."	SSI provides students with real-world evidence about problems relevant to their lives and helps students gain appreciation and understanding of the nature of scientific inquiry, including its complexity and interdisciplinary nature.
"The *NGSS* focus on deeper understanding of content as well as application of content."	SSI teaches content in context through extended inquiries that reinforce learning and encourage application of learning through numerous methods, including lab and field investigations, internet research, debates, and writing activities.
"Science and engineering are integrated in the *NGSS*."	SSI often relates to engineering concepts and allows students to deliberate on the impacts of technology on society and of society on technology.
"The *NGSS* are designed to prepare students for college, career, and citizenship."	SSI prepares students to be critical consumers of scientific information and lets them rehearse skills necessary for college and beyond, including argumentation and discourse, evaluation and analysis of primary sources, understanding of diverse perspectives, and informed decision making.
"The *NGSS* and *Common Core State Standards* (in English language arts and mathematics) are aligned."	SSI is interdisciplinary in nature, reinforcing language, literacy, and math skills through evidence-based argumentation, research and writing, debate, and data analysis, all in a manner that emulates real-world scientific applications.

Source: Adapted from Zeidler and Kahn (2014).

of thoughtful solutions. Extensive research has suggested that the SSI approach supports increases in students' science content knowledge, understanding of the nature of science, quality of argumentation abilities, and characteristics for global citizenship, including empathy and perspective taking (Zeidler 2014). And SSI's emphasis on evidence-based argumentation and civic engagement makes integration with the *Common Core State Standards* in English language arts and mathematics (NGAC and CCSSO 2010), as well as the National Council for Social Studies (NCSS 2010) curriculum standards, seamless. This type of integrated curriculum supports what we know about how young children learn best. It is also an efficient way of teaching, a critical consideration given how little time elementary teachers are typically allocated for science and how much is expected to be covered across disciplines in a given day.

Most important, SSI makes science real for students; they are invested in their learning because they feel real connections to the questions and want to make a difference in the world, and in their own lives. This book uses a model for elementary SSI (see Figure 1.1) that recognizes the importance of social skills and discourse, interdisciplinary connections, character development, and of course, science learning. In the following chapter, SSI is connected to two other pedagogical models to form a framework that will help ensure that inquiry and equity are emphasized in your classroom.

FIGURE 1.1.
Key Elements of Elementary SSI Model

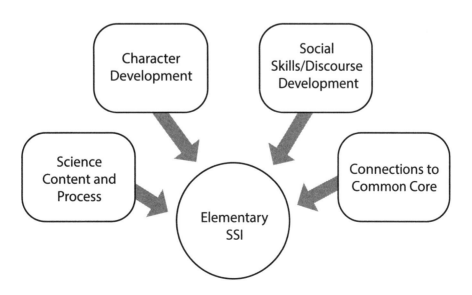

References

National Council for the Social Studies (NCSS). 2010. *National curriculum standards for social studies: A framework for teaching, learning and assessment*. Silver Spring, MD: NCSS.

National Governors Association Center for Best Practices and Council of Chief State School Officers (NGAC and CCSSO). 2010. *Common core state standards.* Washington, DC: NGAC and CCSSO.

National Research Council (NRC). 2012. *A framework for K–12 science education*. Washington, DC: National Academies Press.

NGSS Lead States. 2013. *Next Generation Science Standards: For states, by states*. Washington, DC: National Academies Press. *www.nextgenscience.org/next-generation-science-standards*.

Zeidler, D. L. 2014. Socioscientific issues as a curriculum emphasis: Theory, research, and practice. In *Handbook of research on science education,* vol. 2, ed. N. G. Lederman and S. K. Abell, 697–726. New York: Routledge.

Zeidler, D. L., and S. Kahn. 2014. *It's debatable! Using socioscientific issues to develop scientific literacy, K–12*. Arlington, VA: NSTA Press.

Unit II
Framework for This Book

This book weaves together three research-based pedagogical approaches to develop a cohesive framework that supports inquiry-based, equitable lessons that emphasize scientific literacy and citizenship: (1) socioscientific issues (SSI), (2) BSCS 5E Instructional Model (5Es), and (3) Universal Design for Learning (UDL). These approaches are described in more detail below.

Socioscientific Issues (SSI)

As discussed in the introduction, SSI is a pedagogical framework that uses debatable, socially related scientific issues as the context for science teaching. Some examples of SSI are fluoride in water, space colonization, animals in circuses, offshore oil drilling, vaccine safety, bicycle helmet regulation, and even whether chocolate is good for you. The main idea behind SSI is that providing students with opportunities for argumentation about scientific issues prepares them for informed, participatory citizenship as adults. SSI emphasizes science learning in a highly student-centered manner

with a goal of promoting character and civic virtue (Zeidler et al. 2005). Through SSI curriculum, students are prompted to collect and examine evidence through multiple research experiences, develop arguments, and question their existing beliefs. Through discourse and debate, they become exposed to different thought processes and are encouraged to approach decisions in an open, unbiased way while respecting and acknowledging different perspectives and ways of knowing (NSTA 2016). This mode of learning is aligned with constructivist pedagogy in that it respects the fact that students come to school with preconceived notions of how the world works. SSI deliberately presents challenges to those preconceptions, thereby triggering cognitive dissonance and creating opportunities for students to examine and reflect on new ideas. Through SSI, teachers help students acquire flexibility, open-mindedness, and perspective-taking abilities so that they can integrate content knowledge with real-world deliberation. In short, SSI prepares students for science-related decision making in an ever-changing global society. Table 2.1 describes more specifically what SSI curriculum is and isn't.

TABLE 2.1.

Features of Socioscientific Issues (SSI) Curriculum

What SSI Curriculum *Is*	What SSI Curriculum Is *Not*
A research-based, interdisciplinary approach that enlists higher-order problem solving, argumentation, and research skills to analyze challenging, contextualized scientific concepts and issues.	A "cookbook" approach to scientific exploration that emphasizes "one right method" and predictable outcomes.
A method that uses real-world scenarios and real data to prepare students for their future roles as societal decision makers.	Simplistic use of hypothetical scenarios that are irrelevant to students' lives.
A conduit for scientific argumentation and discourse skills that mimic the manner in which real scientists research, discuss, debate, and deliberate on scientific issues.	Emphasis on esoteric debates that allow students to contribute opinions rather than evidence.
A relevant and meaningful context for probing students' moral/ethical beliefs on controversial issues while guiding them to become tolerant and open to conflicting opinions and perspectives.	Reliance on "safe" subjects that avoid emotional connections and moral/ethical dilemmas.
A logical approach for modeling nature of science, including the tentativeness of scientific conclusions, the importance of rational argument and skepticism, the role of creativity, and the distinction between science and pseudoscience.	A traditional approach to scientific methodology that fails to recognize the varying social, contextual, and personal influences that contribute to scientific progress.

Source: Adapted from Zeidler and Kahn (2014).

Framework for This Book

While the SSI framework does not specify any particular instructional model for its implementation, SSI's emphasis on constructivist learning and student-centered inquiry lends itself perfectly to the BSCS 5E Instructional Model for lesson planning.

BSCS 5E Instructional Model (5Es)

For many years, science curriculum relied on "cookbook" approaches to scientific investigations. Teachers instructed students on the steps to follow, and students followed these steps to confirm what their teacher (and often they) already knew. More contemporary notions of inquiry-based science teaching became prominent in the late 20th century, representing a shift away from teacher-centered instruction and toward student-centered learning driven by student questions, interests, and approaches to learning. One model that can support inquiry in the science classroom is the BSCS 5E Instructional Model, often referred to as the 5Es, which was developed by the Biological Sciences Curriculum Study and more recently applied to the *Next Generation Science Standards* (*NGSS*) (Bybee 2015). This model follows a learning cycle that consists of the following five stages: Engage, Explore, Explain, Elaborate, and Evaluate. The five stages are described in more detail in Table 2.2.

TABLE 2.2.

The BSCS 5E Instructional Model (5Es)

Engage	Students engage in short activities that prompt curiosity and elicit prior knowledge.
Explore	Students collaborate on hands-on activities that are designed to explore new ideas.
Explain	Students develop explanations of the phenomena they have been engaging with and exploring, while the teacher clarifies and extends learning on the topic.
Elaborate	Students engage in activities that deepen understanding and extend what they have learned to new contexts.
Evaluate	Students assess their understanding, and teachers evaluate student progress toward learning goals.

Source: Adapted from Bybee (2015).

Each lesson in this book is presented in the 5Es format. At times, certain stages of the model are repeated in a cyclical manner to ensure that students have sufficient opportunities to explore and engage in their investigations.

Universal Design for Learning (UDL)

Universal Design for Learning (UDL) (Rose and Meyer 2002) is an educational framework that considers the variability of learners in the classroom and rejects the "one size fits all" approach to teaching and learning. In sum, UDL attempts to meet the learning needs of as many students as possible in advance, thereby reducing the need for specific accommodations for individual students. Whereas teachers must often "retrofit" existing curriculum to accommodate a variety of learners, UDL provides a structure for proactive planning and implementation that anticipates the range of learners and uses research-based approaches from the fields of education, psychology, and neuroscience to make lessons as accessible to as many students as possible. UDL is based on an architectural approach known as universal design (UD), which attempts to create environments that are accessible to as many as possible. Chances are, you are already familiar with some forms of UD. Curb cuts, the small ramps at street corners, and closed-captioning on television are both examples of UD, as they increase access to larger numbers of people. Interestingly, although both of these examples were originally designed for people with disabilities, they are helpful for many purposes. For example, curb cuts help not only those who use wheelchairs but also the elderly, people with baby strollers or bicycles, and young children, while closed captioning is used not only by people with hearing impairments but also by those learning a new language or to read, as well as on televisions in public spaces where audio is difficult to hear or muted. When applied to learning, this approach translates to flexibility in the way information is presented to students, the ways students can show what they have learned, and the ways students are engaged in the learning itself. There are three guidelines of UDL (CAST 2018):

- Provide multiple means of engagement (the "why" of learning). The teacher should provide options for the way students are engaged by and sustained in learning.

- Provide multiple means of representation (the "what" of learning). The teacher should provide options for the way information is presented.

- Provide multiple means of action and expression (the "how" of learning). The teacher should provide options for the ways students interact with instructional materials and communicate their learning.

Framework for This Book

Figure 2.1 is a graphic organizer that helps communicate some of the many ways in which UDL can be implemented. Throughout the lesson plans (pp. 56–480), information boxes identify places where UDL was incorporated.

FIGURE 2.1.
UDL Guidelines Graphic Organizer

Source: CAST (2018), reprinted with permission.

Putting it all together, the framework for this book (Figure 2.2) provides the structure for research-based, age-appropriate, and equitable science learning. In combination with explicit alignments with the *NGSS*, the *Common Core State Standards* in English language arts and mathematics, and the *National Curriculum Standards for Social Studies* (from the National Council for the Social Studies), this book represents a powerful resource for interdisciplinary standards-based learning.

FIGURE 2.2.

Framework for *It's Still Debatable!*

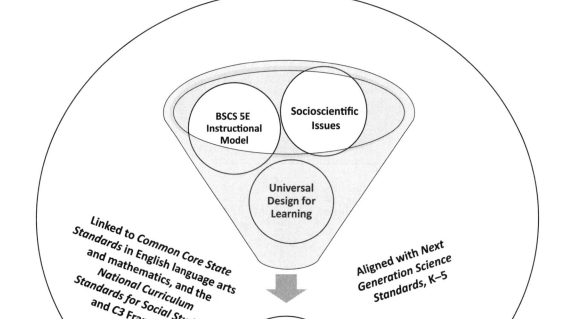

References

Bybee, R. W. 2015. *The BSCS 5E instructional model: Creating teachable moments*. Arlington, VA: NSTA Press.

CAST. 2018. Universal design for learning guidelines, version 2.2. *http://udlguidelines.cast.org*.

National Science Teachers Association (NSTA). 2016. NSTA position statement: Teaching science in the context of societal and personal issues. *www.nsta.org/about/positions/societalpersonalissues.aspx*.

Rose, D. H., and A. Meyer. 2002. *Teaching every student in the digital age: Universal design for learning*. Alexandria, VA: Association for Supervision and Curriculum Development.

Zeidler, D. L. 2014. Socioscientific issues as a curriculum emphasis: Theory, research, and practice. In *Handbook of research on science education,* vol. 2, ed. N. G. Lederman and S. K. Abell, 697–726. New York: Routledge.

Zeidler, D. L., and S. Kahn. 2014. *It's debatable! Using socioscientific issues to develop scientific literacy, K–12*. Arlington, VA: NSTA Press.

Zeidler, D. L., T. D. Sadler, M. L., Simmons, and E. V. Howes. 2005. Beyond STS: A research-based framework for socioscientific issues education. *Science Education* 89 (3): 357–377.

Unit III
Strategies for Promoting Inquiry, Argument, and Inclusion

Within each of the lesson plans in this book are research-based tools and strategies for supporting all students' understanding of inquiry and argument. Some of these strategies, particularly graphic organizers like Venn diagrams and KWL charts, will likely be quite familiar, while others, such as choice boards and sentence frames, may be less so. I have intentionally tried to use a variety of tools and strategies in this book to provide you with a range of options that you can use throughout your teaching day. This chapter serves as an introduction to these tools and strategies.

Research-Based Tools and Strategies

OWL and KWL Charts

The Observe, Wonder, Learn (OWL) chart (Figure 3.1, p. 16) is a variation of the classic Know, Want to Know, Learned (KWL) graphic organizer (Ogle 1986) that helps assess students' prior knowledge, primes and organizes their learning on a particular topic, encourages inquiry by allowing their questions to guide learning, and serves as a tool for formative assessment. With both of these graphic organizers, instruction begins by asking students what they Know (or Observe) about a topic, followed by asking what they Want to Know about it. As students proceed through investigations, they revisit the organizer to add new Wonderings and answers that they have Learned. The OWL chart was introduced into science instruction in

response to the situation that frequently occurs when we ask students what they know about a subject, only to learn that what they *think* they know isn't actually true. This is a common situation in science, where students hold many misconceptions about phenomena, often because of making incorrect inferences from their observations. By focusing on observation, you can not only hone students' observational skills but also help students distinguish between observation and inference. Once while using a KWL chart during a unit on birds, I showed my class of third graders several large tail feathers and asked what they knew about feathers. One student excitedly responded as he pointed to the quill, or calamus, "They're made of plastic!" This is, of course, incorrect because feathers are made of keratin, a similar but harder version of the keratin protein that makes up human hair and nails. Using the KWL chart, I faced the awkward situation of having to correct the student and not add his contribution to the organizer. Yet, the student actually was making an astute observation because the quill looks and feels like plastic. The problem rested with his inference that something that looks and feels like plastic *is* plastic.

By using Observe in place of Know, the OWL chart keeps students focused on making observations and prompts thoughtful discussions of observation versus inference. If I had asked my class, "What do you observe about the feathers?" the same student would likely have said, "They look like [or feel like] plastic!" If the student said, "I observe it's made of plastic," there is a clear trajectory to then clarifying, "Is that an observation—information you gather directly by using your senses—or an inference—an explanation of your observations that requires you to fill in missing information with prior knowledge?" In sum, OWL charts are very useful when there is something (an object or a phenomenon) for students to observe or that they have observed, while KWL charts are helpful for more abstract concepts (e.g., the water cycle) or when objects or pictures of a concept aren't available to observe.

An OWL chart is used in Lesson 1, and a KWL chart is used in Lessons 6 and 11.

FIGURE 3.1.

OWL Chart

Observe	Wonder	Learn

KLEW Chart

The KLEW chart (Hershberger, Zembal-Saul, and Starr 2006) is another variation of the KWL chart that adds Evidence to the graphic organizer (Figure 3.2). The acronym KLEW stands for what we think we Know, what we are Learning, our Evidence, and what we are still Wondering. The other innovation in the KLEW chart is that it makes clear that learning isn't the final step; in science, we continually collect data that contribute to the development of claims (what we've learned), which then need to be supported by evidence. The KLEW chart, like the KWL and OWL charts, can be kept available throughout the lesson or unit so that students can track their learning, maintain focus on the question or questions being investigated, and become familiar with the language and process of evidence-based argument. KLEW charts can be created with teacher-written student responses or by students writing on sticky notes and posting them to the chart. They can also be used by students individually or in small groups. I like to use KLEW charts to introduce investigations after students have had experience with KWL and OWL charts; that way, I can focus a bit more on the new Evidence element of the chart.

KLEW charts are used in Lessons 2, 7, and 14.

FIGURE 3.2.

KLEW Chart

What do we think we KNOW?	What have we LEARNED?	What is our EVIDENCE?	What are our new questions or WONDERINGS?

RAN Chart

The Reading and Analyzing Nonfiction (RAN) chart (Stead 2004, 2014) is yet another variant of the KWL chart that was conceived for nonfiction read-alouds (Figure 3.3). It focuses on inquiry and asks students to evaluate what they *think* they know, what they have learned, and what they want to know, in addition to providing students with a template for identifying misconceptions and confirming understandings as they read. The RAN chart can be used during whole-class instruction, with students being given sticky notes to add their ideas (and moving them around as appropriate), or it can be used individually by students as a map of their thinking and learning. In this book, a RAN chart is used as part of a jigsaw reading activity (see p. 426) so that students can be exposed to a large amount of text through peer reporting, and then record their learning on their RAN chart.

The RAN chart is used in Lesson 13.

FIGURE 3.3.
RAN Chart

Topic	What I Think I Know	Confirmed	Misconceptions	New Information	Wonderings

Venn Diagram

Venn diagrams are highly versatile tools used for showing similarities and differences between concepts (Figure 3.4). They can be used in the science classroom to compare, contrast, and recognize relationships between objects or ideas. Venn diagrams consist of two or more overlapping circles each representing a concept, with the overlapping portion representing commonalities between the concepts represented by those circles. The concepts can range from fairly simple, such as comparing dogs and cats or plants and animals, to more complex, such as photosynthesis and respiration or evaporation and precipitation. Venn diagrams can be two-dimensional drawings or projections, or they can be made three-dimensional by using manipulatives such as Hula-Hoops, strings, Wikki Stix (wax-covered string), or paper plates to represent the circles. Using

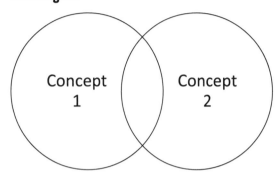

FIGURE 3.4.

Venn Diagram

three-dimensional Venn diagrams can ensure that students receive both visual and tactile cues. Venn diagrams can be used at any point in a lesson by either the teacher or students.

Venn diagrams are used in Lessons 1 and 12.

Concept Map

Concept maps (Novak 1998) are graphic organizers that help students analyze, visualize, and find relationships among concepts (Figure 3.5). Concept mapping can be a particularly powerful tool when used with brainstorming, as students can add terms that they associate with the given topic and then organize them in a way that is meaningful to them. Concept maps are typically developed by starting with the topic or term of focus, usually within a circle or other geometric figure, and then inviting students to suggest related terms or concepts, which are added to the map. These concepts are then organized and connected to each other and the key concept

FIGURE 3.5.

Concept Map

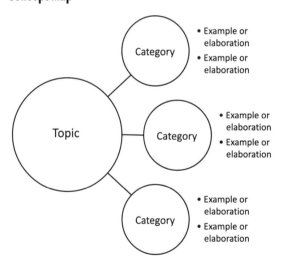

with lines or arrows that are drawn to show relationships. For example, if students were studying frogs, the teacher would begin by writing the word *frogs* in the center of the map and invite students to share words or ideas associated with frogs. Students could write contributions on sticky notes and post them, or the teacher could write them on the board. Then the class could work together to develop sensible categories within which to sort the concepts, such as "what it eats," "where it lives," and "what it looks like." These categories are then connected to each other and to the main topic. Concept maps work well as preassessments that can be built on as students progress in their unit study. They are also excellent for use in small groups as students read through new material and work together to develop a collaborative concept map.

Concept maps are used in Lessons 4 and 12.

Four Square Writing Method

The Four Square Writing Method (Gould and Gould 1999) is a framework or template that helps students organize their writing (Figure 3.6). It is typically used for persuasive writing but can be used in a variety of contexts. Similar to the Frayer Model (Frayer, Frederick, and Klausmeier 1969), the Four Square Writing Method uses a graphic organizer consisting of a series of four rectangles surrounding a central rectangle. Unlike the Frayer Model, which focuses on a vocabulary word and provides definitions, characteristics, examples, and non-examples, the central rectangle in the four square framework contains a topic sentence such as "Insects are very interesting." The surrounding rectangles then support the topic sentence with reasons or rationales. Many variations of this method's template exist, and each can be adapted to students' ages and abilities. For example, for younger students, you might choose to have the first three rectangles simply include reasons for the student's topic sentence (e.g., "Insects are very interesting because they have six legs."), with the fourth box dedicated to a feeling or a picture. For older students, setting up the upper left-hand box as an opening sentence (e.g., "I think that insects are very interesting because they are different from other animals in several ways."), followed by two supporting sentences (e.g., "For one thing, insects have a hard outside skeleton called an exoskeleton.") and a final summary sentence in the other three boxes, can be helpful in organizing an opinion piece or advocacy letter.

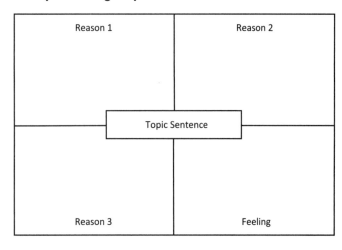

FIGURE 3.6.

Four Square Writing Template

The four square template can be easily adapted for developing an evidence-based argument by arranging the boxes as in Figure 3.7.

This template requires students to support their claim with both evidence and sources for the evidence and is used in Lessons 4, 8, and 10.

Claim, Evidence, Reasoning (CER) Framework

Developing scientific explanations is a key practice in the *Next Generation Science Standards*

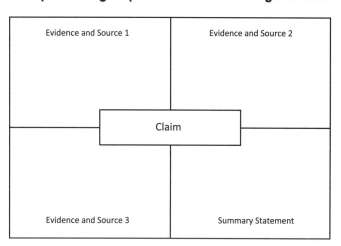

FIGURE 3.7.

Four Square Writing Template for Evidence-Based Argumentation

(*NGSS*). Scientific explanations are the answers to the questions that are asked in scientific investigations. Too often, students respond to questions like "What do plants need to grow?" with answers such as "They need light," but they aren't able to connect their answer to evidence or explain why the evidence they can provide justifies their answer. The Claim, Evidence, Reasoning (CER) framework (McNeill et al. 2006; McNeill and Martin 2011) addresses this challenge by simplifying a complex set of processes into a three-part model:

Scientific Explanation = Claim + Evidence + Reasoning

- A *claim* is a statement or conclusion that answers the question in an investigation. For example, "Plants need light to grow" is a claim.

- *Evidence* is data that support the claim. Data can be numerical, written, recorded, or pictorial and can be taken from many sources, including observations, experimental results, texts and other media, personal experiences, and others' experiences. For example, "In our experiment, we found that the plants near the window grew taller than the plants in the cabinet."

- *Reasoning* connects the evidence to the claim and justifies its use in this situation. For example, "The plants near the window and in the cabinet had everything else the same (cup, soil, water) except light. Since the plant in the light grew taller, plants must need light."

Supporting students' ability to develop explanations in this manner forms the basis for engaging in argument from evidence, a related *NGSS* practice. A template for using CER is found in Figure 3.8.

FIGURE 3.8.

CER Template

Question:

⬇

Claim:

⬇

Evidence:	➡	Reasoning:

A related approach, specifically for evidence-based argumentation in socio-scientific issues (SSI), is to frame students' arguments as Claim, Evidence, Source (CES) (Zeidler and Kahn 2014). In SSI, the claim is often debatable in part based on the credibility of the sources of evidence. Therefore, it is critical for students to identify their sources so that those sources can be examined and evaluated. A template for using CES is found in Figure 3.9.

The CER template is used in Lesson 11, and the CES template may be used in Lessons 7, 9, and 13.

FIGURE 3.9.

CES Template

Claim	Evidence	Source

Agree/Disagree or Yes/No T-Chart

An agree/disagree or yes/no T-chart is another graphic organizer for helping students organize their thinking, particularly on controversial issues (Figure 3.10). On this type of chart, students develop lists of arguments based on evidence to support their claim. The evidence that they supply can be in either written or pictorial form. While the T-chart doesn't quantitatively determine a student's final opinion or conclusion on an issue (as a single compelling argument on one side of the T-chart can outweigh several weaker or simply less compelling arguments on the other side) it gives students the opportunity to record and organize their thinking.

An agree/disagree or yes/no T-chart is used in Lessons 4, 11, and 14.

T-Chart for Evaluating Models

Developing and using models is a science and engineering practice that even the youngest students must employ. Models help scientists and engineers predict and explain systems based on evidence and theories. Modeling can be used in the classroom for many purposes, such as to reveal students' thinking about systems (e.g., having students draw a food web to communicate their understanding of how food webs work) and to serve as predictive tools (e.g., having students simulate a forest food web through either role-playing or a computer simulation and predicting how the food web might be affected by a wildfire). Because models are simplified representations of the

FIGURE 3.10.

Agree/Disagree T-Chart

Claim: _____.

Agree	Disagree
(write or draw *evidence* here)	(write or draw *evidence* here)

real phenomena that they are modeling, they necessarily have strengths and weaknesses (or ways that they accurately or inaccurately depict the real thing). Helping students evaluate models is a critical step in supporting their understanding of how to interpret their research (e.g., asking whether a particular model is reliable) as well as the tentativeness of scientific knowledge (e.g., knowledge can change as models are revised based on new evidence), and even that using different models may yield different but reliable results. To scaffold students' evaluation of models, teachers can use T-charts that elicit students' ideas about the quality of their model by asking, "How well does our model represent (the) real _____?" (filled in with terms like *wetlands*, *tornadoes*, *water cycle*, or *rocket launches*), and "How was our model like the real system?" "How was our model not like the real system?" or "In what ways was our model accurate?" "In what ways was our model not accurate?"

This can even be simplified to a thumbs-up or thumbs-down question. In Lesson 8, for example, students create egg-sized sports helmets and test their effectiveness by dropping eggs with the helmets on. In some ways, this model seems like a reasonable representation of the real thing: Heads and eggs are hard on the outside, soft on the inside. They are also both somewhat fragile and can be damaged by impact. But they also differ: People aren't as fragile as eggs, and they aren't generally dropped during sports. Rather than dismiss models that aren't perfect, scientists and engineers evaluate the strengths and weaknesses of each, always revising and striving for more accuracy. After evaluating a model like the egg-helmet experiment, you can ask your students, "How could we make our model stronger?" "How could we make it a more accurate representation (or more like the real thing)?" This is a wonderful practice for your students to engage in, and a T-chart like the one in Figure 3.11 can be a helpful tool.

T-charts are used for evaluating models in Lessons 7 and 8.

FIGURE 3.11.

T-Chart for Evaluating a Model

How well does our model represent (the) real _____?

Ways the Model Is Accurate	Ways the Model Is Not Accurate

How could we improve our model?

Weighing the Evidence Template

Another tool students can use to help organize their thinking is the Weighing the Evidence graphic organizer (Figure 3.12). In this graphic organizer, students list evidence on either side of an issue and determine which side they think outweighs the other. Similar to the T-chart but a bit more concrete, the Weighing the Evidence template reinforces the idea that evidence has weight in deciding one's position on an issue.

The Weighing the Evidence template is used in Lesson 12.

FIGURE 3.12.

Weighing the Evidence Template

Weighing the Evidence

Question

Choice 1   Choice 2

Write and explain your decision here:

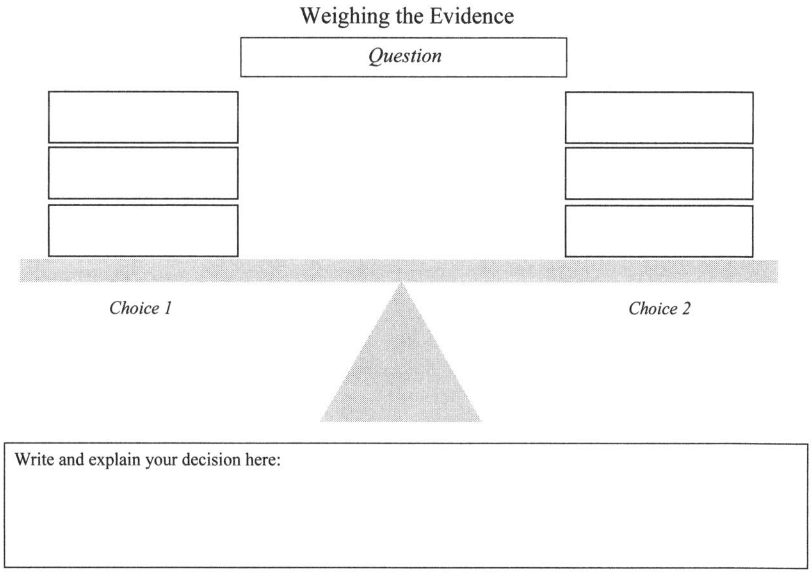

Sentence Frames for Arguments

One powerful way of scaffolding or supporting students' ability to write or discuss arguments is by providing sentence frames (Fulwiler 2007; Ross, Fisher, and Frey 2009). Sentence frames help organize and externalize students' conceptual thinking. The ability to externalize or communicate one's thinking is an essential process in science and engineering, as well as across all disciplines. Scaffolding this process is critical for all students but can be particularly helpful with students who have learning disabilities or are English language learners. The sentence frames included in this graphic organizer are specifically designed to help students articulate claims, evidence, and reasoning, as well as polite ways of disagreeing with others (Figure 3.13). This helps students enhance social skills while engaging in discourse.

Sentence frames are used in most lessons but are particularly highlighted in Lessons 4, 8, and 11.

FIGURE 3.13.

Sentence Frames for Arguments

Making a Claim	Providing Evidence
I think that _____. I know that _____. My claim is that _____. *I think that we should invest in wind energy.*	I think this because _____. I know this because _____. The evidence I have to support this claim is _____. *I think this because wind energy is a renewable resource.*
Explaining My Reasoning	**Disagreeing With Others**
My evidence supports my claim because _____. According to _____ (source), *According to our reading, renewable resources never run out so they are a good choice.*	I disagree with your claim because _____. I don't think your evidence is accurate because _____. I think you should also consider _____. One difference between my idea and yours is _____. *I think you should also consider solar and water power because they are also renewable.* *I disagree with your claim because we don't have enough wind in our area.*

Source: Adapted from Ross, Fisher, and Frey (2009).

Sources of Evidence Template

Evidence is data that support a claim. Students often hold the misconception that data necessarily mean numbers. In actuality, data, and therefore evidence, can be in any number of forms, including numerical, written, recorded, physical, and pictorial. Moreover, evidence can be obtained from many sources, such as observations, results of experiments, texts, other media, or one's own or others' experiences. This is an important concept for students, as they will often need to integrate evidence from a number of sources to develop a sophisticated argument. Helping students organize their evidence while reinforcing the variety of sources available to them can be accomplished through the use of this template in Figure 3.14.

The Sources of Evidence template is used in Lessons 7 and 13.

FIGURE 3.14.

Sources of Evidence Template

What Are Your Sources of Evidence?

Texts	Videos
Our Experiments	Other People's Experiments
Personal Experience	Other People's Experiences

Source: Adapted from Shim, Thompson, Richards, and Vaa (2018).

Evaluating Media Sources Template

One of the most difficult tasks for educators interested in promoting students' ability to engage in evidence-based argument is helping them evaluate the quality and trustworthiness of their media sources, such as books, websites, and videos. While the information age has certainly made accessing information easier than ever before, it has not necessarily improved the quality of that information. Consumers of information still need to be savvy about evaluating the sources and ensuring that they are reliable, accurate, and reasonably free from bias. Conducting a comprehensive evaluation of a resource is often time-consuming and quite complex. While young students may not be able to perform extensive research on their media sources, they can gain sophistication by asking key questions as they decide how or whether to use a resource. I have found the CARS rubric (Harris 2000) to be extremely helpful in guiding students' thinking on this. CARS stands for Credibility, Accuracy, Reasonableness, and Support. Figure 3.15 (p. 28) has a template for Evaluating Media Sources in which I have simplified the original rubric to make it appropriate for young children and help them ask key questions as they evaluate media sources.

The Evaluating Media Sources template is used in Lessons 9 and 14.

FIGURE 3.15.
Evaluating Media Sources Template

Evaluating Media Sources

It's always important to determine whether your media sources such as books, videos, and websites are trustworthy. Here are some questions to ask using the CARS rubric:

	Questions to ask	My answers
Credibility	• What do I know about the author(s), reporters, or website hosts? • Do they have expertise and strong credentials, such as education or jobs in the field? • Have any organizations recommended them or given them awards?	
Accuracy	• Does the information make sense? • Does the information seem outdated? • Are there other sources that have updated information?	
Reasonableness	• Does the information seem to be fair and free from bias, or does the source seem to be promoting a particular viewpoint?	
Support	• Are there other sources that support (agree with) the information presented? • Does the source include references or a bibliography to show where they found their information?	

Opinion Letter

Persuasive writing is an essential skill as students progress to increasingly sophisticated levels in language arts. Helping students write in a persuasive manner not only supports their use and organization of evidence but also provides teachers with a powerful assessment tool, as students' reasoning and integration skills are clearly evident in opinion letters. Moreover, giving students multiple opportunities to communicate their opinions on different issues empowers them to take action and voice their concerns. I use two different opinion letter templates, a simpler one-reason version and a more complex three-reason letter (Figure 3.16).

The opinion letter is used in Lessons 4 and 11.

FIGURE 3.16.

Opinion Letter Template

```
I think that _____.

Here are my reasons.

┌─────────────┐  ┌─────────────┐  ┌─────────────┐
│ Reason #1   │  │ Reason #2   │  │ Reason #3   │
│             │  │             │  │             │
│             │  │             │  │             │
│             │  │             │  │             │
│             │  │             │  │             │
│             │  │             │  │             │
│             │  │             │  │             │
└─────────────┘  └─────────────┘  └─────────────┘

That is why I think that _____.
```

Town Hall or Congressional Meeting

Town hall meetings and congressional meetings are excellent culminating events for SSI lessons, as they allow for students to extend their research from a variety of sources, collaborate with teammates, and engage in evidence-based argument. The typical format of such meetings involves challenging students to take on the role of a townsperson or member of Congress to address some debatable issue. Students then work with others to research the topic, using books, the internet, or other provided sources to prepare for the event. The teacher then explains the format to students, which for elementary-age students generally consists of some variation of the following:

1. Opening statements: Students state their claims and evidence-based rationales.

2. Questioning: Students have an opportunity to question and rebut (respond to) other teams' claims.

3. Closing statements: Students summarize their claims and rationales.

Depending on the scenario, students may be asked to vote on or negotiate a final position. The teacher then debriefs the meeting by asking what students found challenging or interesting, what new information they learned, and how their personal positions on the issue compare with the one that they argued. Assessing these types of meetings can be done using a rubric such as the one in Figure 3.17.

FIGURE 3.17.

Rubric for Town Hall or Congressional Meeting

	Early (1 pt.)	Emerging (2 pts.)	Sophisticated (3 pts.)	Points and Comments
Use of Evidence	Students use opinion without evidence to back their claims.	Students use tenuous or incomplete evidence to back claims.	Students demonstrate complete and accurate use of evidence to back claims.	
Source and Quality of Evidence	Students are unable to identify sources of evidence.	Students demonstrate some effort in identifying and evaluating the sources of evidence.	Students thoughtfully identify and evaluate the sources of evidence.	
Science Content Understanding	Students demonstrate minimal understanding of science content.	Students demonstrate a moderate degree of understanding of science content.	Students demonstrate strong understanding of science content and consistently apply it to their arguments.	
Clarity and Organization of Presentation	Presentation is unclear and disorganized.	Presentation is somewhat clear and organized.	Presentation is clear, organized, and compelling.	
Response to Questions	Students are unable to respond to questions.	Students respond in inappropriate manner or with inaccurate information.	Students respond appropriately, thoughtfully, and accurately.	

Total__/15 pts.

Source: Adapted from Zeidler and Kahn (2014).

Strategies for Promoting Inquiry, Argument, and Inclusion

You can choose to assess groups and individual students during the meeting, or you can assess groups during the meeting and have students create an individual product, such as an opinion letter or CER template (see p. 22) to share their own position on the issue, rather than the one they are role-playing. Town hall and congressional meetings are used in Lessons 1, 7, 9, 11, and 13.

Choice Board

Choice boards (Wormeli 2006) are a set of options given to students to provide them with choice and autonomy. While the concept of learning styles has been widely discredited, the importance of maintaining student interest is key, and choice is among the most powerful ways that we can maintain student interest and promote autonomy. To create a tic-tac-toe choice board, you begin with a set of boxes and put the main assessment task that all students will complete in the center box (Figure 3.18). Then, identify two or three learning objectives that you would like students to demonstrate. Design options that reflect students' interests for these

FIGURE 3.18.

Sample Tic-Tac-Toe Choice Board

Make tic-tac-toe (three in a row) using the center box.

Write a **story** that describes how a seed becomes a plant.	Create a **painting** that shows what part of a plant three (3) different foods come from.	Write a **song** that describes the parts of plants and what they do.
Create a set of **flash cards** that teach what part of a plant three (3) foods come from.	Complete **My Plant Journal.**	Draw a **comic strip** about how a seed becomes a plant.
Write a **play** that teaches what part of a plant three (3) different foods come from.	Perform a **dance** that shows how a seed becomes a plant.	Make a **mobile** that shows the parts of plants and describes what they do.

objectives, and place them in the boxes such that completing three in a row using the center box will address two different learning objectives. Students mark the center box with an X when they've completed the main assessment task, and then choose two other activities to make three in a row, marking them off as they are completed. Figure 3.18 (p. 31) shows the choice board from Lesson 3, where My Plant Journal is the main assessment that all students complete. The learning objectives (of which I'd like students to demonstrate two of three) are to (1) identify the parts of plants and their functions, (2) identify which parts of plants at least three foods come from, and (3) describe how a seed becomes a plant. If you would like a simpler choice board (not tic-tac-toe style), use fewer options and have students complete two out of three or three out of four.

This strategy is included in Lesson 3.

Cooperative Learning Strategies

Elementary science class, and SSI in particular, presents an ideal opportunity to support students' social skills. The practice of science requires communication and collaboration, so ensuring that students practice and master these skills is key to their future science (or any) career and personal goals. Cooperative learning (Johnson and Johnson 1989; Slavin 1980) is a research-based approach to classroom organization that fosters cooperative skills. It is particularly effective in helping promote understanding among diverse student populations with different backgrounds and skill sets (Putnam 1997). SSI lends itself to the implementation of several cooperative learning strategies, as it is a highly interactive framework that is enhanced by students' ability to communicate freely and productively together.

Cooperative learning consists of five main elements: (1) individual and group accountability, (2) positive interdependence, (3) face-to-face interaction, (4) promotion of social skills, and (5) group processing. Each of these elements can be integrated into the negotiation of SSI by using the following tips:

- During activities, assign and rotate team roles, such as recorder, reporter, timekeeper, materials manager, and principal investigator, to ensure full participation.

- Ensure that the group task (e.g., book or online research, laboratory investigation, presentation) requires the input of all students. In other words, make each role essential.

- Assess students on both individual and group work.

- Consider distributing "C-points" to teams for demonstrating desired social skills. Teams might tally points to earn a privilege, or the points can count

Strategies for Promoting Inquiry, Argument, and Inclusion

toward their group project grade. Here is a list of some "C" skills that can be targeted:

- Cooperation
- Collaboration
- Communication
- Consensus building
- Care of materials
- Completion

Be sure to model and describe the social skill you are targeting so that students understand expectations.

- Process group activities by having students provide input on their group's performance, including areas of strength and challenges. Highlight the processes groups followed to work through challenges.

- Emphasize team building and class building (Kagan, Kagan, and Kagan 1997) to foster trust and positive communication among team members.

Following are descriptions of some common cooperative learning strategies used to organize and manage lessons.

Four Corners

Four corners (Kagan, Kagan, and Kagan 1997) is a wonderful strategy that strengthens students' argumentation and listening skills and helps increase their confidence in class discussions. To implement four corners, you can simply pose a debatable question and offer four options for the answer. Post the four options on signs in the four corners of the classroom (see Figure 3.19). After letting students think about their answers, have them jot down their choice (which commits them to their answer), then have them move to their chosen corner. Once students have formed groups at the corners, allow them to discuss with their group why they chose that corner (i.e., answer),

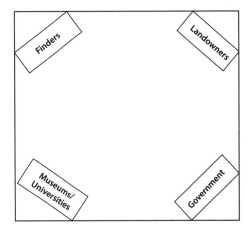

FIGURE 3.19.

Sample Four Corners

Who should own the dinosaur bones?

then have a spokesperson for each group present to the class. Allow students to change corners after hearing different positions on the question or issue. Process the activity by asking students whether they heard anything that made them rethink their position, why some reasons were particularly persuasive, and what else they might want to know about the issue. Four corners can be used as an assessment tool at any stage in a lesson or unit or simply for review and reinforcement.

Four corners is used in Lessons 5 and 10.

Yes/No or Agree/Disagree Argument Lines

Yes/no or agree/disagree argument lines (Kagan, Kagan, and Kagan 1997) are another powerful strategy for debate and discussion in the science classroom. To implement this strategy, you simply post signs on which you've written "Yes" and "No" (or "Agree" and "Disagree") on opposite sides of the room. Pose a debatable question to the class, then have students move to the side of the room that represents their response. This can be set up as a spectrum line, where those wishing to express strong agreement or disagreement stand near the ends and those with moderate opinions stand closer to the center, or as a bright line, where students stand on either side of the line, a strategy used to force students to make a decision rather than straddle the center. (See Figure 3.20.) As in four corners (p. 33), allow students to discuss their position with others in their group, then have a representative from each side present an argument for this position to the class. Again, students should be given a chance to move around after hearing the arguments. You can use argument lines when you are introducing a new topic to gauge students' initial reactions to an issue or at the end of a lesson or unit as an assessment or to see how students' thinking has changed.

Argument lines are used in Lessons 4, 7, 8, and 11.

FIGURE 3.20.
Argument Lines

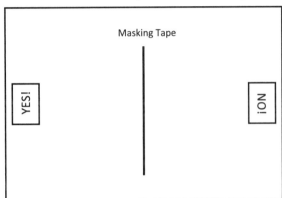
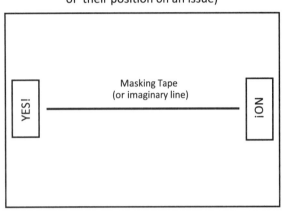

Think-Pair-Share

In the think-pair-share strategy (Kagan, Kagan, and Kagan 1997), you allow students to individually reflect on a question posed, then give them time to discuss the question in pairs before engaging in whole-class discussion (Figure 3.21). This strategy gives students additional time to process questions, reduces stress that can result from having to respond in front of the whole class without testing ideas with classmates first, and fosters confidence in responding to questions.

Think-pair-share is used in Lessons 13 and 14.

FIGURE 3.21.
Think-Pair-Share

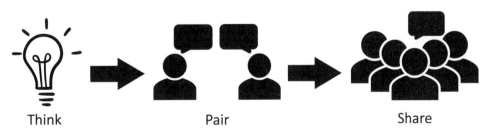

Jigsaw

In the jigsaw strategy (Slavin 1995), team members become experts on a certain research topic by working with members from other teams ("expert groups"), then returning to teach what they've learned to their original teams ("home groups"). Students learn by teaching, and the team is reliant on each member becoming an expert. To implement this strategy, begin by assigning students to home groups. Then, assign one member of each group to a reading selection. Next, create expert groups consisting of students from the various home groups who have been assigned to the same selection. Be sure to provide expert groups with questions that help guide their understanding of the readings. Finally, have experts return to their home groups where they will teach what they learned in their expert groups to their home groups. (See Figure 3.22.)

The jigsaw strategy is used and described in greater detail in Lesson 13.

FIGURE 3.22.

Jigsaw

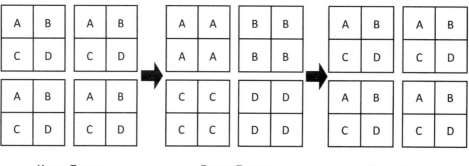

Home Teams   Expert Teams   Home Teams

Talking Chips or Talking Sticks

The talking chips and talking sticks strategies (Kagan, Kagan, and Kagan 1997) both help ensure that everyone on a team gets a chance to participate in discussions. With talking chips, you give each student two or three chips of a particular color, with each member of a team receiving a different color. When students wish to speak, they place a chip on the table. When they run out of chips, they must wait and listen to others. With talking sticks, you give each team a special stick that designates the speaker (Figure 3.23). As an alternative, you might have teams create their own "microphones" out of aluminum foil or paper. Students may speak only when they are holding the stick or microphone. I found this to be a great tool for helping young students (kindergarten and first grade) with turn taking during discussions. Though I didn't specifically include these strategies in the lesson plans in this book, you can use talking chips or talking sticks freely during any team activity.

FIGURE 3.23.

Kindergarten Team Talking Stick/Microphone

References

Frayer, D., W. Frederick, and H. Klausmeier. 1969. *A schema for testing the level of cognitive mastery*. Madison: Wisconsin Center for Education Research.

Fulwiler, B. R. 2007. *Writing in science: How to scaffold instruction to support learning.* Portsmouth, NH: Heinemann.

Gould, J. S., and E. J. Gould. 1999. *Four square: Writing method for grades 1–3: A unique approach to teaching basic writing skills.* Dayton: Teaching & Learning Company.

Harris, R. 2000. *WebQuester: A guidebook to the web*. Boston: McGraw-Hill.

Hershberger, K., C. Zembal-Saul, and M. Starr. 2006. Evidence helps the KLW get a KLEW. *Science and Children* 43 (5): 50–53.

Johnson, D. W., and R. T. Johnson. 1989. *Cooperation and competition: Theory and research*. Edina, MN: Interaction Book.

Kagan, L., M. Kagan, and S. Kagan. 1997. *Cooperative learning structures for teambuilding*. San Clemente, CA: Kagan Cooperative Learning.

McNeill, K. L., D. J. Lizotte, J. Krajcik, and R. W. Marx. 2006. Supporting students' construction of scientific explanations by fading scaffolds in instructional materials. *Journal of the Learning Sciences* 15 (2): 153–191.

McNeill, K. L., and D. Martin. 2011. Claims, evidence, and reasoning: Demystifying data during a unit on simple machines. *Science and Children* 48 (8): 52–56.

Novak, J. D. 1998. *Learning, creating, and using knowledge: Concept maps as facilitative tools in schools and corporations.* Mahwah, NJ: Erlbaum.

Ogle, D. M. 1986. K-W-L: A teaching model that develops active reading of expository text. *Reading Teacher* 39 (6): 564–570.

Putnam, J. W. 1997. *Cooperative learning in diverse classrooms.* Englewood Cliffs, NJ: Merrill/Prentice Hall.

Ross, D., D. Fisher, and N. Frey. 2009. The art of argumentation. *Science and Children* 47 (3): 28–31.

Shim, S-Y., J. Thompson, J. Richards, and K. Vaa. 2018. Agree/disagree T-charts: Supporting young students in scientific argumentation and modeling. *Science and Children* 56 (1): 39–47.

Slavin, R. E. 1980. Cooperative learning. *Review of Educational Research* 50 (2): 315–342.

Slavin, R. E. 1995. *Cooperative learning: Theory, research, and practice.* Boston: Allyn & Bacon.

Stead, T. 2004. *Reality checks: Teaching reading comprehension with nonfiction, K–5.* Portsmouth, NH: Stenhouse Publishers.

Stead, T. 2014. Nurturing the inquiring mind through the nonfiction read-aloud. *Reading Teacher* 67 (7): 488–495.

Wormeli, R. 2006. *Fair isn't always equal: Assessment and grading in the differentiated classroom.* Portland, ME: Stenhouse Publishers.

Zeidler, D. L., and S. Kahn. 2014. *It's debatable! Using socioscientific issues to develop scientific literacy, K–12.* Arlington, VA: NSTA Press.

Unit IV
A Guide for Reading and Implementing the Lesson Plans

The lesson plans in this book represent a range of science content areas and societal issues, as well as interdisciplinary connections to English language arts (ELA), social studies, and math. Moreover, the lessons are designed to be fairly self-contained, in the sense that you can find background information on the topic, guiding questions, readings for students, assessments, and tools for implementation all within a single lesson plan. The lessons assume a 45- to 50-minute period but can easily be adjusted to your classroom time frame. Each lesson will take multiple class periods that can be adjusted to the pace and schedule of your class. Many elementary teachers find they have little time to teach science in their busy day. Something to keep in mind is that many of the activities in this book connect so closely to both social studies and ELA that they can be integrated into those time blocks. To help ensure that lessons are used to their fullest extent, this chapter gives descriptions and justifications for each component within the lessons.

Lesson Title

Each lesson title includes a question that encapsulates the SSI. As discussed earlier, a socioscientific issue, or SSI, is a societal issue that is informed by science but is not decided by science alone. Deliberations about these issues, including the way scientific information is interpreted and integrated, are influenced by ethical, political, economic, religious, and philosophical beliefs. This makes the issues debatable and ripe for consideration of different perspectives and consequences.

A Guide for Reading and Implementing the Lesson Plans

Suggested Grade Levels

The lessons in this book are organized around the *NGSS* disciplinary core ideas (DCIs) and divided by grade bands (e.g., K–2 or 3–5), rather than single grade levels, following the progression matrices in *The NSTA Quick-Reference Guide to the NGSS, Elementary School* (Willard 2014). This approach was chosen in recognition of the fact that teachers may be working with state standards that may or may not be aligned with the *NGSS*. The DCIs present clear statements of the science content addressed in the lesson, thereby simplifying alignment with any set of standards. The grade band approach also recognizes that science learning occurs over time and through multiple experiences. (More on alignment with the *Next Generation Science Standards* [*NGSS*] is found in the Connecting to the *NGSS* sections described below.)

Driving Questions

The driving questions serve as focal points for teaching, as they summarize the questions that are answered by the DCIs on which each lesson is based. In some cases, the societal question that is addressed by the lesson may be included in the driving question if the lesson's title doesn't adequately summarize it.

Lesson Overview

The lesson overview is a brief synopsis of the lesson. In some cases, specific activities are mentioned to apprise the teacher.

Connecting to the *NGSS*

The *NGSS* represent a bold departure from earlier science standards in that they were developed as performance expectations carefully crafted to integrate three dimensions of science learning: disciplinary core ideas (DCIs), science and engineering practices (SEPs), and crosscutting concepts (CCs). These dimensions are designed to capture the breadth and depth of science as practiced by scientists and engineers. Though at first glance you might find the *NGSS* to be quite complicated (and a bit overwhelming), further examination reveals an extraordinary interweaving of content, practice, and conceptual understandings, all within age-appropriate learning progressions that are designed to help students master complex concepts and ways of thinking about science and apply them to real-world scenarios.

As mentioned earlier, I decided to emphasize the DCIs in the lesson plan to make the science content transparent; that said, each lesson has been aligned to

A Guide for Reading and Implementing the Lesson Plans

support significant progress toward meeting at least one grade-specific performance expectation, as described for each lesson in the appendix. There, you will find the performance expectation (e.g., 3-LS2-1) along with the three-dimensional learning foundation boxes (DCIs, SEPs, and CCs) on which the performance expectation was built. It is my hope that presenting the alignments in this way will make the lessons accessible for non-*NGSS* teachers and also provide flexibility in the way lessons could be used (and be useful) for multiple grade levels. This is important to note because, although the performance expectations in the *NGSS* are grade specific, these expectations represent summative assessments and should reflect learning across a range of activities and experiences. The specific DCIs, SEPs, and CCs on which each performance expectation was built were not intended to take the place of rich, diverse combinations of experiences that integrate DCIs, SEPs, and CCs in different ways. Therefore, lessons may reflect, for example, SEPs that go beyond the ones on which a performance expectation was built.

Societal Issues

While science is at the heart of the lessons, it is important to keep in mind that SSI are societal issues that are related to science but are debatable because science alone doesn't determine the resolution of the issues; other societal factors, which may include economic, political, moral/ethical, religious, and historical considerations, come into play. These factors can (and do) influence how people integrate, incorporate, and weigh scientific evidence that is presented. This section of the lesson plan identifies some of the overarching societal issues that relate to the main SSI question addressed in the lesson, such as animal rights, property rights, roles of government, environmental issues/stewardship, land use, individual versus societal interests, consumer protection, poverty, allocation of resources, justice, and equity. These are very big concepts that may seem daunting to address, but the beauty of the SSI lessons is that students will begin to grapple with them without much prompting. The lessons are designed with these issues built in. The important thing to remember is that the teacher's role is not to tell students *what* to think about these issues, but rather *how* to maintain an open mind, integrate information from different sources, and consider the consequences of their actions (or inactions) in various scenarios. In some cases, the societal issue is introduced at the beginning of the lesson, while in other cases, the issue is introduced after students have a bit more content background, or schema.

Nature of Science

Among the most important things we as science teachers can impart to students is that science is a unique way of knowing about the world. Those aspects of science that together differentiate it from other ways of understanding the world are referred to as the nature of science (NOS) (Lederman 1992; Lederman et al. 2002). NOS concepts, such as the understanding that science is based on empirical evidence or that scientific knowledge is tentative and open to revision when new evidence is found, are critical to helping students differentiate science from nonscience and understand that contradictory evidence doesn't mean that science is arbitrary or untrustworthy, but rather that multiple approaches to and interpretations of scientific evidence are precisely what makes science rigorous. NOS understandings are described more fully in Appendix H of the *NGSS* (NGSS Lead States 2013; the *NGSS* appendixes can be found at *www.nextgenscience.org/resources/ngss-appendices*). In addition, NOS understandings that are connected to specific *NGSS* performance expectations can be found in the *NGSS* foundation boxes (either with the CCs or SEPs) underneath the performance expectations. In this book, I've articulated the NOS understandings that relate most closely to the lesson in this section of the lesson plan. I've also noted specifically within many lesson plans where the particular teachable moment is for NOS. Making NOS as explicit as possible in your lessons helps students incorporate it into their thinking about science.

CCSS Connections

The *Common Core State Standards* (*CCSS*; *www.corestandards.org*) initiative aimed to develop rigorous and coherent standards in ELA and mathematics that would prepare all K–12 students across the United States for college or to enter the workforce. Over 40 states have adopted the *CCSS* and are working to fully implement them. The *NGSS* are fully aligned with the *CCSS*; teachers can look at the *NGSS* performance expectations and find the ELA and mathematics connections to that performance expectation in the connection boxes, which appear below the foundation boxes (e.g., DCIs, SEPs, and CCs). The ELA standards prepare students to read, write, speak, listen, and use language effectively in a variety of contexts. Specifically, the ELA standards are designed to prepare students to use these skills in science, social studies, and technical subjects. You can find a full description of the interplay between the ELA standards and the *NGSS* in Appendix M of the *NGSS* (NGSS Lead States 2013). The standards for mathematics also attempt to provide students with deep conceptual understanding of key math topics and prepare students to apply math in a variety of contexts. The interplay between the *CCSS Mathematics* and the *NGSS* is described in Appendix L of the *NGSS* (NGSS Lead States

A Guide for Reading and Implementing the Lesson Plans

2013). There are many common practices among ELA, mathematics, and science, a fact brilliantly communicated by NSTA in Figure 4.1.

FIGURE 4.1.

Venn Diagram Illustrating Common Practices Among ELA, Mathematics, and Science

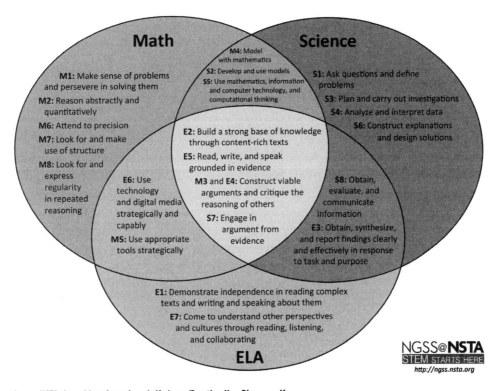

Source: NSTA, http://nstahosted.org/pdfs/ngss/PracticesVennDiagram.pdf.

Because of the interdisciplinary nature of SSI, many of the activities in this book are aligned with both the ELA and math core standards. This is especially true given that the intersection of all three disciplines focuses on evidence-based argumentation. The *CCSS* for each lesson are included in this section of the lesson plan.

NCSS Connections

The *National Curriculum Standards for Social Studies: A Framework for Teaching, Learning, and Assessment* (*NCSS*; NCSS 2010; www.socialstudies.org/standards) consists of a set of learning expectations designed to reflect rigorous core content and disciplinary habits of mind. The content is divided into 10 broad themes: (1) culture; (2) time, continuity, and change; (3) people, places, and environments; (4) individual development and identity; (5) individuals, groups, and institutions; (6) power, authority, and governance; (7) production, distribution, and consumption; (8) science, technology, and society; (9) global connection; and (10) civic ideals and practices. Each theme is followed by a series of questions for exploration within that theme, as well as sections that include (1) knowledge (the core ideas), (2) processes (the skills that students should possess), and (3) products (suggestions of classroom activities and artifacts that serve as evidence of understanding).

C3 Framework

A complementary resource in social studies education is the *College, Career, and Civic Life (C3) Framework for Social Studies State Standards* (www.socialstudies.org/c3), which was released in 2013. Its purpose is to help states and educators implement the *NCSS* in a manner that emphasizes the skills and practices necessary for applying content in real-world settings as active and engaged citizens (Herczog 2013). Unlike the *NCSS*, which focus more heavily on the *what* of student learning, the *C3* emphasizes the *how* with a strong emphasis on inquiry. Among the framework's guiding principles are a commitment to skills and practices that prepare students for democratic decision making; a focus on social studies as preparation for college, career, and civic engagement; an affirmation of the need for interdisciplinary applications; and close connection to the *Common Core State Standards for English Language Arts*. The *C3* is designed around an "inquiry arc" consisting of four dimensions (see Table 4.1).

TABLE 4.1.

Four Dimensions of the *C3* Framework

Dimension 1: *Developing Questions and Planning Inquiries*	**Dimension 2:** *Applying Disciplinary Tools and Concepts*	**Dimension 3:** *Evaluating Sources and Using Evidence*	**Dimension 4:** *Communicating Conclusions and Taking Informed Action*
Developing questions and planning inquiries	Civics Economics Geography	Gathering and evaluating sources	Communicating and critiquing conclusions
		Developing claims and using evidence	Taking informed action

Source: Adapted from NCSS (2013, p. 12, Table 1).

It is clear that many of the priorities of contemporary social studies educators are shared by science education; the focus on inquiry, investigation, evidence-based argument, evaluation of sources, and communication of ideas is key to both disciplines. Perhaps not surprisingly, SSI is a natural fit for social studies practices, with its emphasis on societal issues related to science. Moreover, the C3 focus on informed action can be seen as a vehicle for extending the SSI framework from debate and discourse to action. For this book, I chose to emphasize the *practices* from the *NCSS* to highlight the overlaps in inquiry and argument as well as the habits of mind that these disciplines share. In addition, I have included C3 connections to dimensions 3 and 4, which have grade band progressions as outlined in Tables 4.2–4.6 (pp. 45–47).

TABLE 4.2.

Dimension 3: Gathering and Evaluating Sources

By the End of Grade 2	By the End of Grade 5
Individually or with other students:	Individually or with other students:
D3.1.K-2. Gather relevant information from one or two sources while using the origin and structure to guide the selection.	D3.1.3-5. Gather relevant information from multiple sources while using the origin, structure, and context to guide the selection.
D3.2.K-2. Evaluate a source by distinguishing between fact and opinion.	D3.2.3-5. Use distinctions among fact and opinion to determine the credibility of multiple sources.

Source: Adapted from NCSS (2013, p. 54, Table 25).

TABLE 4.3.

Dimension 3: Developing Claims and Using Evidence

By the End of Grade 2	By the End of Grade 5
Individually or with other students:	Individually or with other students:
Begins in Grades 3–5	D3.3.3-5. Identify evidence that draws information from multiple sources in response to compelling questions.
Begins in Grades 3–5	D3.4.3-5. Use evidence to develop claims in response to compelling questions.

Source: Adapted from NCSS (2013, p. 55, Table 26).

TABLE 4.4.

Dimension 4: Communicating Conclusions

By the End of Grade 2	By the End of Grade 5
Individually or with other students:	Individually or with other students:
D4.1.K-2. Construct an argument with reasons.	D4.1.3-5. Construct arguments using claims and evidence from multiple sources.
D4.2.K-2. Construct explanations using correct sequence and relevant information.	D4.2.3-5. Construct explanations using reasoning, correct sequence, examples, and details with relevant information and data.
D4.3.K-2. Present a summary of an argument using print, oral, and digital technologies.	D4.3.3-5. Present a summary of arguments and explanations to others outside the classroom using print and oral technologies (*e.g.*, posters, essays, letters, debates, speeches, and reports) and digital technologies (*e.g.*, internet, social media, and digital documentary).

Source: Adapted from NCSS (2013, p. 60, Table 28).

TABLE 4.5.

Dimension 4: Critiquing Conclusions

By the End of Grade 2	By the End of Grade 5
Individually or with other students:	Individually or with other students:
D4.4.K-2. Ask and answer questions about arguments.	D4.4.3-5. Critique arguments.
D4.5.K-2. Ask and answer questions about explanations.	D4.5.3-5. Critique explanations.

Source: Adapted from NCSS (2013, p. 61, Table 29).

TABLE 4.6.

Dimension 4: Taking Informed Action

By the End of Grade 2	By the End of Grade 5
Individually or with other students:	Individually or with other students:
D4.6.K-2. Identify and explain a range of local, regional, and global problems, and some ways in which people are trying to address these problems.	D4.6.3-5. Draw on disciplinary concepts to explain the challenges people have faced and opportunities they have created, in addressing local, regional, and global problems at various times and places.
D4.7.K-2. Identify ways to take action to help address local, regional, and global problems.	D4.7.3-5. Explain different strategies and approaches students and others could take in working alone and together to address local, regional, and global problems, and predict possible results of their actions.
D4.8.K-2. Use listening, consensus-building, and voting procedures to decide on and take action in their classrooms.	D4.8.3-5. Use a range of deliberative and democratic procedures to make decisions about and act on civic problems in their classrooms and schools.

Source: Adapted from NCSS (2013, p. 62, Table 30).

UDL Toolkit

Every lesson has a UDL (Universal Design for Learning) Toolkit that points out where and how the multiple means of engagement, multiple means of representation, and multiple means of action and expression are employed. In addition, some lessons may include accommodations for students with special needs if the UDL Toolkit does not adequately meet the needs of most children. It should be noted that additional accommodations may still be required for students with disabilities in your classroom. An outstanding resource is the National Science Teaching Association web page on "Science for Students with Disabilities" at *www.nsta.org/disabilities*.

Suggested Schedule and Sequence

In each lesson, I have outlined a suggested sequence that follows the BSCS 5E Instructional Model (5Es) and could be completed over 4–6 days, except for a few lessons that require long-term observations or extended learning experiences.

The lessons in *It's Still Debatable!* assume a 45–50 minute period but can easily be adjusted to your classroom time frame.

Materials

I have tried hard to keep materials as simple (and inexpensive) as possible. Depending on the activity, materials are listed per class, per group (of a specified number), or per student. In addition to hands-on materials, the lists also include student handouts, which are found at the end of each lesson.

Safety Notes

Science teaching necessarily involves working with different materials, and at times, this can pose safety hazards. Safety *always* needs to be the first concern in all our teaching. You need to be sure that your classroom and other spaces are appropriate for the activities being conducted. This means that engineering controls such as proper ventilation, a fire extinguisher, and an eye wash station must be available and used properly. In addition, students must wear personal protective equipment (PPE) such as sanitized indirectly vented chemical-splash goggles or safety glasses with side shields as appropriate, nonlatex aprons, and vinyl gloves during all components of investigations (setup, the hands-on investigation, and cleanup) when they are using potential harmful supplies, equipment, or chemicals. At a minimum, the eye protection PPE you provide for students to use must meet the ANSI/ISEA Z87.1 D3 standard. Be sure to also review and comply with all safety policies and procedures, including but not limited to appropriate chemical management, that have been established by your place of employment. It is imperative to properly dispose of all materials, even common items like baking soda and vinegar, as well as to properly maintain all equipment. NSTA maintains an excellent web page on "Safety in the Science Classroom" (*www.nsta.org/safety*), which provides guidance on safety for teachers at all levels and includes a safety acknowledgment form (sometimes called a "safety contract") (*http://static.nsta.org/pdfs/SafetyAcknowledgmentForm-ElementarySchool.pdf*) specifically for elementary students to review with their teachers and have signed by parents or guardians. It cannot be overstated that safety is the single most important part of any lesson.

The Safety Notes provided for each lesson highlight specific concerns that might be associated with that particular lesson. These precautions are based in part on the use of the recommended materials and instructions, legal safety standards, and better professional safety practices. Use of alternative materials or procedures for these investigations may jeopardize the level of safety and therefore is at the user's own risk. Remember that an investigation includes three parts: (1) setup, what you do to prepare the materials for students to use; (2) the actual investigation, which involves students using the materials and equipment; and (3) cleanup, which includes cleaning the materials and putting them away for later use. The safety procedures and PPE stipulated for each investigation apply to all three parts.

Media

Many of the lessons use picture books to engage students and make content accessible. Following in the footsteps of the fantastic NSTA Press *Teaching Science Through Trade Books* and *Picture Perfect* series, I often use combinations of fiction and nonfiction texts to create a conceptual storyline that I hope is coherent and enjoyable. In choosing books, I am particularly concerned about two features: that the books are accurate and do not introduce or perpetuate misconceptions, and that they reflect the diversity of today's classrooms.

Some of the lessons in this book use videos that are available on YouTube at *www.youtube.com/playlist?list=PLv-4JxAyjFq07VehCRQtq-LmU-6o0Zhh2*, while other lessons use media such as posters that are available at stable links on NSTA Connections. The URLs for each lesson are provided in the Media section.

Background for Teachers

The background for teachers section includes information on the content area covered in the lesson, as well as some information on the lesson sequence and *NGSS* connections. The content goes beyond what students need to know; it is meant to give you a quick brushup on topics that you may not have taught or learned about in some time. The section also provides information on the societal issue that is addressed and in some cases includes tips for managing the lesson in your classroom.

Additional Resources

Additional resources include books, websites, and other media for supporting your teaching. These are not required for the lesson, but they can help you extend the lessons and gain confidence in the content.

Misconception Alert

At times in the book, you will see a Misconception Alert box that contains descriptions of common misconceptions that students might have about certain relevant content. Science is fraught with misconceptions for many reasons, including the fact that sometimes science may seem counterintuitive and counter to preconceived notions. For example, while it might seem to make sense to students that heavy objects always sink in water or that the Sun moves up and down in the sky each day, these examples demonstrate that sometimes the explanations students develop about phenomena aren't accurate, even though they may seem sensible based on their initial (but often limited) observations. In addition, at times our common language can convey the wrong information. For example, neither inchworms nor mealworms are actually worms. I have tried to provide tips on addressing the misconceptions in several lessons. For example, Lesson 6, Bee-ing There for Bees, includes a discussion about the misconception that bees visit flowers for the purpose of pollinating them, rather than to collect nectar and pollen for their own (and their hive's) benefit. This is understandable, because children are told from an early age that bees are important because they are pollinators, so confusion arises over bees being beneficial versus bees *intending* to be beneficial. This and several other misconceptions are addressed in this book. For extensive resources on identifying and addressing misconceptions in your classroom, I strongly suggest you take a look at the outstanding *Uncovering Student Ideas* series from NSTA Press.

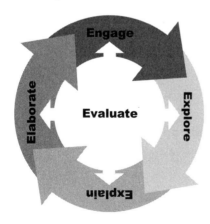

FIGURE 4.2.
The 5E Cycle

5E Lesson Plan

The 5E lesson plans follow a cycle of Engage, Explore, Explain, Elaborate, and Evaluate, as described on pages 9–10. However, it should be noted that Evaluate can come at any point in the lesson, as it is critical to have ongoing formative assessment (to help you determine the next steps in teaching) and summative assessment (to help you determine whether students have met the learning objectives). Therefore, Evaluate sections often appear throughout the lessons, or lessons may include suggestions on how informal evaluation may be conducted at particular points. Figure 4.2 illustrates this cycle.

Going Deeper

The Going Deeper sections suggest extensions that will allow you to continue the learning in your classroom. These extensions involve further research, family involvement, and actions in the community.

A Note About Assessment, Including High-Stakes Testing

It's Still Debatable! uses a variety of assessment types and formats. Formative assessments, which are meant to help teachers gauge student understanding and guide ongoing instruction, are embedded throughout the book and take various forms, including anticipation guides, class discussions, graphic organizers that elicit prior and developing knowledge, yes/no argument lines, observations during hands-on investigations, and so on. Summative assessments, on the other hand, are meant to provide students with opportunities to show what they know and give teachers opportunities to objectively measure student understanding of the lesson objectives. In this book, summative assessments also take many forms, including debates, written products such as persuasive letters and informational pamphlets, performance projects such as commercials or class presentations, and so on. These assessments typically include rubrics to guide teachers in assessing student work in an equitable manner and may also be available to students to support their understanding of expectations and self-assessment strategies.

However, it should be said that *It's Still Debatable!* was not specifically written as a preparation resource for high-stakes testing. While the activities and assessments reinforce key concepts and practices commonly assessed on such tests, the assessment formats in this book, for the most part, do not replicate those of elementary high-stakes tests. With that caveat in mind, there are still many ways for you to maximize the benefit of this book for your students' testing success:

1. Reinforce key vocabulary words throughout the day. Try to work science vocabulary into all of your daily routines including morning meetings/greetings, closing circles, read-alouds, snack and recess times, and so on. For example, if you are teaching Lesson 2, which focuses on forces, you could begin the day with a morning message that reads, "Good morning! Yesterday we learned that forces are pushes or pulls. What are some examples of forces you saw on your way to school today?" You could allow students to write or draw pictures of their examples on a chart or whiteboard and discuss them in your morning meeting. Reinforcing this vocabulary on the playground and in PE would be helpful as well. Similarly, if you were teaching Lesson 3, which explores plants,

a fun morning greeting activity would be for you to create cards out of the Edible Plant Parts Sort chart on page 116 and distribute one card (either a plant part name, a picture, or a function) to each student. Then challenge students to find their plant partners! Reinforcing plant part vocabulary during lunch and incorporating plant books during read-aloud and choice times would reap tremendous benefit as well. Any opportunities you can find to creatively reinforce vocabulary is helpful in demystifying language and reducing anxiety that can feel overwhelming for students during testing periods.

2. Reinforce science concepts when you teach other subject areas. *It's Still Debatable!* lessons integrate ELA, math, and social studies quite seamlessly. That means that you are promoting content and skills from these other subject areas every time you implement the lessons. But this cross-pollination works in the other direction as well; you can reinforce many of the science concepts that students learn about in these lessons during those subject times as well. For example, if you are working on measuring length or distance in math, why not connect it to building and testing puff mobiles in Lesson 13? Or perhaps you are teaching area or perimeter. In this case, you can provide context with Lesson 9, in which students calculate the areas or perimeters of Bullfrog Town and its various structures. If persuasive writing is your focus in ELA, think about using topics such as whether football is too dangerous for children (Lesson 8), whether we need a space program (Lesson 11), or whether distracted walking should be against the law (Lesson 14) as the context for students' writing. By thinking this way, you will find that you suddenly have more time for science, and your students will reap the benefits from extended opportunities for science and greater integration of subject areas, both of which are quite helpful for standardized test taking.

3. Challenge students to apply learning to novel situations. A key goal of science education is to prepare students to apply science in their everyday lives and in society. Of course, everyday life and society are always changing, so we want to ensure that children can generalize what they learn and apply it to new situations. A common challenge on standardized tests is that students learn science content in a particular context in class; when confronted with a different context on tests, they don't even recognize that the content is familiar. Giving students opportunities to see or at least discuss different applications of science learning objectives is extremely helpful. For example, Lesson 5 is about developing fair

tests that evaluate how well certain materials work for their intended purpose. In this context, students are evaluating paper towels, cloths, and sponges for properties such as absorbency, wet strength, environmental impact, and so on. The same objectives could be applied to any consumer product you might wish to have students test (e.g., shampoos, backpacks, toys). Challenge students to select other products that they'd like to test, emphasizing the relationship between properties and functions as well as the trade-offs that are inherent to any design (which is a critical science and engineering practice). Emphasizing the considerations for designing fair tests is critical. If time is short, you can simply pose questions/problems to students such as, "I got so wet coming into school today. My umbrella just didn't do a good job. What qualities do you think go into a good umbrella? How could I test them?" This type of conversation models scientific thinking aloud for your students and extends the learning beyond the initial lesson context.

4. Pay extra attention to reading and interpreting charts, graphs, and maps. A critical practice that is tested on standardized assessments is reading and interpreting data presented in different formats. In several *It's Still Debatable!* lessons, students have an opportunity to create, read, and/or interpret charts (Lesson 13), graphs (Lesson 2), and maps (Lesson 7). Several of the Going Deeper sections also emphasize this skill because it is so critical for academic success. Taking class opinion surveys and creating bar graphs with sticky notes, for example, like the one done in Lesson 2, is a wonderful practice that you can use regularly during morning meetings or closing circles. Ongoing monitoring of weather (e.g., the wind measurements in Lesson 7), plant growth (e.g., the seedling measurements in Lesson 3), or wildlife sightings (e.g., the bee observations in Lesson 6), and having students record observations in different formats, helps them to become comfortable with the many ways that scientists record and interpret data. This, in turn, prepares students for such items on standardized tests.

5. Read (and reread) nonfiction passages with your students. Although standardized science tests are designed to assess science learning, they are also in many ways reading tests—if students have difficulty reading the information provided, they are less likely to be able to show what they know about science. It is, therefore, critical for you to provide students with many opportunities to read scientific language and organize their thinking about what they have read. Many of the lessons in *It's Still*

Debatable! include nonfiction readings and ask students to use graphic organizers such as T-charts, four square writing templates, RAN charts, and others to identify the big ideas and organize key concepts. Following up with writing about what they have learned, as is the task in many of the summative assessments in the book, promotes students' mastery of science concepts and advances interdisciplinary literacy—all of which can help your students to succeed on high-stakes testing and beyond.

References

Herczog, M. M. 2013. The links between the *C3* framework and the NCSS national curriculum standards for social studies. *Social Education* 77 (6): 331–333.

Lederman, N. G. 1992. Students' and teachers' conceptions of the nature of science: A review of the research. *Journal of Research in Science Teaching* 29 (4): 331–359.

Lederman, N. G., F. Abd-El-Khalick, R. L. Bell, and R. S. Schwartz. 2002. Views of nature of science questionnaire: Toward valid and meaningful assessment of learners' conceptions of nature of science. *Journal of Research in Science Teaching* 39 (6): 497–521.

National Council for the Social Studies (NCSS). 2010. *National curriculum standards for social studies: A framework for teaching, learning, and assessment*. Silver Spring, MD: NCSS.

National Council for the Social Studies (NCSS). 2013. *The college, career, and civic life (C3) framework for social studies state standards: Guidance for enhancing the rigor of K–12 civics, economics, geography, and history*. Silver Spring, MD: NCSS. *www.socialstudies.org/C3*.

NGSS Lead States. 2013. *Next Generation Science Standards: For states, by states*. Washington, DC: National Academies Press. *www.nextgenscience.org*.

Willard, T. 2014. *The NSTA quick-reference guide to the NGSS, elementary school*. Arlington, VA: NSTA Press.

Unit V
Lesson Plans

Lesson Plans

Lesson 1

Leave It to Beavers

Should We Relocate the Beaver Dam?

Suggested Grade Levels

K–2

Driving Questions

- How do animals change their environment to meet their needs?
- How can we use drawings and models to communicate our ideas for solving problems?

Lesson Overview

Students examine the manner in which beavers change their environment to survive by building dams and lodges. After learning about beaver anatomy and behavior through a variety of media, students engage in an engineering design challenge in which they work together as a beaver family to build and test a dam that can raise water to a certain level using limited resources. Students then engage in a debate about whether a beaver dam that is causing flooding in a town should be left alone or moved. By engaging in the controversial question, students apply their knowledge of beaver anatomy, behavior, and survival needs, as well as their engineering experiences, to support their arguments.

Connecting to the *NGSS*

(See full alignment in Table A.1 on p. 502.)

- ESS2.E: Biogeology
    - Plants and animals can change their environment. (K-ESS2-2)

- ETS1.B: Developing Possible Solutions
  - Designs can be conveyed through sketches, drawings, or physical models. These representations are useful in communicating ideas for a problem's solutions to other people. (K-2-ETS1-2)

Societal Issues

Human-Animal Conflicts, Animal Rights, Environmental Stewardship

Nature of Science

- Science Models, Laws, Mechanisms, and Theories Explain Natural Phenomena
  - Scientists use drawings, sketches, and models as a way to communicate ideas.
  - Scientists search for cause-and-effect relationships to explain natural events.
- Science Addresses Questions About the Natural and Material World
  - Scientists study the natural and material world.

CCSS Connections

- English Language Arts
  - R.K.1. With prompting and support, ask and answer questions about key details in a text.
  - W.K.2. Use a combination of drawing, dictating, and writing to compose informative/explanatory texts in which they name what they are writing about and supply some information about the topic.

NCSS Connections

- Theme 8: Science, Technology, and Society
  - Research and evaluate various scientific and technological proposals for addressing real-life issues and problems.
- Theme 10: Civic Ideals and Practices

- Evaluate positions about an issue based on the evidence and arguments provided, and describe the pros, cons, and consequences of holding a specific position.

C3 Framework

- Dimension 4: Communicating Conclusions
  - D4.1.K-2. Construct an argument with reasons.
  - D4.3.K-2. Present a summary of an argument using print, oral, and digital technologies.
- Dimension 4: Critiquing Conclusions
  - D4.4.K-2. Ask and answer questions about arguments.

UDL Toolkit

Multiple Means of Engagement	Multiple Means of Representation	Multiple Means of Action and Expression
Students are engaged using a riddle, observation of a beaver photo, and a video.	Information is presented orally and through writing and pictures.	Students express their ideas through talking, writing, and drawing during dam-building activity.
The dam-building activity allows for additional challenges (holding water for longer times) and choices to maintain motivation.	Graphic organizers like Venn diagrams (dams versus lodges) and sequencing organizers like the yearly cycle activity help students organize information.	Students can assume different roles to accomplish tasks of dam building and testing, and they are given assessment options to show what they know.

Suggested Schedule and Sequence

- Day 1: **Engage** with Who Am I? riddle, student observations written on Beaver OWL Chart, and Animal Planet's "Fooled by Nature—Beaver Dams" video
- Day 2: **Explore** with Build a Beaver Dam Engineering Design Challenge

Lesson 1: Leave It to Beavers

- Day 3: **Explain** with *The Beavers' Busy Year* read-aloud and create Busy Beavers' Yearly Cycle
- Day 4: **Elaborate** with Riverville Town Hall Meeting, and **Evaluate** with My Beaver Proposal and Gallery Walk

Materials

(per team of 4)

- Long plastic planter box (available at hardware or home improvement stores)
- Craft sticks
- Pipe cleaners
- Toothpicks
- Small amount of clay
- Sand (optional)
- Yarn or string
- Tape
- Measuring cup
- Water source
- Timer or clock (optional)
- Block or book to place under one side of planter box
- Pencils, crayons, and markers
- Set of Build a Beaver Dam Team Role Cards (p. 71), precut

(per student)

- Paper plate
- Safety glasses or goggles and a nonlatex apron

Student Handouts

- Beaver OWL Chart (optional)

- Build a Beaver Dam
- My Beaver Proposal

Safety Notes

1. All students must wear safety glasses or goggles and nonlatex aprons during the setup, hands-on, and takedown phases of the activity.

2. Use caution when working with sharp tools or materials to avoid cutting or puncturing skin.

3. Immediately wipe up any spilled water on the floor to avoid a slip-and-fall hazard.

4. Wash hands with soap and water after completing this activity.

Media

Website

- PBS infographic on beavers
  www.pbs.org/wnet/nature/leave-it-to-beavers-infographic-beavers-101/8868

Video

- Animal Planet's "Fooled by Nature—Beaver Dams"
  www.youtube.com/watch?v=Na2HYq11yuM

Book

- *The Beavers' Busy Year,* by Mary Holland

Background for Teachers

Beavers are incredibly fascinating animals that change their environment to meet their needs. Beavers are mammals, which means that they are warm-blooded, have fur, and provide milk for their babies, which are known as kits. The kits are live-born (as opposed to eggs) and nurse from their mothers. About 3 feet long and 60 pounds when fully grown, beavers are the largest rodents in North America. Rodents, which include squirrels, mice, hamsters, and guinea pigs, are mammals that have large front teeth (incisors) used for gnawing. These teeth never stop growing, so it is important for rodents to continually gnaw on food to keep their incisors

Lesson 1: Leave It to Beavers

from overgrowing. Beavers are herbivores that feed on leaves, bark, twigs, roots, and aquatic plants. Their life span is approximately 24 years. They are adapted for life on both land and water. Their adaptations or special features include thick waterproof fur, large webbed back feet, a large paddle tail that acts like a rudder for steering, and transparent eyelids that help them see underwater. They can stay underwater for 15 minutes before coming up to the surface to breathe.

Beavers are considered to be among the best engineers in the animal kingdom because they create ponds by building dams out of logs, twigs, rocks, mud, and leaves. A beaver dam slows the flow of water enough to create a pond behind it. The beavers live in a lodge surrounded by the pond, which protects them from large predators like coyotes and wolves. The lodges can only be entered through underwater openings. The deep water also allows beavers to store food in cool areas underneath their lodges.

To build dams and lodges, a single beaver can cut down nearly 200 trees per year! This can cause problems for landowners who do not want their trees cut down or have their land flooded by the creation of a pond, and commercial forest owners claim that beavers cost them millions of dollars in lost timber. Beavers also can build dams across culverts, structures that allow water to flow under roads, railroads, or trails, which can result in flooding over surrounding areas. When beaver dams cause floods or when beavers cut down trees, people often call for trapping or killing of the beavers and destruction of their dams and lodges. But supporters of beavers point to the diversity of wildlife that live in the wetlands that beavers produce. Beavers are considered keystone species because so many plants and animals depend on the ponds that they create. Beaver dams also contribute to cleaner water and help prevent flooding after storms by slowing water down. Some ways that people have attempted to prevent damage by beavers include the use of tree guards around tree trunks and "beaver deceivers" to keep them away from culverts.

Although you may not have beaver issues in your area, urban, suburban, and rural areas all have issues where wildlife and human interests seem at odds. Deer foraging in yards, raccoons rooting through garbage cans, and even coyotes wandering through cities create situations where people make decisions about how to best manage or coexist with wildlife. These human-animal conflicts are increasingly common as human populations grow and natural habitats are reduced. This is a global concern, as many species around the world, such as Asian leopards, African and Asian elephants, and bears and wolves in North America and Europe, are often at the center of controversies. The beaver controversy is a first step in introducing this issue.

Lesson Plans

In this lesson, students learn about beaver anatomy and behavior through various media to prepare them for designing and testing a beaver dam. This activity addresses the *NGSS* standards for kindergarten, which expect students to be able to articulate the manner in which animals change the environment to suit their purposes. The activity also focuses on students' ability to convey their ideas through drawings and the physical model of the dam, to assess the engineering standard that emphasizes students' ability to describe how the shape and structure of an object (in this case, the dam) can make it suitable for solving a problem (stopping the water flow).

Additional Resources

Tree for All's "Living With Beavers"

- *www.jointreeforall.org/beavers*

"Beavers: Wetlands & Wildlife"

- *www.beaversww.org*

World Wildlife Fund's "Human-Wildlife Conflict"

- *http://wwf.panda.org/our_work/wildlife/problems/human_animal_conflict*

National Geographic Kids' "Awesome Animals: American Beaver" (an excellent extended video)

- *https://kids.nationalgeographic.com/videos?videoGuid=american-beaver-7-kids*

5E Lesson Plan

Engage: Who Am I? Riddle, Beaver OWL Chart, and "Fooled by Nature—Beaver Dams" Video

1. Begin by inviting students to help solve a riddle: "Who am I?" The following list of clues should be read aloud and exposed one at a time, on a projector or written on paper. Allow students to guess the answer to the riddle, pausing after each clue to allow students time to think and process answers.

   Who Am I?

   - I am an animal.
   - I have fur.

- I am an herbivore (I eat plants).
- My babies are called kits.
- I weigh around 60 pounds when I'm fully grown.
- I am a good swimmer.
- I am the largest rodent in North America.
- I have big front teeth.
- I have a big paddle tail.
- I build dams.

2. When students have guessed that it's a beaver, take a moment to review the facts that they learned in the riddle. Ask, "Which facts were giveaways?" (multiple answers)

3. Show a picture or a projection of a beaver (a poster of one is available at *www.pbs.org/wnet/nature/leave-it-to-beavers-infographic-beavers-101/8868*). Ask, "What are some things you observe about the beaver?" Use think-pair-share to allow students to discuss with a partner before reporting to the class. After two to three minutes, ask pairs to share with the class. Record student observations on the Beaver OWL organizer. You can project the OWL chart that is provided on page 70, print it on large paper, or simply create your own on the board. You can also provide each student with a copy of the handout so that students can record their own information.

4. Ask students, "What else do you wonder about beavers?" Record answers on the OWL chart.

5. Tell the class, "Let's see if we can find out any answers in the video." Watch Animal Planet's "Fooled by Nature—Beaver Dams" video (*www.youtube.com/watch?v=Na2HYq11yuM*), then complete the "L" column with students.

Explore: Build a Beaver Dam Engineering Design Challenge

Advance Preparation: Before students arrive, use a permanent marker to mark the waterlines on one end of each of the plastic "river" boxes. Draw a small line approximately 6 inches in from the end and 2 inches above the bottom of the box.

1. Show students the OWL chart and review their Observations, Wonderings, and things that they have already Learned. Ask, "How do dams work?" (responses will vary)

2. Explain to students that they will have the opportunity to be the beavers today as they design and build a dam on a simulated river. Distribute copies of the Build a Beaver Dam student handout (p. 72).

3. Explain that each beaver family (team of four students) will work together to design, build, and test a dam that can hold enough water to reach the waterline mark for 10 seconds.

4. Show students the materials that they will have available for building (e.g., craft sticks, clay, yarn, tape), and explain that they must first plan their designs by discussing ideas with their beaver families and drawing the plans for their dams. Students should be encouraged to label their drawings (Figure 5.1). You may wish to create a word bank of available materials on the board to support students' writing.

5. Once beaver teams have created their designs, they can collaboratively build their dams (Figure 5.2). Dams should be built adjacent to the waterline marks, which will leave approximately 6 inches of open area at the base of the river. (This will allow for collection of any water that gets through the dam.)

6. When teams have completed their dams, have students begin testing. Remind students that the challenge is to have the water held back behind the dam reach the waterline mark and stay there for 10 seconds. Students may add enough water to the river to reach the waterline mark, but once it's reached, they should stop adding water and begin timing or counting slowly to 10. Randomly assign cooperative team roles using the role cards on page 71.

7. Provide beavers teams with time to repair their dams if they do not hold back the water for 10 seconds. Encourage students to record their new designs on their data sheets. If teams are successful on their first attempt, challenge them to reach a higher water mark for 30 seconds or to see if their dam can hold for 1 minute.

8. Allow beaver families the opportunity to share their findings with each other. Ask students:

    - "How did the shape of the dam affect its success?" (it was tall enough to keep rising water back, being a little curved helped it stop water from going around the sides, it was wide enough to reach the sides)

    - "How did the materials you used affect the success of the dam?" (the materials were packed tightly, used clay for waterproofing)

- "How is your model similar to a real river?" (water is flowing downhill, it is long)

- "How is your model different from a real river?" (it is made of plastic, it is small, we added the water, it doesn't change shape)

- "In a real river, what would form behind the dam?" (a pond)

- "How are beavers like engineers?" (beavers plan and use materials to solve a problem, they change their environment)

FIGURE 5.1.
Build a Beaver Dam Data Sheet

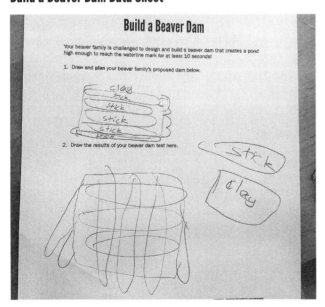

FIGURE 5.2.
Students Building a Beaver Dam

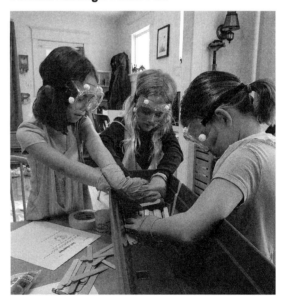

Explain: *The Beavers' Busy Year* Read-Aloud and Busy Beavers' Yearly Cycle

1. Show the cover of the book *The Beavers' Busy Year*, by Mary Holland. Ask, "What types of things do you think keep beavers busy during the year?" (accept different answers)

2. Read the book aloud. Encourage students to show "connection" hand signals if they hear a fact that they can connect to their study so far.

3. After the read-aloud, ask:

   - "How are dams and lodges alike? How are they different?" Develop a Venn diagram from student responses to visually depict these similarities and differences (see Figure 5.3).

   - "What are some of the different activities that beavers do during the year?" (looking for food, making dams and lodges, having babies, sleeping)

   - "How do these different activities help beavers survive?" (answers will vary)

FIGURE 5.3.

Sample Venn Diagram Comparing Beaver Dams and Lodges

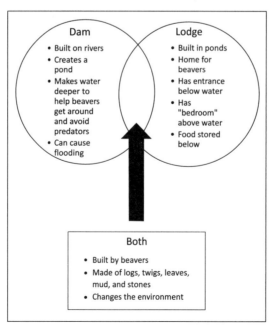

4. Now, have each student create a Busy Beavers' Yearly Cycle on a paper plate. Instruct students to draw a large plus sign (+) on the paper plate to divide it into four sections. Have students write the seasons on the four sections, as shown in Figure 5.4. Then, instruct students to draw what the beavers do during each season of the year.

FIGURE 5.4.

Busy Beavers' Yearly Cycle

Elaborate: Riverville Town Hall Meeting

1. Inform students that they are now citizens of a town called Riverville that is experiencing a big problem. It seems that a beaver family has built a dam on the river that runs through the middle of town, creating a pond that sometimes overflows onto roads and into some backyards. The town leaders, who are members of Riverville's Town Council, have called a meeting to find out what the town's citizens think should be done about the beavers. Should they be allowed to stay or should they be moved? Several different groups are attending the meeting and have different opinions about the beavers:

   - River Road Homeowners. The homeowners are very unhappy about the flooding in their yards and want the beavers trapped and removed.

   - Riverville Nature Club. The club members want the beavers to stay because the new pond that has been created attracts birds, deer, and other wildlife.

   - Lovely Lumber Business Owners. The business owners want the beavers gone because the beavers are cutting down the lumber company's trees to build the dams.

   - Riverville Elementary School Students. The students want the beavers to stay because they think that watching the beavers is fun and educational.

   - Town Council. The council members are neutral and want to listen to the different arguments that the various groups make before coming to a decision.

2. Randomly assign Town Hall Meeting roles from the above list to students. Inform them that for now, they are going to be taking the position of the

group they received, but they will have a chance to share their own personal positions on this issue later.

3. Have students representing the same group come together at tables. Explain the Town Hall Meeting rules:

   - Groups will spend five minutes working together to discuss their positions.

   - Each group will then have one minute to explain their position to the Town Council. They must include at least one fact about how beavers change the environment.

   - The Town Council members can ask questions. After the four groups have presented and the Town Council has completed the questioning, the council members will take a few minutes to deliberate and then announce their decision and reasons.

   - Everyone must listen carefully to the other town members, and all students must be respectful in their words and actions.

4. After the Town Council announces its decision, ask:

   - "Were you surprised by the Town Council's decision?" (responses will vary)

   - "How did you feel when another group shared a position that was the same as yours?" (students may express different feelings, including happiness and relief)

   - "How did you feel when another group shared a position that was different from yours?" (students may express different feelings, including anger and frustration)

   - "Why are meetings like this important for communities?" (they allow different ideas and information to be shared)

Evaluate: My Beaver Proposal and Gallery Walk

1. Inform students that they now have the opportunity to share their own opinions on beavers. Ask, "Should the beaver family be allowed to stay in Riverville? Should the dam be removed? Are there other solutions?"

2. Have students create a proposal in the form of a picture, poem, comic strip, or story using the My Beaver Proposal handout (p. 73). Each proposal must include *at least* the following:

    - One fact about how beavers' bodies are adapted for survival
    - One fact about how beavers change their environment to survive
    - The student's own opinion about whether the beavers should be moved to another town or should remain in Riverville

3. Allow students to share their proposals through a Gallery Walk. Have half of the class stand in a large circle around the room, holding their proposals, while the other students walk around the circle to visit the students in the first group as they present. Then have the two groups switch places.

4. Use the rubric on page 74 to assess each student's proposal.

Going Deeper

- Contact local park and wildlife management organizations to find out about human-wildlife conflict issues in your area. Involving students in these issues, through either readings or interviews with park management and local residents, can give them insights into how they can become involved.

- Students can explore historical maps of your region to examine the changes over time to natural habitats.

- Students can interview their families and neighbors about interactions they may have had with "nuisance" wildlife.

Name: _____ Date: _____

Beaver OWL Chart

Observe	Wonder	Learn

Build a Beaver Dam

Team Role Cards

Beaver Dam Builder:_____   Date: _____

Build a Beaver Dam

Your beaver family is challenged to design and build a beaver dam that creates a pond high enough to reach the waterline mark for at least 10 seconds!

1. Draw and **plan** your beaver family's proposed dam below.

2. Draw the results of your beaver dam test here.

Name: _____ Date: _____

My Beaver Proposal

My Beaver Proposal Is a (circle one):

    Picture   Poem   Comic Strip   Story

          By: _____

Rubric for My Beaver Proposal

Criteria	Not Yet (1 pt.)	Emerging (2 pts.)	Secure (3 pts.)
Describes how beavers' bodies are adapted for survival	Not included or unclear.	Identifies body part but does not clearly describe connection between form and function.	Identifies relevant body part and makes clear connection between form and function.
Describes how beavers change their environment to survive	Not included or unclear.	Identifies dams, lodges, or treefalls but does not clearly describe changes in environment.	Identifies dams, lodges, or treefalls as ways that beavers change their environment and makes connection to survival.
States opinion on beaver plan	Not included or unclear.	Provides an opinion but is unclear or not backed by any rationale.	Provides a thoughtful opinion that is backed by a clear rationale.

Total: ___/ 9 pts.

Lesson 2: Swingy Thingy

Lesson 2

Swingy Thingy

What Makes a Great Playground?

Suggested Grade Levels

K–2

Driving Question

- How do pushes and pulls affect the speed and direction of objects?

Lesson Overview

Students explore the way pushes and pulls affect the speed and direction of objects using both playground swings and model swings that they build and test. After a read-aloud about a girl's dream playground, students vote to determine the features of a class dream playground and collaboratively develop blueprints, a class model, and a list of playground rules. By engaging in collaborative planning and rule making, students apply their knowledge of forces to support their arguments for safe and enjoyable playground design while modeling civic engagement.

Connecting to the *NGSS*

(See full alignment in Table A.2 on p. 503.)

- PS2.A: Forces and Motion
    - Pushes and pulls can have different strengths and directions. (K-PS2-2)
    - Pushing or pulling on an object can change the speed or direction of its motion and can start or stop it. (K-PS2-2)
- ETS1.A: Defining and Delimiting Engineering Problems

- A situation that people want to change or create can be approached as a problem to be solved through engineering. Such problems may have many acceptable solutions. (K-2-ETS1-1)

Societal Issues

Land Use, Accessibility, Fair Negotiations

Nature of Science

- Science Models, Laws, Mechanisms, and Theories Explain Natural Phenomena
    - Scientists use drawings, sketches, and models as a way to communicate ideas.
    - Scientists search for cause-and-effect relationships to explain natural events.
- Science Addresses Questions About the Natural and Material World
    - Scientists study the natural and material world.

CCSS Connections

- English Language Arts
    - R.K.1. With prompting and support, ask and answer questions about key details in a text.
    - W.K.2. Use a combination of drawing, dictating, and writing to compose informative/explanatory texts in which they name what they are writing about and supply some information about the topic.
- Mathematics
    - MP.2. Reason abstractly and quantitatively. (K-PS2-1)
    - K.MD.A.1. Describe measurable attributes of objects, such as length or weight. Describe several measurable attributes of a single object. (K-PS2-1)

- K.MD.A.2. Directly compare two objects with a measurable attribute in common, to see which object has "more of"/"less of" the attribute, and describe the difference. (K-PS2-1)

NCSS Connections

- Theme 3: Individual Development and Identity
  - Explore factors that contribute to personal identity, such as physical attributes, gender, race, and culture.
- Theme 6: Power, Authority, and Governance
  - Examine issues involving the rights and responsibilities of individuals and groups in relation to the broader society.
- Theme 10: Civic Ideals and Practices
  - Ask and find answers to questions about how to plan for action with others to improve life in the school, community, and beyond.

C3 Framework

- Dimension 4: Taking Informed Action
  - D4.8.K-2. Use listening, consensus-building, and voting procedures to decide on and take action in their classrooms.

UDL Toolkit

Multiple Means of Engagement	Multiple Means of Representation	Multiple Means of Action and Expression
Students are introduced to the topic through a charades game, KLEW chart, and an outdoor activity.	Playground Charades involves auditory, visual, and kinesthetic presentation of information.	Students choose the type of playground apparatus they'd like to build.
Students are provided with a series of challenges in their STEM engineering activity that can be extended.	Information on forces is presented through pictures, words, and hands-on experiences.	Students express their individuality by drawing their ideal playground.
Students engage in small groups, whole-class activities, and individually to maintain interest and provide a variety of experiences.	Use of a KLEW chart, playground equipment bar graph, and playground blueprint map presents information in an interactive and organized way.	Students express their learning through writing, talking, creating models, and drawing.

Suggested Schedule and Sequence

- Day 1: **Engage** with Playground Charades and Swingy KLEW Chart and **Explore** and **Evaluate** with Swingy Science
- Day 2: **Explain** with *Move It! Motion, Forces, and You* read-aloud and **Explore, Explain,** and **Evaluate** with Model Swing STEM Challenge
- Day 3: **Elaborate** with *My Dream Playground* read-aloud and Our Dream Playground graph
- Day 4: **Elaborate** with Build a Class Playground Model
- Day 5: **Evaluate** with My Dream Playground scene

Materials

For Swingy Science (outdoor activity)

(per team of 2–3)

- Playground swings
- Clipboards (optional)
- Pencil

For Model Swing STEM Challenge

(per team of 2–3)

- 7 paper straws (uncut)
- 3 paper straws cut in half (6 half-sized straws)
- 5 pipe cleaners cut in half (10 half-sized pipe cleaners that will serve as connectors)
- 2 cupcake liners
- Hole punch
- Scissors
- Small toys to place in the seats
- Straw swing set instructions from the Craft Train: *www.thecrafttrain.com/straw-swing-set*

Lesson 2: Swingy Thingy

For Class Playground Model Equipment (e.g., swings, slides, monkey bars, merry-go-rounds)

(per team of 2–3)

- Craft supplies such as craft sticks, pipe cleaners, string or yarn
- Cardboard
- Paper straws
- Aluminum foil
- Marbles
- Scissors
- Paper towel or toilet paper tubes
- Tape
- Chart paper
- Sticky notes

(per student)

- Safety glasses or goggles

Student Handouts

- Swingy Science
- Model Swing STEM Challenge
- My Dream Playground

Safety Notes

1. All students must wear safety glasses or goggles during the setup, hands-on, and takedown phases of the activity.

2. Use caution when working with sharp tools or materials to avoid cutting or puncturing skin.

3. Immediately pick up any items dropped on the floor to avoid a slip-and-fall hazard.

4. Wash hands with soap and water after completing this activity.

Media

Books

- *Move It! Motion, Forces, and You*, by Adrienne Mason
- *My Dream Playground*, by Kate M. Becker

Background for Teachers

Playgrounds are not only great places for children to exercise and socialize, but they are also excellent laboratories for studying forces and motion. Swings, slides, seesaws, and merry-go-rounds all demonstrate *motion*, which is the movement of something from one place to another. Motion requires *forces*, or *pushes and pulls*, to occur. Forces can change the speed or direction of an object: more force (bigger pushes or pulls) leads to faster movement, while applying force from a different direction can change the direction of an object. Children can observe these basic principles by pushing and pulling a swing. By applying more force to the swing, the swing moves back and forth for a longer period of time. If the person on the swing pumps his or her legs, this applies even more force to keep the swing moving. Of course, if the person on the swing gets tired and stops pumping his or her legs, the swing will eventually stop. Why? Because there is *friction* between the swing and the air (and even the point where the swing's chain is attached to the swing frame). Friction is a force that slows things down when they rub together. Another force at work when observing a swing is *gravity*, which is a force that pulls objects toward the center of Earth. When someone pushes you on a swing, you go up, but then you come back down because of gravity. Another important concept students can observe with swings is that they can push a swing sideways to change the direction of the swinging motion. They can even twist swings around to make them spin!

    Although this lesson focuses on swings, you can apply many of these same concepts to other playground equipment. For example, gravity is the reason that you go down a slide. Slides have polished surfaces to reduce the friction so that children can slide down quickly. Monkey bars provide opportunities for children to pump their feet in order to swing, pull themselves up by exerting enough force to overcome gravity, or drop down to the ground, again thanks to gravity. Seesaws are a bit more complicated but also demonstrate these concepts well. Seesaws show how forces can be balanced if the riders are roughly the same weight and seated the same distance from the center pivot point, called a *fulcrum*. In this situation, gravity is pulling the same amount on each side. But if one rider pushes down on the ground with his or her feet (exerting a force), this rider can disrupt the equilibrium, raising himself or herself and lowering the other rider. This position doesn't last long, though, because gravity soon pulls the rider back down toward Earth. Children intuitively figure out that if one person is significantly heavier than the other, the heavier rider can easily keep the lighter rider's side of the seesaw up in the air. However, moving the heavier person closer toward the fulcrum and the lighter person away from the fulcrum allows the lighter rider to keep the heavier rider up in the air because of a turning force known as *torque*. One final concept explains why none of the equipment on a playground moves unless someone (or a very big wind) moves it: *Inertia* is the tendency of things to stay in one place unless a force acts on them.

In this lesson, students apply forces (pushes and pulls) to make swings move. They test how the strength of the force they exert makes swings move for a longer time and observe that they can change the direction of swings by pushing or pulling from a different direction. They experiment with swings on the playground and model swings that they build in a STEM challenge. The STEM challenge has the added criterion of working with given materials to develop a successful design that can swing in concert with a friend's swing; this requires students to adjust their swings and their pushes and pulls to meet the challenge. Finally, the development of a class playground model requires students to apply their knowledge of forces and motion to create working models of playground equipment. This demands thoughtful planning, collaboration, discussions about accessibility, and negotiation to create a whole-class playground display, along with playground rules, that models real-world community projects.

Additional Resources

Book on forces and motion for teachers

- Robertson, W. C. 2002. *Force and motion: Stop faking it! Finally understanding science so you can teach it.* Arlington, VA: NSTA Press.

Article on teaching playground science to students with visual impairments

- Fast, D., and T. Wild. 2018. Traveling with science. *Science and Children* 55 (5): 54–59.

Article on designing playground equipment using 3-D printing

- Wendt, S., and J. Wendt. 2015. Printing the playground. *Science and Children* 52 (5): 43–47.

5E Lesson Plan

Engage: Playground Charades and Swingy KLEW Chart

1. Begin by asking students to imagine that they are on a playground, playing on their favorite playground equipment. Explain to students that instead of telling the class what their favorite playground equipment is, they are going to act out their activity on the equipment, allowing the class to guess, like a game of charades. Call on four or five volunteers to come up one at a time and act out their favorite playground activity. As students correctly guess, write or draw the playground equipment on the board (e.g., swing, slide, seesaw, monkey bars). When all the favorite activities on playground equipment have been named, explain to students that they're going to focus on swings today.

2. Draw a KLEW chart like the one in Figure 3.2 (see p. 17) and ask, "What do we think we Know about swings?" List their responses under the "K." The idea of using pushes to make swings move will likely come up, but if students don't volunteer it, prompt them with the question "How do swings move?" Once all answers have been added to the chart, inform students that they will be building on this knowledge by going out to the playground to explore swings!

Explore: Swingy Science

1. Distribute the Swingy Science handout (p. 92) to students, along with clipboards and pencils. Assign partners (or groups of three if needed) and head out to the playground.

2. Explain to students that they are going to test the number of times their partners swing with (1) a pull, (2) a pull and a push, and (3) a pull, a push, and pumping their legs. Model the investigation for students with a student volunteer who is seated on a swing. Explain that Partner 1 (the student) sits on the swing without pumping legs while Partner 2 gives one pull. They should count the number of times the swing comes back to Partner 2 before it stops. Show students where on the sheet they should enter their data. Also demonstrate a pull and a push, then a pull, a push, and leg pumping. (Demonstrate the start of each, but don't count the number of swings.) After they have counted using each method, partners should switch roles.

3. Allow students to work on this activity for several minutes. Students will quickly notice that when their partners are pumping their legs, the swings don't stop. Have students record that in their own words (e.g., does not stop, goes forever, keeps swinging).

Lesson 2: Swingy Thingy

4. When students have completed the handout, allow them a few extra minutes to explore other ways of moving the swing, such as changing directions or stopping it with their feet.

5. When students are done, ask:

   - "How does pushing and pulling affect the number of swings?" (the more pushes and pulls, the higher the number of swings)

   - "Why do you think this is so?" (a push and a pull is stronger than just a pull, so there is more motion)

   - "What happened when your partner pumped his or her legs?" (the swing didn't stop)

   - "Why do you think pumping keeps the swing moving?" (pumping keeps adding pushes and pulls; answers may vary and students will learn about forces in the next activity)

   - "How do you stop swinging?" (putting feet on the ground)

   - "Were you able to find out anything else about moving the swings?" (answers will vary, but students may mention changing directions, going diagonally, or spinning)

6. Revisit the KLEW chart and allow students to contribute what they have Learned in the "L" column. Students will say, for example, that pumping legs helps swings move longer and bigger pushes make them go higher. Introduce the "Evidence" ("E") column and ask, "What Evidence (or proof) do you have for what you've Learned?" Write answers such as "When we pumped our legs, the swing went longer."

Evaluate: Swingy Science

1. Evaluate student work by ensuring that they have recorded their data and correctly interpreted these data in the final question. Students should observe that they are able to change the motion of the swings by changing the direction of pushes and pulls and by adding leg pumping.

Explain: *Move It! Motion, Forces, and You* Read-Aloud

1. Read aloud *Move It! Motion, Forces, and You*, by Adrienne Mason, using the guided reading prompts listed below. (*Note:* Some pages in this book describe activities to demonstrate certain concepts. You can clip these pages in advance

Lesson Plans

so that they aren't part of your read-aloud, but you can always choose to do them at another time.)

- On pages 6–7, note that the picture shows children playing on a playground and asks students to identify where forces are being used. Ask students, "Do you see any connections between this page and our playground investigation?" (students should recognize that the pushes and pulls that move the swing are forces)

- Similarly, on pages 12–13, the book describes how big forces make things move farther. Ask students, "Do you see any connections between this page and our playground investigation?" (students should recognize that when they gave pushes and pulls, or pushes, pulls, and leg pumping, the swing kept moving)

- On pages 16–17, the book discusses how pushes and pulls can change directions. Ask students, "Do you see any connections between this page and our playground investigation?" (students should recognize that pushing moved the swing in one direction and pulling moved it in another; students may have also noticed that pushing sideways produced a sideways swinging pattern)

- On pages 18–19, the book discusses stopping motion. Ask, "Do you see any connections between this page and our playground investigation?" (students used their feet to stop the swinging motion, and this required force)

Explore and Explain: Model Swing STEM Challenge

1. Explain to students that they are going to have a chance to build and test their own model swings. Show your prebuilt model, demonstrating that it really swings. Ask students to identify any shapes they notice. (the base is a rectangle, the sides are triangles, the swings are rectangles)

2. Show students the materials that they will have available for building (paper straws, pipe cleaners, and cupcake liners). Demonstrate how the pipe cleaners (halves) can be used as connectors between straws by simply pushing a pipe cleaner into one end of a straw, leaving enough outside to add the other straw; then the pipe cleaner can be bent, forming a joint, so that the straws can go in different directions. You can also put two pipe cleaners into one end of a straw so that three straws can be joined.

Lesson 2: Swingy Thingy

3. Challenge students to build a swing that is able to hold a small figure or toy in the seat and swing back and forth at least four times with only a pull (only pulling the swing back and releasing it, not pushing). They will count exactly the same way they did on the playground, by counting each time the swing comes back to the starting side.

4. Allow students time to work on their models. *Tip:* If students are struggling, encourage them to create the base (a rectangle) first, and then use the short sides of the base to make triangles for the sides of the model. Then, place one straw across the top. Encourage students to help each other as they finish.

5. When students are finished building, it is time to test the swings. Students may need to make adjustments in how tightly the pipe cleaner is wrapped around the top bar. Encourage students to keep testing until they have a well-functioning swing that is able to accomplish the challenge.

6. Distribute the Model Swing STEM Challenge handout (p. 93), and have students complete the set of challenges on the sheet.

7. When students are done with the challenges, ask:

    - "How did you make your swing move?" (by pushing and pulling)
    - "How are pushes and pulls alike?" (they make things move)
    - "How are pushes and pulls different?" (they move things in a different direction; pushes move things away from us, and pulls move them toward us)
    - "How did you make your swing move the same as your friend's swing?" (start the swings at the same time from the same height)
    - "How did you make the swing move back and forth at least four times with only one pull?" (pull it back very far)

Evaluate: Model Swing STEM Challenge

1. Evaluate students' work on the STEM Challenge through observation of their building and testing, as well as their success with the challenges. Questions 4 and 5 on the STEM Challenge handout require testing and data collection (e.g., counting the number of swings), as well as possibly revising their designs to get the coordinated swinging with their friend's swing and to move back and forth with only one pull.

Elaborate: *My Dream Playground* Read-Aloud and Our Dream Playground Graph

1. Explain to students that you are going to read *My Dream Playground*, by Kate M. Becker, a true story about a girl who dreamed about playgrounds because she lived in a place where there weren't any. Note that although the girl's name isn't included in the book itself, the author's note in the back of the book tells us that the story is based on the experiences of a girl named Ashley who lived in Washington, DC. After reading, ask:

    - "Why do you think there weren't any playgrounds in Ashley's neighborhood? The book doesn't tell us so we have to infer." (answers will vary; perhaps the city she lived in didn't have much space or money for playgrounds)

    - "What were some of the things that Ashley wanted to have in her playground?" (slides, swings, monkey bars, trampolines)

    - "What advice did her mom give her?" (never stop dreaming)

    - "Who were some of the people who helped Ashley's playground dream come true?" (her brothers helped her plan, Darell managed the project, Mr. Sid brought sandwiches, Ms. Gonzalez brought the tent, Gregory played music, and hundreds of volunteers helped build)

    - "What lessons do you think the author might want us to learn from this book?" (never stop dreaming, much can be accomplished when people work together for a common goal, everyone can contribute in his or her community, children can make a difference)

2. Announce to students that they are going to design and build a model of a class dream playground. Using think-pair-share, ask students, "What are some questions we need to think about before we can build a model playground?" Give students a few moments to think quietly, then allow them to turn to a partner to discuss, and finally, have them share out. Elicit and prompt questions such as the following:

    - Where would our playground be?

    - How big will it be?

    - What kinds of equipment will it have?

    - Who would use it?

Lesson 2: Swingy Thingy

- Should it be *accessible* for children who use wheelchairs or walkers?
- What materials can we use to build it?
- How long do we have to build it?
- Do the equipment pieces have to work?
- Can we have other things, like people or trees, in it?

Write student questions on the board or on chart paper so that they remain accessible throughout the activity.

3. Show students a dedicated area in the classroom for the playground, such as a tabletop or counter space, and place on it a piece of chart paper labeled "Blueprint." Explain to students that this will be the space where they will create their playground; the blueprint is going to be a layout of their plan.

4. Ask, "How can we decide on what equipment should be in our playground?" (accept all answers, which may include voting for different pieces of equipment, having everyone contribute one idea, or having everyone draft his or her own playground design and voting on the best one) Tell students, "Let's begin by voting to see what kinds of equipment we might like to have in our playground model."

5. On chart paper or a board, draw an Our Dream Playground graph, labeling the x-axis "Playground Equipment" and the y-axis "Number of Votes." With the class, brainstorm some examples of playground equipment and write them on the x-axis. Give each student three sticky notes, and have them write their names and a small drawing of the playground equipment type they want. Then, have students attach their sticky notes to the graph to create a bar graph that looks something like the one in Figure 5.5 (p. 88).

FIGURE 5.5.

Sample Our Dream Playground Graph for Sticky Note Voting

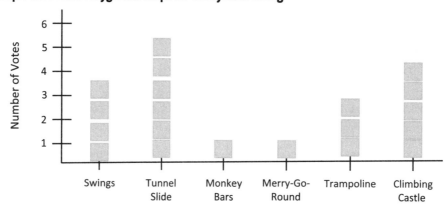

6. Ask students to *interpret the data,* or information, on the graph. "What does this graph tell us?" (in the sample shown in Figure 5.5, tunnel slide and climbing castle got the most votes, monkey bars and merry-go-round the least) Ask, "How could we use this information in planning our playground?" (e.g., we might want to have two tunnel slides and castles and just one merry-go-round, or perhaps we want to have a really big castle that would accommodate the interests of the group)

7. Have students respond to the following questions:

    • "How many pieces of equipment can we fit in our playground model?" (allow them to estimate the number and sizes of pieces of equipment)

    • "What materials do we have for building?" (show students the craft materials available)

    • "How much time do we have to build?" (give students time constraints)

    • "Can we make a background and model people for it?" (yes, but you can also allow students to use figurines and other classroom supplies)

    • "Does the equipment have to work?" (yes, equipment that moves, such as merry-go-rounds, swings, and seesaws, must actually move, and equipment like climbing castles and slides have to be able to hold the weight of the model people and demonstrate the way the equipment is used)

- "Should the playground be accessible for children who use wheelchairs or walkers?" (Explain to students that there are laws that require playgrounds to have accessible pathways and equipment that can allow all children, including children with different physical challenges, to play and have fun. Ask students if they have ever seen playground equipment like swings that accommodate wheelchairs or have high backs to help support children, ramps for accessing bridges or castles, or merry-go-rounds that accommodate wheelchairs. You can show photos of such equipment by searching for "accessible playground" on the internet. Encourage students to think about how they can make pieces of equipment that can allow the greatest number of children of all physical abilities to enjoy them.)

- "How do we decide what goes where?" (Explain to students that they will be working in groups and will need to *negotiate* with each other's work groups to devise a blueprint, or plan, that works for the space available and the equipment that is desired. Teams will draw their pieces of equipment on sticky notes that approximate the size and shape of their equipment. For example, a slide might be two sticky notes long. The team would put the two notes together and draw their slide on them. A climbing castle might be four sticky notes put together into a large square, with the castle drawn on it. Teams will then place the sticky note equipment where they think it should go on the blueprint; other teams can respectfully question and discuss the placement, and teams can come to an agreement on where equipment will go. No building will take place until the class has reached a consensus on the blueprint.)

- "What are some ways we can work out disagreements about what equipment goes where, or who works on what equipment?" (accept all answers; discuss options such as "give and take," where one team might shift the placement of its playground equipment but gets a bigger space; "all in," where all teams get a piece of something they want, such as a part of the space that they want; or "all out," where both teams that want the same space find another spot)

8. Explain to students that they will work in small groups on one *type* of playground equipment, but they may decide to make more than one of that type,

especially if several students are interested in making it. For example, they may decide to have two tunnel slides or two castles. Assist students in dividing into work groups by interest so that all students are working on something that engages them.

9. Give each student group a large piece of chart paper, and explain that they will have to plan their equipment before they start and will need to check in with you when they are ready. They will need to include a list of materials to pick up at the materials table when their plan and the playground blueprint are approved. Distribute some sticky notes to each team so that the teams can position their equipment on the blueprint.

10. As teams draw out their plans and begin to place their ideas on the blueprint, monitor the discussions to ensure that negotiations are productive. When the blueprint includes each team's sticky note equipment drawing, ask the class if they are in agreement on the design. If there is any disagreement, help students compromise on equipment placement and design. Once the blueprint is approved, congratulate students on their good collaboration and negotiation!

11. Allow students to begin building. Remind them that their playground equipment will have to work. This means that moving parts will have to work similarly to a real piece of playground equipment. Nonmoving equipment, such as a climbing castle, will have to withstand the weight of model people. Teams will have to test their equipment, and revise if necessary, until it is a working model. As students are working, ensure that they are collaborating and discussing the material and space constraints.

12. As teams finish, allow them to place their equipment on the playground. If teams finish early, allow them to work on additional aspects of the playground area, such as grass, trees, signage, and paths. Once all teams have finished (or the time is up), have teams present their pieces of playground equipment and demonstrate how they work. Ask teams:

    - "What went well in your planning or building process? What didn't?"

    - "Did you need to revise your plan as you started building? If so, why?"

    - "How well did your team work together? What is one thing that your team did really well? One thing that you could improve on?"

13. Once all teams have presented, tell them that there is still something missing from their playground. Allow them to guess. Then, explain that they need

to develop some rules to post on the playground to make sure that it is a great community place for all children. As a group, brainstorm a list of rules that might be appropriate for the playground, such as "Be kind to everyone," "Be safe: no jumping off the castle," "No pushing: keep your hands to yourselves," "Take turns," and "Say yes if someone asks if he or she can play with you." Write the rules or have students write them as they are suggested.

Evaluate: My Dream Playground Scene

1. Distribute the My Dream Playground handout (p. 94) to each student. Explain to students that they can draw any kind of dream playground they can imagine, but it must include the following two elements:

    - A push that is changing the direction or speed of an object
    - A pull that is changing the direction or speed of an object

    Assess the drawings based on the presence of these two elements or students' ability to explain their drawing in a way that conveys these two elements.

Going Deeper

- Students can survey playground equipment at school and in their neighborhood and create class maps and graphs of what they've found.

- Students can interview parents and other adults about their favorite playground equipment and childhood memories.

- Students can interview people with disabilities about playground access.

- Students can contact school, town, or city administrators about getting new playgrounds or accessible equipment in a playground.

- Students can research the history of playgrounds (see "How We Came to Play: The History of Playgrounds" at *https://savingplaces.org/stories/how-we-came-to-play-the-history-of-playgrounds/#.W3B_POgzo2*).

- Students can make connections between this lesson and simple machines.

Name: _____   Date: _____

Swingy Science

How many times does Partner 1 swing with …

1. a pull? _____

2. a pull and a push? _____

3. a pull, a push, and pumping legs? _____

How many times does Partner 2 swing with …

1. a pull? _____

2. a pull and a push? _____

3. a pull, a push, and pumping legs? _____

We found that _____ gives the most swings (circle one):

a pull   a pull and a push   a pull, a push, and pumping legs

Name: _____ Date: _____

Model Swing STEM Challenge

1. Can you build a swing that can hold a figure or toy in the seat?

    ☐ yes   ☐ no

2. Can you make your swing move back and forth when you **push** it?

    ☐ yes   ☐ no

3. Can you make your swing move back and forth when you **pull** it?

    ☐ yes   ☐ no

4. Can you make your swing move back and forth at the same time as a friend's swing?

    ☐ yes   ☐ no

5. Can you make your swing move back and forth at least four times with only one pull? (Count one for every time the swing comes back to the starting side.)

    ☐ yes   ☐ no

My Dream Playground

By: _____

Use your imagination to draw your dream playground. Remember to include (1) a push that is changing the speed or direction of an object and (2) a pull that is changing the speed or direction of an object.

Lesson 3: Take a (Farm) Stand

Lesson 3

Take a (Farm) Stand

Can Plants Help Us Fight Hunger?

Suggested Grade Levels

K–2

Driving Questions

- What do plants need to live and grow?
- How do individual plants of the same kind vary, and how are they similar?
- How are humans dependent on plants?

Lesson Overview

Students conduct a series of experiments on bean plants to determine what plants need to live and grow. They also compare and contrast individual bean plants to recognize the variation that can exist in observable traits. After learning about the parts of plants from which various human foods are derived, students brainstorm and take action on ways of addressing food insecurity in their school and in their community.

Connecting to the *NGSS*

(See full alignment in Table A.3 on p. 504.)

- LS1.C: Organization for Matter and Energy Flow in Organisms
  - All animals need food in order to live and grow. They obtain their food from plants or from other animals. Plants need water and light to live and grow. (K-LS1-1)

- LS2.A: Interdependent Relationships in Ecosystems
  - Plants depend on water and light to grow. (2-LS2-1)
- LS3.B: Variation of Traits
  - Individuals of the same kind of plant or animal are recognizable as similar but can also vary in many ways. (1-LS3-1)

Societal Issues

Distribution of Resources, Addressing Poverty

Nature of Science

- Scientific Knowledge Is Based on Empirical Evidence
  - Scientists look for patterns and order when making observations about the world.
- Science Models, Laws, Mechanisms, and Theories Explain Natural Phenomena
  - Science searches for cause-and-effect relationships to explain natural events.
- Science Addresses Questions About the Natural and Material World
  - Scientists study the natural and material world.

CCSS Connections

- English Language Arts
  - R.K.1. With prompting and support, ask and answer questions about key details in a text.
  - RL.1.1. Ask and answer questions about key details in a text.
  - RL.2.3. Describe how characters in a story respond to major events and challenges.
  - W.K.7. Participate in shared research and writing projects (e.g., explore a number of books by a favorite author and express opinions about them). (K-LS1-1)

- Mathematics
  - K.MD.A.2. Directly compare two objects with a measurable attribute in common, to see which object has "more of"/"less of" the attribute, and describe the difference. (K-LS1-1)

NCSS Connections

- Theme 7: Production, Distribution, and Consumption
  - Ask and find answers to questions about the production, distribution, and consumption of goods and services in the school and community.
  - Examine and evaluate different methods for allocating scarce goods and services in the school and in the community.

C3 Framework

- Dimension 4: Communicating Conclusions
  - D4.1.K-2. Construct an argument with reasons.
  - D4.7.K-2. Identify ways to take action to help address local, regional, and global problems.

UDL Toolkit

Multiple Means of Engagement	Multiple Means of Representation	Multiple Means of Action and Expression
Including blank pages in student journals for independent inquiries optimizes choice and autonomy.	Providing a word bank highlights the key vocabulary and activates background knowledge.	Through the Edible Plant Parts Sort and Plant Parts Salad, students show what they know through words, pictures, and tangible objects.
Using a choice board promotes autonomy and self-regulation and fosters motivation.	Providing templates with written and pictorial cues highlights big ideas and promotes comprehension.	Using a choice board for assessment and extension gives students multiple ways to express their learning.

Suggested Schedule and Sequence

- Day 1: **Engage** with *The Ugly Vegetables* read-aloud, and **Explore** with Planting Bean Seeds

- Day 2: **Explain** with All About Beans read-aloud, **Elaborate** with *National Geographic Readers: Plants* mini read-aloud, and **Evaluate** with Edible Plant Parts Sort and Plant Parts Salad

- Day 3 (about one week later): **Explore** and **Explain** with Seedlings Same and Different

- Day 4: **Explore** with What Do Plants Need to Live and Grow? experiment setup and My Plant Journal

- Day 5: **Explain** with What Do Plants Need to Live and Grow? results, **Elaborate** with My Plant Journal, and **Evaluate** with My Plant Journal Rubric and Plant Choice Board (optional)

- Day 6: **Elaborate** with *Maddi's Fridge* read-aloud and Let's Help Friends Who Have Empty Refrigerators brainstorm

- Days 7+: **Elaborate** and **Evaluate** with food drive, food bank, school garden project, or community garden

Materials

For Planting Bean Seeds Activity

(per student)

- Paper cup (9 oz. works well)
- 2–3 bean seeds (pesticide-free; wax, green, and lima all work well)
- Small clear plastic cup with waterline marked at halfway point
- Safety glasses or goggles and a nonlatex apron

(per class)

- 2–3 clear cups or jars (for viewing roots underground)
- Large bag of potting soil (sterilized)
- Markers

Lesson 3: Take a (Farm) Stand

- Water supply
- Window area where seedlings can receive sunlight

For Edible Plant Parts Sort Activity

(per group)

- Scissors
- Glue sticks
- Assortment of edible cut vegetables, fruits, and seeds

For Seedlings Same and Different Activity

(per class)

- Unifix or other stacking cubes (for measuring plants)

Student Handouts

- All About Beans
- Edible Plant Parts Sort
- Seedlings Same and Different
- My Plant Journal
- Plant Choice Board

Safety Notes

1. All students must wear safety glasses or goggles and nonlatex aprons during the setup, hands-on, and takedown phases of the activity.

2. Use caution when working with glassware or plasticware, which can shatter if dropped and cut skin.

3. In this lesson, students will be eating the food items used in the Plant Parts Salad. It's imperative to check with the school nurse and all parents beforehand about student food allergies, and avoid using any of these foods in the activity. Be sure to sanitize the classroom area where food will be eaten. Food should not be eaten in formal science laboratories.

4. Immediately wipe up any spilled water on the floor to avoid a slip-and-fall hazard.

5. Immediately pick up any items dropped on the floor to avoid a slip-and-fall hazard.

6. Wash hands with soap and water after completing this activity.

Media

Books

- *The Ugly Vegetables,* by Grace Lin
- *National Geographic Readers: Plants,* by Kathryn Williams
- *Maddi's Fridge,* by Lois Brandt

Video

- Time-lapse video of bean seed growth
  www.youtube.com/watch?v=w77zPAtVTuI

Background for Teachers

Plants are arguably the most important organisms on Earth. Thanks to their ability to use the Sun's energy to produce their food, they are the first living link in all food chains. This means that all animals, including humans, are dependent on plants for food either by directly eating plants or by eating animals that eat plants. Unlike animals, which must eat other organisms, plants make their own food (glucose, a simple sugar) through the process of photosynthesis. During photosynthesis, plants take in carbon dioxide (a gas in the air) and water, and in the presence of sunlight, they convert it to glucose while releasing oxygen. This is yet another reason that plants are essential to life on Earth: They provide the oxygen that animals need, while reducing excess carbon dioxide in the air. This is particularly essential given the fact that carbon dioxide, a by-product of burning fossil fuels, contributes to climate change. Plants also help maintain moisture in the air and keep soil in place. They are, quite simply, ecosystem superstars! Plants are also classroom superstars because they can be used to teach students about the needs of living things, fundamentals of heredity, and important social skills, including caring and responsibility. This lesson engages students in growing bean seeds to observe variation among

Lesson 3: Take a (Farm) Stand

individuals of the same kind of plant, as well as conducting tests of bean seeds' need for water and sunlight.

The expectations for the *NGSS* standards in this lesson are that kindergartners will recognize patterns in what living things need to live, including plants' needs for light and water. The related second-grade expectation anticipates that students will design and conduct experiments to test that plants need light and water. It should be noted, however, that second-grade assessments should be limited to testing one variable at a time. In other words, assessments should test for students' ability to set up a test for either water or sunlight (or some other variable), but not both together. Because this activity is designed for K–2, the What Do Plants Need to Live and Grow? experiment has teams of students running two separate but collaborative experiments: one that tests sunlight versus no sunlight (both with water) and one that tests water versus no water (both with sunlight). This is admittedly a simpler setup than one that handles two variables at a time, which would involve testing sunlight with water, sunlight without water, water without sunlight, and no water and no sunlight, but it is done intentionally to ensure that students have a clear understanding of a fair test, one that keeps all the variables controlled except for the one variable that you're testing. The lesson also asks students to design their own additional experiment, which could involve such things as using a different type of seed, varying the amount of light or water, or using liquids other than water.

The final part of this lesson involves reading *Maddi's Fridge,* by Lois Brandt. It is a book about friendship, telling a tale of two girls who help each other in different ways. Maddi helps Sophia on the playground with a climbing wall challenge, while Sophia helps Maddi by bringing her food to help make up for her home's empty fridge. Over 13 million children in the United States—nearly one in five—live in "food insecure" homes, which means that the families regularly don't have enough food to eat. Poor nutrition contributes to stress, illness, and difficulties in concentration and performing schoolwork. While school breakfast and lunch programs fill an important need, weeknight and weekend meals are still often unpredictable for children from food insecure homes. One of the ways that schools can help is to support local food banks, which make food available for those in need. While it is critical to check with your local food bank beforehand to see what types of donations it accepts, typically nonperishable, nutritious items are accepted. These items include many plant-based foods, such as canned or dried beans, peanut butter, low-sodium canned vegetables and vegetable soups, canned fruits (in juice or light syrup), rice, pasta, oatmeal, canned bean chili, unsalted nuts and seeds, whole-grain (low-sugar) cereal, dried fruits, and olive or canola oil, as well as canned fish and meats. Having students coordinate a food drive and sort the foods into plant-based and non-plant-based foods can reinforce the importance of plants to humans.

Another way that schools can help is to set up onsite school food pantries that make food available to children and families in need. School and community gardens can also assist with making fresh produce available to all children and families. These gardens have been shown to reduce food insecurity, improve nutrition and vegetable intake, and promote students' understanding of plants and life cycles. School groups can often volunteer at community gardens if no school garden is feasible, although methods such as square foot gardening (see the resources listed below) make it possible for classrooms and schools to grow produce in very small spaces, including rooftops! This lesson concludes with an open-ended opportunity for classrooms to apply their newfound knowledge of plants to help address food insecurity. While it may be tempting to skip this work-intensive piece, I urge you to empower your students to take on this challenge; it will likely be one of the most memorable and meaningful activities in their school careers.

Additional Resources

Book on how children can learn through gardening

- Bartholomew, M. 2014. *Square foot gardening with kids: Learn together—Gardening basics, science and math, water conservation, self-sufficiency, healthy eating.* Minneapolis: Cool Springs Press.

Feeding America website

- *www.feedingamerica.org*

Find your local food bank

- *www.feedingamerica.org/find-your-local-foodbank*

Facts on hunger in the United States

- *www.nokidhungry.org/who-we-are/hunger-facts*

Lois Brandt's site

- *www.loisbrandt.com/take-action*

Article about an elementary school garden that helps feed the hungry

- Associated Press. *U.S. News and World Report.* 2017. School Gardens, Food "Rescues" Are Helping Nevada's Hungry. October 6. *www.usnews.com/news/best-states/nevada/articles/2017-10-06/school-gardens-food-rescues-are-helping-nevadas-hungry.*

Lesson 3: Take a (Farm) Stand

5E Lesson Plan

Engage: *The Ugly Vegetables* Read-Aloud

1. Show and describe the cover of the book *The Ugly Vegetables,* by Grace Lin, which depicts a mother and daughter in a vegetable patch. Tell students that this is a true story of the author's childhood. Ask students, "What do you think the title of the book might be referring to? What could make vegetables 'ugly'?" (accept all answers) Read the book aloud, using the following guided reading prompts. (*Note:* The pages in this book are unnumbered; the numbers below assume that the story begins on page 5.)

    - After reading pages 5–11, ask, "Grace notices several differences between her mother's garden and the neighbors' gardens. What were some of those differences?" (big shovel versus small shovel, hose versus watering can, seed packages versus papers with Chinese vegetable names, seedlings look like grass versus leaves)

    - After reading pages 12–17, ask, "Why did Grace think that flowers were better than vegetables?" (they were colorful, smelled good, had butterflies and bees; the vegetables were lumpy, bumpy, and "icky yellow" and green; her garden was different from the others, which may have made her feel that the vegetables weren't as good as the flowers)

    - After reading pages 25–31, ask, "Why did the neighbors come over to Grace's house?" (they wanted to try the soup, and they brought flowers to trade)

    - Then ask, "What did everyone plant the following spring?" (flowers *and* Chinese vegetables)

    - Finally, ask students, "What were some of the things that Grace's mother and the neighbors did to help their seeds grow?" (prepared soil, watered, planted in spring when weather was warm) "What do you think plants need to live and grow?" (accept all answers) Lead into Explore by saying, "Let's find out!"

Explore: Planting Bean Seeds

1. Distribute a paper cup, two or three bean seeds, and a small cup with water-line mark to each student. Have students use markers to write their names on the sides of the cups. Call students by table to scoop soil out of the bag with

their cups, filling the cups slightly below the top. Then, model how to plant seeds by poking your pointer finger into the soil about an inch deep (or about half the length of your finger), dropping a bean seed into the hole, and gently covering it with soil. Students can plant two or three seeds in a cup, each in its own separate hole. (Plant seeds in a few extra cups in case some children's seeds don't grow.) Show students how to fill the water cup to the line and water their plants. (*Note:* In a later experiment, students will need to control for the amount of water each bean plant receives. Today's activity prepares students for that skill.)

2. Next, have two or three student volunteers fill the clear plastic cups with soil and plant two or three seeds in each, placing them right up against the cups' walls; this will make it possible to view germination over the next several days.

3. Finally, have all of the students place all the cups (the paper cups and the clear viewing cups) in a spot where they will not be disturbed. You can take daily photos of the plants to develop a time-lapse sequence of bean growth. Have students water their seeds each day.

Misconception Alert

Students often think that sunlight is needed for seeds to germinate. While the Sun provides warmth for seeds that can help trigger and possibly accelerate germination, seeds get their food from the cotyledon while they are underground. The seedlings rely on the Sun to make food through photosynthesis only after the leaves have formed. Although this isn't a focus of this particular activity, you can have students place their cups in different spots around the room to see if the timing (or success) of germination differs based on whether they receive sunlight.

Explain: All About Beans Article Read-Aloud

1. Distribute copies of the All About Beans handout (p. 115) to students. Display the article using a projector or document camera to assist students in following the reading. Read the article aloud to model fluency, or invite student volunteers to take turns reading. After reading, ask:

    - "What are legumes?" (plants that have pods with seeds inside of them)

    - "What are some examples of legumes?" (e.g., beans, peas, peanuts, lentils)

- "Why are beans and other legumes nutritious?" (because they are high in protein)
- "What is 'Three Sisters' gardening?" (Native American practice of planting beans with corn and squash, which help each other grow)
- "What does it mean when a seed germinates?" (the embryo breaks through the seed coat and grows roots and shoots)
- "How do seeds get their food?" (it is stored in the cotyledon)

2. Give students a few minutes to observe and water their seeds. If this is only the second day since planting, they won't see any signs of germination above ground, but they will likely see evidence of germination in the seeds that were planted against the walls of the clear cups.

3. Explain to students that they are going to engage in an activity about parts of plants that we eat, but first, they need to learn some vocabulary about plant parts.

Elaborate: *National Geographic Readers: Plants* Mini Read-Aloud

1. Read pages 14–27 of *National Geographic Readers: Plants,* by Kathryn Williams, which describe the functions of roots, stems, leaves, fruits, and seeds.

Evaluate: Edible Plant Parts Sort and Plant Parts Salad

Safety Alert: This activity involves sorting and eating various plant parts. Be aware of student allergies, and avoid common allergens such as peanuts and strawberries.

1. Have students seated in table groups. Distribute copies of the Edible Plant Parts Sort handout (p. 116), and provide scissors and glue sticks. Place a plate of edible cut vegetables, fruits, and seeds in the center of each table.

2. Explain to students that the middle column on the handout describes the jobs of different plant parts. Their challenge is to work with their tablemates to read through what each plant part does, and then cut and paste the correct plant part name in the first column. Next, they should put an edible example of that plant part in the third column.

3. If you feel it would be helpful, do one example with students. For example, point to the bottom of the chart and say, "This plant part absorbs water and nutrients from the soil. What plant part is that?" (root) "Cut and paste 'Root'

into the box in the first column. Which of our edible plant parts is a root?" (e.g., carrot, radish). "Place one of these in the third column."

4. As students are working, check for understanding by asking, "What do the roots do?" and "Do any of the other edible plant parts fit into that category?" When students are done, review the answers (see Figure 5.6). As an extra challenge (and assessment), distribute paper plates to students and see if they can create plant diagrams using some of the edible plant parts. For example, students might use carrots for the roots, celery for the stem, spinach for leaves, broccoli or cauliflower for the flower, and so on. When students have correctly created their plant diagrams, allow them to eat and enjoy the edible plant parts salad (Figures 5.7 and 5.8). Ranch dressing is optional!

FIGURE 5.6.

Edible Plant Parts Sort Answer Key

Plant Part	What This Plant Part Does
Seed	Germinates and grows into a new plant
Fruit	Carries and protects seeds
Flower	Makes fruits and seeds through pollination
Leaf	Uses sunlight to make the plant's food (sugar)
Stem	Carries water and nutrients through plant
Root	Absorbs water and nutrients from the soil

Lesson 3: Take a (Farm) Stand

FIGURE 5.7.
Working on Plant Parts Salads

FIGURE 5.8.
Edible Plant Diagram. All the Plant Parts Are There!

FIGURE 5.9.
Parts of a Plant Diagram

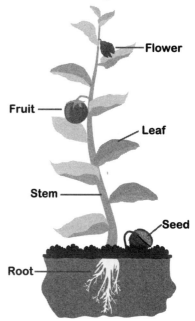

Explore and Explain: Seedlings Same and Different

1. Once most of the seedlings have emerged, approximately one week later, tell students that they are going to be *botanists,* or plant scientists, and have an opportunity to make some careful observations of their seedlings. Review the parts of a plant by drawing a diagram like the one in Figure 5.9 on the board or displaying the figure on poster board or with a document camera (Figure 5.10).

FIGURE 5.10.
Reviewing Plant Parts

2. Review the parts of the seedlings and their functions by asking students about the functions of the plant parts listed in Table 5.1.

TABLE 5.1.

Plant Part Functions

Plant Part	Function
Roots	Absorb water and nutrients; hold plant in the ground
Stem	Carries water and nutrients up to leaves; provides support
Leaves	Make plant's food using sunlight and a gas in the air called carbon dioxide
Cotyledon	Provides food for the seed before its leaves form; falls off afterward

3. Distribute copies of the Seedlings Same and Different handout (p. 117). Assign students to work in pairs so that they can observe each other's seedlings. Read the handout, which asks students to draw and label diagrams of their own and their partner's seedling cups. Stress to students that it is important that their drawings show small details of the seedlings, including the number and shape of leaves, whether the cotyledon is still attached, and the colors. Students can also measure the height of the seedlings using Unifix or other stacking cubes. Model this for students by placing a Unifix cube at the base of a seedling, then stacking the cubes until they reach the highest part of the plant. Have students record the measurements on their diagrams (e.g., 2 cubes high). If more than one seed has sprouted in a cup, have students include them all in their drawings. Allow students to refer to their All About Beans article to help with labeling the parts, such as stems, leaves, and cotyledons. Some possible answers to the questions on the handout are as follows:

- How are the bean seedlings similar? (have stems, leaves, and cotyledons; are green)

- How are the bean seedlings different? (different number of leaves, some have cotyledons still attached and some don't, different heights, different shapes with some more bent than others, slightly different leaf shapes)

4. Ask students, "Does anything surprise you about what you observed here?" (answers will vary, but students may be surprised by how different the seedlings can appear when you look at them carefully; they also may note that

Lesson 3: Take a (Farm) Stand

they look a little different even though they are all bean plants) Remind students that all these seedlings are the same *kind,* or species, because they are all one type of bean plant. But they are each separate *individuals* that may have differences in appearance. Ask, "Do you know any other species of plants or animals where the main body parts or features are similar but the individuals are each different?" (answers will vary, but students may mention that this is true for humans, as well as all plants and animals) Inform students that all species have variability or differences among individuals, and this helps the species survive. They will learn more about this as they progress in school. Have students try to complete the fill-in sentence on the handout using the word bank. The answer is as follows: I observed that <u>individuals</u> of the same <u>kind</u> of plant have the <u>same</u> parts but can look <u>different</u>.

5. To prepare for the next activity, have students thin out the seedlings if more than one is growing in the cup. They can do this by gently removing extra seedlings using their fingers or a spoon and replanting them in a separate cup with soil. These extras will be used for an experiment during the next class. Do not have students write their names on these cups. Have students observe the roots of the transplanted beans, as well as any roots that have emerged from the bean seeds in the clear cups.

Explore: What Do Plants Need to Live and Grow? Experiment Setup and My Plant Journal

1. Divide the class into groups of four. Distribute copies of the My Plant Journal handout (pp. 118–123) to students, as well as their bean plant cups. Ask students, "What do plants need to live and grow?" (accept all answers; students will likely suggest sunlight, water, soil, and possibly plant food) Ask students how they know this. (answers will vary but might include reading about it, seeing it on television, having taken care of plants at home, in a garden, or in the classroom) Explain to students that when scientists want to answer a question, they develop fair tests to find out the answers. "Today, we are going to set up fair tests to see if bean plants need sunlight and water."

2. Show students two bean plant cups and ask, "If I wanted to test whether bean plants need *sunlight* to grow, how could I do it with these two plants?" (guide students to the idea that one plant would be exposed to sunlight, while the other would be placed in the dark) Then say, "OK, let's say I put Plant A in the sunlight and I also give it water, and I put Plant B in the dark and I *don't* give it water. Do you think that this is a good test?" (students will likely recognize that this wouldn't be a fair test; to be a fair test, you would

need to give both plants water or not give either plant water) Ask, "Why is it important to do a fair test?" (because otherwise, you won't know if the plant is growing or not growing because of the sunlight or the water) Say, "When scientists conduct an experiment, they want a fair test where all the conditions are the same except for the thing, or variable, you are testing. In this example, what are we testing?" (whether bean plants need sunlight) "So in this test, what things should be exactly the same between the two plants?" (the kind of plant, amount of water, soil) "What is the one thing that should be different between the two?" (one will be in sunlight, and the other will be in the dark) "Let's think about another variable. What if I wanted to see if bean plants need *water* to live and grow? How would I set up that experiment with these two plants?" (one plant gets water, and the other doesn't) "OK, let's say I give Plant A water and I put it in sunlight, and I don't give Plant B water *and* I put it in the dark. Do you think that this is a fair test?" (students will likely recognize that this wouldn't be a fair test, as both plants would need to be in sunlight or in the dark; the only thing that should be different between the two cups is whether they get water or not—everything else should be the same) "Now we are ready to be botanists and set up fair tests for water and sunlight." (*Note:* In this lesson, I am choosing to have students experiment with the extra seedlings because I have found that some young children become quite attached to their seedlings and will not want to withhold light or water from them. There are several ways around this issue. One is not to give students "ownership" of *any* of the seedlings, and simply have them plant the seeds without putting their names on the cups. Another solution is to have students use their seedlings as the control plants that receive both light *and* water, while the extra seedlings are the test plants that receive either light *or* water. This is a reasonable solution, except that the extra seedlings were transplanted when they were thinned out, which adds another variable into the mix, leading to a question of whether seedlings are doing poorly because they didn't receive light or water or were transplanted—all of which are possible. For these reasons, I choose to use *only* the extra seedlings for this experiment and reassure students that if any of the seedlings do very poorly ["look really sick"] during the experiment, we will attend to them promptly. My classes have set up "plant hospitals" where the seedlings get extra light, water, fertilizer, and TLC!)

3. Distribute four seedling cups to each table, and have students write the numbers 1–4 on the cups. Students with Cups 1 and 2 will set up the test for water, and students with Cups 3 and 4 will set up the test for sunlight. Have students draw their experiments on pages 1–2 of their journals and make their

predictions on page 3. Have them use Unifix or stacking cubes to measure their seedlings and record the heights. Page 3 also asks students to explain how they will know if a plant is doing well. Some indicators might include getting taller, staying green, or developing leaves. Indicators that the plant is not doing well might be drying out or shriveling up, turning yellow or brown, or not growing. Over the next three to four days, have students use the watering cup (with the waterline marked on it) for the plants that receive water so that they will all receive the same amount.

Explain and Elaborate: What Do Plants Need to Live and Grow? Results and My Plant Journal

1. Allow students to check on their plants each day. After three or four days (time will differ depending on the type of beans), have students observe and record their team's results on pages 4–5 of their journals and write their conclusions on page 6.

2. Discuss the results together as a class. Students will likely observe that the plants that didn't get light or water did not grow as well as those that did. The plants may appear paler, smaller, and shriveled. Ask students to propose ways of helping these plants (e.g., providing sunlight and water), and then allow them to do these things. Ask students, "What other tests would you like to do if you had the opportunity?" Brainstorm tests such as: Do sunflowers grow more quickly than beans? Do bean plants do better with plant food? Does the amount of water (a lot versus a little) affect bean plant growth? Have students design their own experiments on page 7 of the journal. Remind them that all the conditions should be controlled. The only difference between their two setups should be the variable that they are testing. (*Note:* There are two blank pages in the journal after page 7. If possible, allow students to conduct the tests they designed, and then record their results on these pages.)

Evaluate: My Plant Journal Rubric and Plant Choice Board (optional)

1. Assess each student's journal using the rubric on page 125.

2. (Optional) Give students copies of the Plant Choice Board handout (p. 124) and have them place an X on the center box when they complete their plant journals. Then, allow them to complete two other tasks of their choice to make tic-tac-toe using the center box.

Elaborate: *Maddi's Fridge* Read-Aloud and Let's Help Friends Who Have Empty Refrigerators Brainstorm

1. Show and describe the cover of *Maddi's Fridge,* by Lois Brandt, which depicts two girls standing in front of a refrigerator. Ask students, "What do you think might be special about Maddi's fridge?" (accept all answers) Read the book aloud to students. After reading, ask:

   - "What did Sophia notice about Maddi's fridge that seemed to trouble her?" (there was very little food in it—only some milk)

   - "Why is having food in the fridge or house important?" (because we all need food to live)

   - "Why do you think Maddi asked Sophia to promise not to tell anyone about it?" (because she was embarrassed)

   - "How did Maddi help Sophia on the playground?" (she helped her get better on the climbing wall)

   - "How did Sophia try to help Maddi?" (she brought different foods to school, including a fish, eggs, and a burrito)

   - Show students pages 13 and 26 (with page 5 being the first page in this unnumbered book) and ask, "What are some of the foods in Sophia's fridge? Which ones come from plants?" (e.g., carrots, broccoli, pineapple, lettuce, pickles, tea, coffee, apples, oranges, grapes, bread, hummus, tofu, grape juice, scallions)

   - Turn and talk: Ask students, "Why did Sophia tell her mother about Maddi's empty fridge? Do you think it was the right thing to do? Turn to a partner and share your thinking. Then we'll share all together." (she realized that she needed help from her mother; sometimes it can be OK to break a promise if it's to help someone)

   - "How did Sophia's mother help?" (she brought food to Maddi's mother)

   - "In what ways were Sophia and Maddi good friends to each other?" (they helped each other in different ways: Maddi helped Sophia become a better climber, and Sophia helped Maddi get more food; she also slowed down when running)

   - "Why do you think Maddi's family didn't have much food?" (they didn't have enough money to buy it)

2. Turn to the last page of the book, which is titled "Let's Help Friends Who Have Empty Refrigerators." Ask, "What do you think are some ways to help friends with empty fridges?" Have students brainstorm, and accept all answers. Read or elicit the following ideas:

   - Tell a parent or trusted adult if you notice a friend doesn't have enough food. Offer friends food when they come over to play.

   - Make posters encouraging people to give money or food to a local food bank. (Discuss that a food bank is a place where people can go to get free or low-cost food.)

   - Volunteer with your class at a local food bank.

   - Ask a local food bank what foods it needs. Then organize a school or community food drive for those items. (Let students know if you've already checked with a local food bank or pantry to see what foods it needs. Emphasize that many of the foods are plant-based, such as beans, peanut butter, canned vegetables and fruits, rice, pasta, and other grains that come from plants.)

   - Talk to parents and teachers about childhood hunger.

3. Ask students to brainstorm other ideas, including those that tie into growing plants (working in or starting a school garden and food pantry; volunteering in a community garden that grows plants for food). As students discuss their ideas, choose one (perhaps through hands-up voting or using sticky notes on the board) that your class can do. Then have students help organize it!

Elaborate and Evaluate: Food Drive, Food Bank, School Garden Project, or Community Garden

1. Elaborate and informally evaluate by having your class do any of the following activities:

   - Have students sort donated foods into plant and nonplant boxes or specifically in plant part boxes labeled as "fruits," "seeds," "roots," and so forth.

   - Have students create posters that encourage people to donate to the local food bank and post them in local stores and markets (with permission). The posters should explain why food is so important to everyone.

- Have students create and tend to a "square foot garden," either outdoors or indoors, and put the foods that are harvested out for children to take home. Lettuce, radishes, and carrots all work well!

- Have students volunteer at a community garden or food bank, and ask them to share their experience through drawing, writing, singing, acting, or cartooning.

Going Deeper

- Students can interview family members and other adults about gardening and farming experiences.

- Classes can learn about (and grow) vegetables and herbs that are common in different cultures, using *The Ugly Vegetables* as a launching point. (There is a recipe for soup in the back of the book.)

- Classes can visit a farmers' market or supermarket to survey and record (drawings, photos) what types of plant parts they see.

- Students can create maps of regions where particular plants are grown for food.

All About Beans

How Have You "Bean"?

Baked beans, string beans, green beans, and chili beans (but **not** jelly beans!) ... these are just a few examples of beans that people eat. Beans are a special kind of plant called **legumes** because they have **pods** that contain the bean **seeds.**

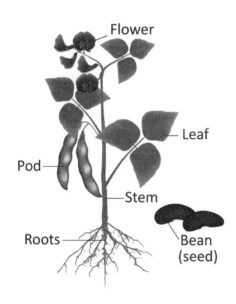

Bean Types and History

There are around 19,000 different **species,** or types, of bean plants. Some familiar types are green beans, string beans, fava beans, lima beans, kidney beans, soybeans, peas, chickpeas (garbanzo beans), lentils, and peanuts. Beans have been grown around the world as food for thousands of years. Beans are very **nutritious,** as they are high in **protein.** Native Americans traditionally grew beans alongside corn and squash, an arrangement referred to as "Three Sisters" gardening because the three species help each other grow.

Planting Bean Seeds

Bean seeds are very easy to plant. Simply poke your finger into the soil about an inch deep (about half the length of your pointer finger). Drop in a bean and gently cover with soil. Water it to keep the soil moist. The seeds will begin to **germinate,** or sprout, within a few days. When bean seeds germinate, the baby plant, or **embryo,** begins to grow as it gets nutrition from the **cotyledon.** The embryo is soon strong enough to break through the tough **seed coat** and form **roots** below the ground and **shoots** (stem, leaves, flowers, and pods) above the ground.

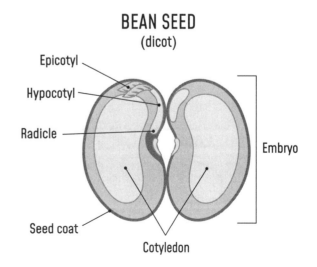

Name: _____ Date: _____

Edible Plant Parts Sort

Glue Plant Part Name Here ↓	What This Plant Part Does	Place Your Edible Plant Part Here ↓
	Germinates and grows into a new plant	
	Carries and protects seeds	
	Makes fruits and seeds through pollination	
	Uses sunlight to make the plant's food (sugar)	
	Carries water and nutrients through plant	
	Absorbs water and nutrients from the soil	

| Leaf | Flower | Root |
| Stem | Seed | Fruit |

Botanist Name: _____   Date: _____

Seedlings Same and Different

Observe your own and your partner's bean seedling cups. Draw and label detailed pictures of the two seedlings in the boxes below. Then answer the questions.

_____'s Cup	_____'s Cup

1. How are the bean seedlings **similar**?

   _____

2. How are the bean seedlings **different**?

   _____

3. Complete the following sentence using each term in the word bank only once:

I observed that _____ of the same _____ of plant have

the _____ parts but can look _____.

Word Bank

kind	different
individuals	same

IT'S STILL DEBATABLE! USING SOCIOSCIENTIFIC ISSUES TO DEVELOP SCIENTIFIC LITERACY, K–5

My Plant Journal

Botanist: _____

Today we set up an experiment to test whether bean plants need water to grow. These are my drawings of the plants.

Water
Cup #1

Height: _____ cubes

No Water
Cup #2

Height: _____ cubes

1

We also set up an experiment to test whether bean plants need sunlight to grow. These are my drawings of the plants.

Sunlight
Cup #3
Height: _____ cubes

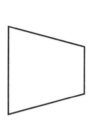

No Sunlight
Cup #4
Height: _____ cubes

In my next experiment, I'd like to test _____
_____
_____
_____

My experiment would look like this:
(Remember to include everything your plants would get and not get in the picture.)

Cup A   Cup B

I predict that the plants in cup(s) _____ will grow best because _____

_____

_____

_____

_____

I will know a plant is growing well by _____

_____

_____

_____

_____

My results showed that the plants in cup(s) _____ grew best.

Therefore, my conclusion is that plants need _____ _____ to grow.

After several days, our water test plants looked like this:

Water
Cup #1

Height: _____ cubes

No Water
Cup #2

Height: _____ cubes

After several days, our sunlight test plants looked like this:

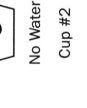

Sunlight
Cup #3

Height: _____ cubes

No Sunlight
Cup #4

Height: _____ cubes

Record the results from your own experiments on these two pages.

Name: _____ Date: _____

Plant Choice Board

Make tic-tac-toe (three in a row) using the center box.

Write a **story** that describes how a seed becomes a plant.	Create a **painting** that shows what part of a plant three (3) different foods come from.	Write a **song** that describes the parts of plants and what they do.
Create a set of **flash cards** that teach what part of a plant three (3) foods come from.	Complete **My Plant Journal.**	Draw a **comic strip** about how a seed becomes a plant.
Write a **play** that teaches what part of a plant three (3) different foods come from.	Perform a **dance** that shows how a seed becomes a plant.	Make a **mobile** that shows the parts of plants and describes what they do.

Lesson 3: Take a (Farm) Stand

Rubric for My Plant Journal

Criteria	Not Yet (1 pt.)	Emerging (2 pts.)	Secure (3 pts.)
Records and communicates observations	Student is unable to record or communicate observations.	Student records or communicates observations, but they are unclear or insufficient.	Student records and communicates observations in a clear and thorough manner using words, pictures, or both.
Makes claims from evidence	Student is unable to make claims.	Student makes claims that are unsupported by evidence.	Student makes claims about plants' needs from observing patterns in data.
Demonstrates understanding of fair test	Student is unable to demonstrate understanding of a fair test.	Student recognizes that the test needs to be fair but cannot identify factors to control.	Student demonstrates understanding of how to design a fair test and can identify controls.

Total: ___ / 9 pts.

Lesson 4

Monkey Business

Do We Need Zoos?

Suggested Grade Levels

K–2

Driving Questions

- How do animal parents care for their young?
- How do young animals take care of themselves?
- How are young animals similar to and different from their parents?

Lesson Overview

Students take virtual tours of zoos and wild spaces through texts and internet research to learn how animal parents care for their young, how young animals take care of themselves, and how young animals are like and different from their parents. After creating a project that communicates how animal parents care for their young from the perspective of the young animal, students learn about the arguments for and against zoos through readings and classroom discourse, then develop a position poster on the issue.

Connecting to the NGSS

(See full alignment in Table A.4 on p. 505.)

- LS1.B: Growth and Development of Organisms
  - Adult plants and animals can have young. In many kinds of animals, parents and the offspring themselves engage in behaviors that help the offspring to survive. (1-LS1-2)

- LS3.A: Inheritance of Traits
  - Young animals are very much, but not exactly, like their parents. Plants also are very much, but not exactly, like their parents. (1-LS3-1)

Societal Issues

Animal Rights, Concepts of Freedom and Altruism

Nature of Science

- Scientific Knowledge Is Based on Empirical Evidence
  - Scientists look for patterns and order when making observations about the world.
- Science Addresses Questions About the Natural and Material World
  - Scientists study the natural and material world.

CCSS Connections

- English Language Arts
  - RI.K.7. With prompting and support, describe the relationship between illustrations and the text in which they appear (e.g., what person, place, thing, or idea in the text an illustration depicts).
  - RI.1.7. Use the illustrations and details in a text to describe its key ideas.
  - W.K.1. Use a combination of drawing, dictating, and writing to compose opinion pieces in which they tell a reader the topic or the name of the book they are writing about and state an opinion or preference about the topic or book (e.g., My favorite book is ...).

NCSS Connections

- Theme 8: Science, Technology, and Society
  - Use diverse types of media technology to research and share information.
- Theme 10: Civic Ideals and Practices

- Evaluate positions about an issue based on the evidence and arguments provided, and describe the pros, cons, and consequences of holding a specific position.

C3 Framework

- Dimension 4: Communicating Conclusions
    - D4.1.K-2. Construct an argument with reasons.
    - D4.3.K-2. Present a summary of an argument using print, oral, and digital technologies.
- Dimension 4: Taking Informed Action
    - D4.8.K-2. Use listening, consensus-building, and voting procedures to decide on and take action in their classrooms.

UDL Toolkit

Multiple Means of Engagement	Multiple Means of Representation	Multiple Means of Action and Expression
Engaging in topics (animals and zoos) that are personally and socially relevant to students adds meaning and motivation.	Using text readings, read-alouds, websites, and pictorial guides provides alternative means of presenting information.	Providing sentence starters, four square organizer, and opinion letter templates supports composition and information management.
Allowing students to choose the animals they "visit" during internet research and the way they present their research project support autonomy and motivation.	Using T-charts, concept maps, and checklists maximizes transfer and generalization of information.	Allowing students to share their ideas through a yes/no argument line, writing, pictures, and options including drama, music, and art gives multiple opportunities for communication.
Providing an extension activity (researching baby animal names) provides varied demands and optimizes challenge.	Having alternative readings (bullet-pointed zoo article) removes unnecessary barriers to reaching instructional goal.	Providing checklists and multilevel templates for projects aids in scaffolding students' planning and strategy development.

Lesson 4: Monkey Business

Suggested Schedule and Sequence

- Day 1: **Engage** with *My Visit to the Zoo* read-aloud and Animal Parents Concept Map

- Day 2: **Explore** with Animal Parents and Their Young four square internet research and What's My (Baby) Name? extension activity

- Days 3–4: **Explain** with *How Animal Babies Stay Safe* read-aloud and *Born in the Wild: Baby Mammals and Their Parents* mini read-aloud, and **Elaborate** and **Evaluate** with How My Parents Care for Me animal baby perspectives project and presentations

- Day 5: **Elaborate** with Do We Need Zoos? article or list and T-chart, and **Evaluate** with Yes/No Line Debate and Are You for Zoos? position poster or My Opinion on Zoos opinion letter

Materials

(per class)

- Internet access (1 device per 2–3 students recommended)
- Sticky notes
- Pencils
- 1 copy of "Yes" and "No" signs (pp. 152–153)
- Tape

For How My Parents Care for Me project

- Craft materials, paper, crayons or markers as needed, depending on how students choose to complete this activity

Student Handouts

- Animal Parents and Their Young
- What's My (Baby) Name?
- How My Parents Care for Me
- Do We Need Zoos? (choice of narrative article or bulleted list)
- Are You for Zoos? (younger) or My Opinion on Zoos (older)

Media

Books

- *My Visit to the Zoo,* by Aliki
- *How Animal Babies Stay Safe,* by Mary Ann Fraser
- *Born in the Wild: Baby Mammals and Their Parents,* by Lita Judge

Websites

- Zooborns, about animal babies born in zoos
  www.zooborns.com

- San Diego Zoo Kids, with age-appropriate information on zoo animals, including live animal cams
  http://kids.sandiegozoo.org/animals

Background for Teachers

Animals are a favorite topic of study for young children. The study of animals through observation can support students' reasoning, particularly with regard to comparing and contrasting physical attributes and recognizing patterns of development and behavior. It's no surprise that the *NGSS* focus on the importance of comparing and contrasting animal parents and their offspring to provide a basis for understanding heredity (that animals can have offspring, and those offspring are like, but not exactly the same as, their parents). The *NGSS* also expect students to recognize patterns of behavior that parents exhibit to ensure the survival of their young (and the behaviors of the young to ensure their own survival). These two standards taken together reinforce a critical aspect of the nature of science—that scientists learn about the natural world by observing patterns.

In this lesson, students take two virtual visits to a zoo: one by reading a book and the other by conducting online research. Students are asked to use their observational skills to spot the ways that parents and offspring are alike and different and to look for those protective behaviors that support the survival of young animals. This study provides a context for reminding students that humans, too, are animals who display many of the same behaviors they observe in other animals.

This lesson also raises the controversial question of whether we need zoos. For some, zoos are remarkable places that bring people close to wild animals, allow for educative experiences, and save wildlife through protection and breeding programs. Others view zoos as cruel or sad places that infringe on animals' rights to freedom and, in some cases, treat animals inhumanely. While some might ask whether such an emotionally charged issue is appropriate for young children, research suggests that it is precisely these types of issues that are successful in engaging students in critical discourse and understanding different perspectives. As someone who taught K–4 students for more than a decade, I found this issue to be one that even the youngest students hold strong intuitions about, yet they have rarely considered both sides of the issue. The activity in this lesson asks students to consider various arguments for and against zoos to help them recognize the complexity of such issues and develop their own informed, evidence-based decisions.

Additional Resources

Articles discussing the debate surrounding zoos

- *www.theatlantic.com/news/archive/2016/06/harambe-zoo/485084*
- *www.wgbh.org/news/2016/06/02/local-news/do-we-really-need-zoos*

5E Lesson Plan

Engage: *My Visit to the Zoo* Read-Aloud and Animal Parents Concept Map

1. Ask students, "Have you ever been to a zoo? If you have, what was it like? If you haven't, what do you think it would be like?" Allow students to turn and talk to a partner briefly. Say, "Today we're going to take a trip to the zoo in a book." Show students the cover of *My Visit to the Zoo*, by Aliki, which portrays a tiger parent and cub in the foreground and an elephant parent with a calf in the background, with children observing. Ask, "I wonder … why do you think the author chose to put these particular scenes of animal parents and their young on the cover?" (accept all answers; students will likely suggest that zoos are places that want animals to have families, perhaps because they are endangered)

2. Read the first two sentences on page 4 of the book: "I didn't really want to visit the zoo. I had been to one that made me feel sad." Ask students, "Why do you think a zoo might have made the author feel sad?" (accept all answers; students will likely indicate that it might have made the author feel sad that animals were in cages and not free) Continue reading the book, which describes a much more pleasant experience for the author because the zoo had habitats where the animals could live similarly to when they are in the wild. As you

read, pay careful attention to the scenes and descriptions of animal parents and their young as follows:

- Page 7 shows a gibbon carrying its baby. Ask, "How is the baby similar to the adult?" (same face, body parts) "How is it different?" (much smaller)

- Page 8 shows macaques and baboons holding and giving a ride to their young. Ask, "How is the macaque baby like its parents?" (same color and body parts) "How is it different?" (no mane or fur around its head, much smaller) "Why do you think the baby baboon rides on its parent?" (it's tired, for safety)

- On page 12, note that the docent told the author that the zoo had recently celebrated over 1,000 births!

- Page 13 shows and reads, "Families of chimps playing and grooming each other affectionately." Ask, "What does the author mean by 'grooming'?" (the chimps pick parasites off each other) "What does 'affectionately' mean?" (with care, not harshly)

- Pages 14–15 show orangutans and gorillas with babies. Ask, "How are these parents caring for their young?" (holding, caressing, playing)

- Pages 16–17 show a cheetah with cub and a lioness with cub. Ask, "How are these parents and their young similar?" (same coloring: cheetah cub has spots like its parent, lion cub has same coloring as well) "How are they different?" (both young are much smaller than their parents but otherwise look very much like them)

- Page 18 shows a dorcas gazelle and an Arabian oryx. Ask, "What is this gazelle fawn doing?" (nursing, getting milk from its mother) "How does this Arabian oryx baby look like its parent?" (same coloring) "How is it different?" (no horns, much smaller)

- Page 19 shows a giraffe, with a marabou stork in the distance. Ask, "What do you think this stork is bringing to the nest?" (food for its young) "How does this giraffe resemble its parent?" (it looks exactly the same except much smaller)

- Page 21 shows African elephants and reads, "Everyone loved the playful, affectionate elephants—especially the baby guarded by its mother." Ask, "How is the mother 'guarding' its baby calf?" (keeping its trunk on it, holding it close, keeping an eye on the people) "The

Lesson 4: Monkey Business

author uses the word 'affectionate' again. Why do you think she does this?" (to show that animal parents are loving and caring toward their offspring)

- Page 24 shows a red kangaroo. Ask, "How is the mother kangaroo caring for its joey?" (she carries her joey in her pouch where she can protect it)

- Page 25 shows a polar bear. Ask, "How is the polar bear parent caring for her cub?" (cuddling and protecting it) (*Note:* On page 25, the author states that "unlike other animals, pandas have not bred successfully in captivity so far." This is no longer true. In addition to successful captive breeding programs in China, several zoos have had panda births. They are more difficult to breed than other animals, though, and there has been limited success in releasing captive-bred pandas into the wild. This is a great opportunity to share with students the fact that scientific information changes over time!)

- Page 28 shows a mute swan. Ask, "How are the cygnets, or baby swans, similar to their parent?" (they swim, have beaks and feathers) "How are they different?" (they are brown, not white, and much smaller) "Why do you think the cygnets may be brown?" (it provides camouflage from predators, especially in the nest)

- Page 31 shows the nursery with a zoo employee feeding a leopard cub with a bottle. Ask, "Why do you think some zoo babies are in the nursery?" (they may need extra care)

- After reading page 32, ask, "What do you think the author thinks about zoos now?" (that they are good places because they breed and protect animals) Ask students to give a thumbs-up, thumbs-down, or thumbs-in-between to show how they feel about zoos now. Allow for brief discussion and assure students that they will have more time to discuss this important topic soon. For now, you'd like to focus on the ways animal parents cared for their young in the book.

3. Write the question "How do animal parents care for their young?" on the board. Ask students to share ideas and write them on the board, trying to organize ideas into big categories such as "Feed," "Protect," "Groom," and "Show Affection." Your class Animal Parents Concept Map might look similar to the one in Figure 5.11 (p. 134).

FIGURE 5.11.

Sample Animal Parents Concept Map

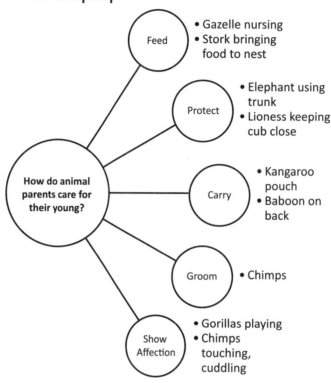

Keep the concept map available during the remainder of the lesson, as your class will build on it as students engage in more research and reading.

Explore: Animal Parents and Their Young Four Square Internet Research and What's My (Baby) Name? Extension Activity

1. Explain to students that they will now have a chance to virtually visit several zoos and learn about how different animals care for their young. Distribute the Animal Parents and Their Young student handout (p. 144), which is a four square graphic organizer. Set up internet access so that there are two or three students per device (e.g., laptop, PC, or iPad).

2. Use a projector to show students the website *www.zooborns.com*. Inform students that they are now zoologists, scientists who study animals, and they are doing research on how animal parents care for their young. When scientists try to understand information about the world, they look for patterns. Your students' job as scientists will be to find examples of animal parents caring

Lesson 4: Monkey Business

for their young in different ways. Show students the home page, scrolling down to look at some of the most recent births. Demonstrate for students how to choose specific animals by selecting a name on the right-hand side of the screen. (*Note:* Although this website has a lot of verbiage, students can scroll through photos of animals to observe behaviors of parents, such as a tiger parent licking its cub or an aardvark nursing its baby. Several of the animals also have videos associated with them.)

3. Instruct students that when they identify a photo of a parent caring for its young, they should draw a picture of it in the square that has that type of behavior and write the animal's name. The squares have the topics "Protect," "Feed," and "Carry," as well as a wild card. The wild card is for a behavior of the student's choice. Challenge students to try to find a different type of behavior, such as grooming, showing affection, or even communicating with their offspring.

4. As students are working, clarify any questions and evaluate informally by checking to make sure students are using the graphic organizer correctly.

5. When students have completed their research, have them switch partners and present their research to each other. After a few minutes, ask students, "Were you able to find any new examples for 'Protect,' 'Feed,' or 'Carry' that we hadn't included on our Animal Parents Concept Map?" (Add any examples to the concept map.) Ask, "Was anyone able to find any examples of 'Grooming' or 'Showing Affection'?" (Add these to the concept map.) Finally, ask, "What did you have in your wild card square? Was anyone able to find any new behaviors that we haven't seen yet?" (students may mention parents building homes or communicating with their offspring; if no new behaviors are mentioned, it is fine, as new ones will be introduced in the next activity)

6. Extension activity: If students have extra time, distribute copies of the What's My (Baby) Name? handout (p. 145). Have them scroll down to the "Baby Animal Names" link (or visit *www.zooborns.com/zooborns/baby-animal-names.html*) and learn the names for the various babies they have found. An answer key for the handout is in Figure 5.12 (p. 136).

FIGURE 5.12.

What's My (Baby) Name? Answer Key

Animal	Baby Animal Name
Camel	Calf
Coyote	Pup or whelp
Eagle	Eaglet or fledgling
Fox	Kit
Giraffe	Calf
Kangaroo	Joey
Llama	Cria
Turtle	Hatchling
Zebra	Colt or foal

Explain: *How Animal Babies Stay Safe* **Read-Aloud and** *Born in the Wild: Baby Mammals and Their Parents* **Mini Read-Aloud**

1. Have the Animal Parents Concept Map in view. Distribute three or four sticky notes to each student, along with pencils. Show students the cover of the book *How Animal Babies Stay Safe*, by Mary Ann Fraser, which has an illustration of a tiger parent and her cubs. Ask students, "Does this cover show a way that the tigress is caring for her cubs?" (protecting them by keeping them close, watching for danger) "We're going to learn about some new animal parents today and how they care for their young. Your job is to jot down any new ways that parents do this. When you hear something new and interesting that you'd like shared on our concept map, jot it on a sticky note and we'll add them to our map after reading."

2. Read the book aloud, periodically stopping to allow students to jot down their notes. Then, allow students to share their notes by posting them on the concept map.

3. Below are some guidelines for the new ideas that the book introduces in case students don't mention them in their sticky note postings:

    - Page 10 shows a dog with puppies. The author states that the puppies "cannot keep themselves warm or get their own food." Ask, "What is something that the dog mom does to care for her young that we haven't mentioned before?" (*keeps them warm*)

- Pages 12–13 show various animal parents *building homes* for their families.

- Pages 20–21 show various animal parents alerting their young to danger. Ask, "How do some animals *alert* their young to danger? (howling, barking, scent, or quick fin movements)

- Pages 22–23 show animal parents risking their lives for their young. Ask, "Why do some animal parents *risk their lives* to protect their young?" (because their young are very important to them, they will fight off enemies even if it is risky for them)

- Pages 24–25 show mother raccoons and ask, "How do mother raccoons *trick enemies*?" (they act as decoys and distract the enemy away from the babies)

- Show students pages 26–27 and ask, "How do elephants protect their young?" (they *stay in a herd* and keep the babies in the middle)

4. Ask students, "The book also mentions some ways that baby animals are able to stay safe by themselves. What are some of those ways?" (camouflage with surroundings; huddle together to stay warm and safe, like penguin babies do when they form a crèche)

5. Tell students that there is one more thing that parents do that wasn't mentioned in either of the books, but it is very common in mammals. Ask, "What is a mammal?" (an animal that feeds its babies milk; some examples are dogs, cats, horses, people, lions, and elephants) Open up the book *Born in the Wild: Baby Mammals and Their Parents*, by Lita Judge, to pages 36–37, which show a sea otter mom and pup. The text reads, "The baby learns." Ask students, "What do you think this baby learns?" (accept all answers) Read the next two pages, which explain that some animal parents *teach* their young how to survive.

- Ask, "How do these animal parents teach their young to survive?" (pika parents *teach* their young that a loud chirping scream means danger, and the baby pikas *learn* that they should go back to the den; the sea otter parent *teaches* her pups how to swim and dive, to find clams and urchins, and to open clams with rocks; young orangutans take at least 10 years to *learn* from their mothers where to find food, how to use sticks for tools, and how to build a sleeping nest)

> **Misconception Alert**
>
> With so much discussion of how animal parents care for their young, it is easy for children to develop the misconception that animal babies are helpless. This is not the case and in fact, the *NGSS* standard 1-LS1-2 emphasizes the fact that many baby animals do their part to protect themselves once they are able. Actions such as responding to parents' danger calls, crying when they're hungry, and imitating the parent as they learn different survival tasks are all ways that animal babies contribute to their own growth and survival. Be sure to take time to discuss these behaviors.

6. Add "Teach" to the Animal Parents Concept Map, along with examples. Let students know that they can look through the rest of the book when they are working on the next project.

Elaborate: Animal Baby Perspectives Project: How My Parents Care for Me

1. Tell students that they will now have a chance to become experts on one kind of animal and the way its parents keep it safe. Distribute copies of the two How My Parents Care for Me student handouts (pp. 146–147). Read aloud the instructions on the first handout (p. 146), which includes a planning checklist and rubric.

2. Tell students that the second handout can help them plan and outline their project. It has blanks for students to fill in and a small four square graphic organizer. Animals can be either chosen by students (preferred) or randomly assigned. If students are choosing, encourage them to select different animals from other students so that there is a good mix for their presentations.

3. Provide students with the books that were read in class, as well as others, and be sure that the Animal Parents Concept Map is accessible. Also, provide students with internet access for research using Zooborns (*www.zooborns.com*) and the San Diego Zoo Kids website (*http://kids.sandiegozoo.org/animals*). The latter has very clear information on zoo animals, along with videos on many of the animals. Provide paper and craft materials as needed for the project.

Lesson 4: Monkey Business

Evaluate: How My Parents Care for Me Presentations

1. Have students present their projects to the class. Assess using the handout list of required elements, giving extra points for clarity and creativity.

Elaborate: Do We Need Zoos? Article or List and T-Chart

1. Show students the cover of *My Visit to the Zoo* to remind them of the book. Ask, "Do you recall how the author felt at the beginning of the book?" (she didn't like zoos) Ask, "And what was her opinion after her visit to the zoo?" (she thought they were good places that cared for and protected animals) Explain that people have different opinions on zoos; zoos are *controversial* because some people think they are needed and some people don't. We're going to read about some of the reasons for this controversy, and then you'll have a chance to share your opinions.

2. Distribute copies of the Do We Need Zoos? handout, using either the narrative article (p. 148) or the bulleted list (p. 149). (*Note:* Two versions have been included to give the teacher the choice of how to present the material. One is in narrative form; this can be read aloud by the teacher, by older students in pairs, or by student volunteers. The other version is presented as bulleted points and is less detailed than the narrative version. This can be presented by the teacher using a document reader displaying an enlarged image of each of the points, with students following along or reading chorally with the teacher, or with students taking turns reading bulleted points aloud.)

3. After reading, draw a yes/no T-chart on the board, with the heading "Do We Need Zoos?" at the top. Ask students to share, *in their own words*, some of the reasons people give on each side of the issue. This will allow you to assess student understanding and reinforce some of the key reasons. Remind students that you are not asking for their opinions yet; you would simply like to summarize some of the reasons they learned from the reading. Summarize students' responses on the T-chart, which might look something like Figure 5.13 (p. 140).

FIGURE 5.13.

Sample Yes/No T-Chart

Question: Do We Need Zoos?

Yes	No
Animals are safe in the zoo.	Wild animals should be free.
Zoos breed endangered species (example: California condor).	Animals don't have a choice whether to be in a zoo or not.
People care more about animals when they visit them up close.	The habitats are too small (example: elephants walk long distances in large groups in the wild).
Scientists can study animals in the zoo to learn more about them.	A lot of the animals that are bred in zoos aren't released into the wild.

Evaluate: Yes/No Line Debate and Are You for Zoos? Position Poster or My Opinion on Zoos Opinion Letter

1. Post the "Yes" sign (with the thumbs-up picture) on one side of the room and the "No" sign (with the thumbs-down picture) on the other. Put a 3-foot line of masking tape on the floor in the center of the room so that it divides the room into two sides, with one side as "Yes" and the other side as "No." (For young students, you can put the signs on the floor on either side of the line to provide a closer visual cue.) The room should look similar to Figure 5.14.

2. Write the following sentence frames on the board:

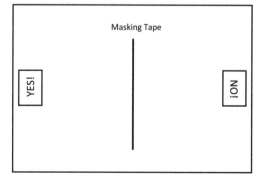

FIGURE 5.14.

Room Setup for Yes/No Line Debate

Lesson 4: Monkey Business

"I think that _____ because _____."

"I disagree with you because _____."

"According to the article/book/website/my personal experience, _____."

"I think you should also consider _____."

3. Tell students that you'd like to hear their opinions on zoos, and we're going to share our opinions in a fun way. Remind students that there is *no wrong answer*. There are good reasons on both sides of the issue. The important thing is that they share their ideas in a thoughtful and respectful manner. Explain that you would like to know whether they think we need zoos. If they think the answer is yes, they should go to the "Yes" side of the line. If they think the answer is no, they should go to the "No" side of the line. They may not straddle the line by putting one leg on either side. For now, you are asking them to choose a side. *Tip:* If you have never done this type of activity before, you may wish to have students write down "yes" or "no" before they get up to go to either side of the line. They can write it on a sticky note and post it on their shirts or simply write it down on paper. This will encourage students to stick to their opinions rather than joining friends or changing their minds because they don't see as many students going to their side of the line. Once your students have done this type of discussion a few times, they will become more confident and will trust that it is a safe environment to have a dissenting opinion.

   As students go to their side of the line, you may notice some of them going far to the edge of the room (to show that they feel strongly about the issue) or coming close to the line (to show that they feel mixed about the issue). This is a wonderful aspect of the activity; another activity in Lesson 8 (p. 249) of this book uses a yes/no spectrum line, which allows for a continuum of opinions. The present activity uses the yes/no line to force students to choose a side, which emphasizes the difficulties in this type of decision making. It is also simpler for children to remember which side is which.

4. When students are on their chosen sides, point to the sentence frames on the board, and ask them to share their opinions in that format. Remind them that they need to take turns and listen hard to other people's opinions. Have students share their opinions, making sure they justify their opinions (claims) using evidence from the books, the article, the websites, or their personal experiences at zoos. Also pay careful attention to whether you see anyone switching sides or scooting toward or away from the line. If you notice that, you can say, "I see that some people may be rethinking their positions on

the issue, and that's fine. Would any of you like to share your thoughts on that?" Ask, "What arguments *persuaded* you to change your mind?" When all students who wish to share their opinions have shared (and all students will have the opportunity to share their opinions in written or pictorial format as well), ask students to move to their final spots so that they can see where their classmates stand on the issue. Ask, "How did it feel to share your opinions this way?" (accept all answers) "What other information might be helpful for you to know about zoos to help you make your decisions?" (accept all answers; students may note that some zoos may be better than others in terms of how they care for their animals, and they may indicate that they would want information on individual zoos rather than making a sweeping decision about all zoos) Encourage students to look at different zoos' websites, as many will indicate whether they have received accreditation from organizations that have more stringent standards than are required by law, such as the Association of Zoos & Aquariums (AZA).

5. A choice of two handouts is available for the next evaluation activity. The first, Are You for Zoos? (p. 150) is the simpler of the two and asks students to fill in the blank as to whether they think we need or don't need zoos, complete the sentence "My top reason is _____," and draw a captioned picture of their top reason. The second, My Opinion on Zoos (p. 151) requires more writing, as it asks students to give three reasons and is in the general form of an opinion letter. For earlier grades or classes with many English language learners, you may wish to use only Are You for Zoos? For older grades, you may wish to use both, with My Opinion on Zoos being used first to elicit more detailed writing, followed by Are You for Zoos? which can be completed afterward and displayed. My Opinion on Zoos can also be used as a homework assignment. Tell students that they have a chance to educate the school community about the issues of zoos by creating opinion posters that will be displayed around the school. Distribute the handouts of your choice, and allow students to display completed Are You for Zoos? posters for the school community.

Going Deeper

- Students can develop booklets or e-books on animal parents and their young.

- Students can observe webcams on zoo websites (such as the San Diego Zoo website included in the list of resources) to watch parenting behavior in real time.

- Students can research information on their local zoo to see how the zoo's work is aligned with the yes/no arguments they read about.

- Students can research zoo breeding programs to add evidence to the yes/no T-chart about breeding and releasing successes and failures.

- Students can interview parents, siblings, and others about their thoughts on zoos.

- Students can research the history of zoos and create a timeline.

Zoologist Name: _____ Date: _____

Animal Parents and Their Young

Draw and describe an example of each of the behaviors in the boxes below. Be sure to include the animal names!

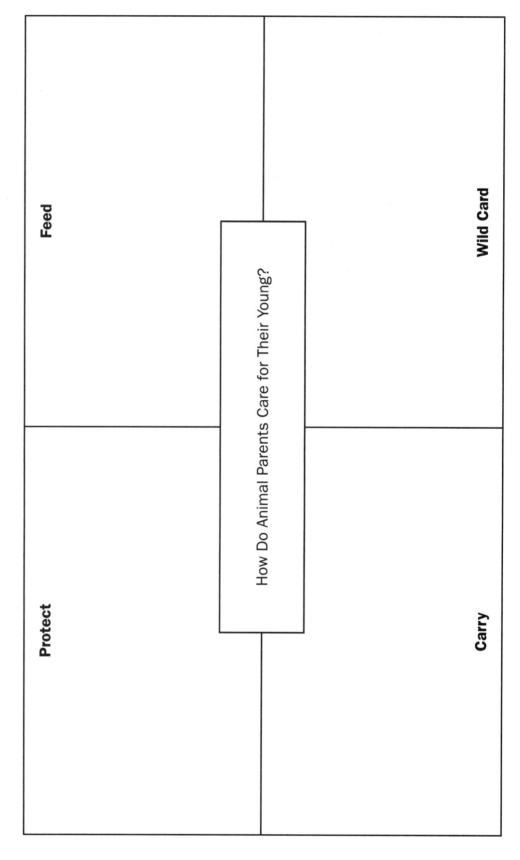

144

NATIONAL SCIENCE TEACHING ASSOCIATION

Name: _____ Date: _____

What's My (Baby) Name?

Using the Zooborns website (*www.zooborns.com*) or other resources, try to find out the names of these amazing animal babies!

Animal	Baby Animal Name
Camel	
Coyote	
Eagle	
Fox	
Giraffe	
Kangaroo	
Llama	
Turtle	
Zebra	

Name: _____ Date: _____

How My Parents Care for Me

Animal Baby Perspectives Project

For this project, research how parents of one type of animal care for their young. Then, teach what you learned to your class *from the perspective of the young animal!* You can write a story, make a poster, create a 3-D diorama or sculpture, perform a play, write and sing a song, or draw a comic strip. Your project must include the following:

- [ ] The type of animal you are
- [ ] Your baby animal name (for example, chick, cub, calf)
- [ ] Where you were born
- [ ] What you eat

- [ ] One way that you are <u>similar to</u> your parents
- [ ] One way that you are <u>different from</u> your parents
- [ ] One way that you take care of yourself or protect yourself
- [ ] One way that your parents take care of you

Check the boxes ✓ as you complete each step. 8 checks = 8 pts. Extra points for clarity and creativity!

Name: _____ Date: _____

How My Parents Care for Me

(Use this form to help you plan your project.)

I am a baby _____. Baby _____s are called
                   type of animal   type of animal

_____. I was born in _____. I like to
  name of baby animal   place where this type of animal lives

eat/drink _____.
            food or drink

One way that I am <u>similar to</u> my parents is …	**One way that I am <u>different from</u> my parents is …**
One way that I can take care of myself (or protect myself) is …	**One way that my parents take care of me is …**

It's Still Debatable! Using Socioscientific Issues to Develop Scientific Literacy, K–5

Do We Need Zoos?

Some Zoo History

Humans have kept wild animals for thousands of years. Records from ancient Egypt show that animals were kept in large **collections** as a sign of wealth and power. Animal collections called **menageries** were common in Europe through the 1800s. The Philadelphia Zoo was the first zoo in the United States. It opened in 1874. At that time, zoos focused on making the animals easy for humans to see, so the animals were kept in small, simple cages.

New and Improved?

As years went by, people became more concerned about the **welfare** of the animals. Today, most zoos focus on **conservation,** which means protecting animals, especially those that are **endangered.** They also try to educate the public so that people will care about animals. But there is a lot of **controversy,** or disagreement, about whether we need zoos.

Woo-hoo for Zoos!

People who support zoos say that they protect animals by keeping them safe. They also argue that many animals that were nearly **extinct** have been bred in zoos. For example, only 22 California condors were alive in the 1980s. Thanks to zoos, over 400 condors exist in zoos and in the wild today. Supporters also argue that people care more about animals when they see them in zoos. Most people would never have the chance to see animals like elephants and tigers in the wild. Many of today's zoos also have large, **realistic** habitats that are healthier for the animals. Zoos argue that their zookeepers care about the animals and that they follow laws that protect animals. Many zoos also have animal **enrichment** activities, such as hiding the animals' food or making toys to keep the animals busy and happy. Scientists also study animals in zoos to learn more about them. Finally, many zoos use money they earn to protect animals in the wild.

Boo for Zoos!

People who are against zoos say that it is cruel to keep animals in **captivity** where they are not free. They also argue that humans don't have the right to keep wild animals, especially because animals don't have the choice of whether they want to live in zoos or not. They also say that people only have zoos for their own **entertainment.** People who are against zoos also feel that the animal habitats are never as real or as large as habitats in the wild. They point to elephant habitats as an example. Elephants in the wild often walk for miles each day and travel in very large packs. This is not possible in most zoos. Some people also argue that many animals that are bred in zoos never get **released,** so it doesn't make sense to keep breeding them. There also are stories about zoos that are unclean and where the animals are not well cared for. Some people say that the laws protecting animals in zoos aren't **enforced** well. Finally, they argue that the money that is spent on keeping zoos could be better spent on protecting the animals' wild habitats.

Do We Need Zoos?

Read what people say for and against zoos to help make your decision.

Woo-hoo for Zoos!

- Zoos protect animals by keeping them safe and by **breeding** them.
- Many animals that were nearly **extinct** have been saved by zoos.
- People care more about animals when they see them and learn about them in zoos.
- Most people would never have the chance to see animals like elephants and tigers in the wild.
- Zookeepers care about the animals.
- Animals are kept in **realistic** habitats.
- Zoos have **enrichment** activities to keep the animals busy and happy.
- Scientists study zoo animals, which helps us understand them better.
- Many zoos use money they earn to protect animals in the wild.

Boo for Zoos!

- It is cruel to keep animals in **captivity** where they are not free.
- Humans don't have the right to keep wild animals.
- Animals don't have a choice about whether they are kept in zoos.
- People keep animals in zoos for their own **entertainment.**
- Zoo habitats are never as real or as large as habitats in the wild.
- Animals that are bred in zoos rarely get **released** into the wild.
- Some zoos are unclean and the animals are not well cared for.
- Laws that protect animals in zoos aren't **enforced** very well.
- Money that is spent on zoos could be better spent on saving wild habitats.

Name: _____ Date: _____

Are You for Zoos?

I think that we _____ zoos.
　　　　　　　need **or** don't need

My top reason is _____.

Here is my picture:

150　　　NATIONAL SCIENCE TEACHING ASSOCIATION

Name: _____ Date: _____

My Opinion on Zoos

I think that we _____ zoos.
                 need **or** don't need

Here are my reasons.

Reason #1

Reason #2

Reason #3

That is why I think we _____ zoos.
                        need **or** don't need

Lesson 5

Soaky Doaky

What's the Best Way to Clean Up Spills?

Suggested Grade Levels

K–2

Driving Questions

- How do properties of materials make them suited for different purposes?
- How can we design a fair test?

Lesson Overview

This lesson empowers students to test whether the claims that advertisers make about their products can be verified through scientific testing. First, students watch paper towel commercials to determine the properties that are promoted by advertisers. They then work collaboratively to design and implement fair tests for paper towel absorbency and wet strength. After determining which is the best paper towel and watching a video about how paper towels are made, students weigh the pros and cons of using paper towels versus cloths and sponges, extending decision making about spill cleanup to include environmental, cost, and hygienic considerations. Students then have the option to create a commercial, develop a presentation, or write a letter to a paper towel company to share their findings. Throughout this lesson, students are introduced to the complexities of consumer decision making and the power of scientific inquiry.

Connecting to the *NGSS*

(See full alignment in Table A.5 on p. 506.)

- PS1.A: Structure and Properties of Matter

- Different properties are suited to different purposes. (2-PS1-2)

Societal Issues

Consumer Product Advertising, Environmental/Sustainability Concerns

Nature of Science

- Scientific Investigations Use a Variety of Methods
    - Science investigations begin with a question.
- Science Addresses Questions About the Natural and Material World
    - Scientists study the natural and material world.

CCSS Connections

- English Language Arts
    - W.K.3. Use a combination of drawing, dictating, and writing to narrate a single event or several loosely linked events, tell about the events in the order in which they occurred, and provide a reaction to what happened.
    - W.1.8. With guidance and support from adults, recall information from experiences or gather information from provided sources to answer a question.
    - W.2.7. Participate in shared research and writing projects (e.g., read a number of books on a single topic to produce a report; record science observations).
- Mathematics
    - MP.2. Reason abstractly and quantitatively. (2-PS1-2)
    - MP.4. Model with mathematics. (2-PS1-2)
    - MP.5. Use appropriate tools strategically. (2-PS1-2)

NCSS Connections

- Theme 7: Production, Distribution, and Consumption

Lesson Plans

- Evaluate how the decisions that people make are influenced by the trade-offs of different options.
- Theme 8: Science, Technology, and Society
    - Identify examples of the use of science and technology in society as well as the consequences of their use.

C3 Framework:

- Dimension 4: Communicating Conclusions
    - D4.1.K-2. Construct an argument with reasons.
    - D4.2.K-2. Construct explanations using correct sequence and relevant information.
- Dimension 4: Taking Informed Action
    - D4.7.K-2. Identify ways to take action to help address local, regional, and global problems.

UDL Toolkit

Multiple Means of Engagement	Multiple Means of Representation	Multiple Means of Action and Expression
Studying a product that is familiar to students and examining advertising and environmental concerns that are relevant to students promote engagement.	Using commercials, hands-on investigations, and a four-corners debate conveys information through visual, auditory, tactile, and kinesthetic modalities.	Using an advantages/disadvantages graphic organizer, a card sequencing activity, a letter template, and a checklist helps students manage information.
Giving students the opportunity to brainstorm and perform their own experiments heightens the salience of goals and objectives.	Providing visual prompts along with text for key vocabulary such as *absorbency* and *wet strength* helps clarify language and symbols.	Providing options of writing a letter, giving a presentation, or developing a commercial gives students maximum opportunity to communicate what they've learned.

Lesson 5: Soaky Doaky

Suggested Schedule and Sequence

- Day 1: **Engage** with Commercial Properties, and **Explore** and **Explain** with Paper Towel Testing for Wet Strength

- Day 2: **Explore** with Paper Towel Testing for Absorbency

- Day 3: **Explain** with Results, Ratings, and Reveal! and **Elaborate** with How Paper Towels Are Made video and sequencing activity

- Day 4: **Elaborate** with Paper Towels Versus Cloths and Sponges Read-Along and Four Corners Activity: What's Most Important to You?

- Day 5: **Evaluate** with Write a Letter to a Paper Towel Company or Make a Presentation or Video for Your School About Paper Towels

Materials

For Paper Towel Testing

(per class)

- 3 rolls of paper towels (different brands), marked A, B, and C
- Set of unit or alphabet blocks (for weights)
- Water supply
- Timer or clock (optional)

(per team of 3–4)

- Tub, basin, or large bowl
- Measuring cup
- 3 identical clear cups marked A, B, and C
- Wide funnel that fits over the cups

For How Paper Towels Are Made Sequencing Game

(per team of 2–3)

- 1 precut set of sequencing game cards in a plastic or paper bag

For Four Corners Activity

(per class)

- 1 set of copied corner cards (pp. 177–180) (*Note:* Copying the pages onto larger paper will help cue students more easily.)
- Tape

(per student)

- Safety glasses or goggles, vinyl gloves, and a nonlatex apron

Student Handouts

- Paper Towel Testing: Wet Strength
- Paper Towel Testing: Absorbency
- Our Own Absorbency Test (copy on the back of the Paper Towel Testing: Absorbency sheet)
- Paper Towels Versus Cloths and Sponges Read-Along
- Letter to a Paper Towel Company or Presentation or Video for Your School About Paper Towels Storyboard (student choice)
- My Paper Towel Project Checklist
- Paper Towel Testing Week Letter to Families (optional)

Safety Notes

1. All students must wear safety glasses or goggles, vinyl gloves, and nonlatex aprons during the setup, hands-on, and takedown phases of the activity.
2. Use caution when working with glassware or plasticware, which can shatter if dropped and cut skin.
3. Immediately wipe up any spilled water on the floor to avoid a slip-and-fall hazard.
4. Immediately pick up any items dropped on the floor to avoid a slip-and-fall hazard.
5. Wash hands with soap and water after completing this activity.

Media

Videos

- Paper towel commercials
  www.youtube.com/watch?v=-y30RsG6DJ8
  www.youtube.com/watch?v=ljUciFgLArA
  www.youtube.com/watch?v=kw9i7vIWxgc

- "This Is How Paper Towels Are Made"
  www.businessinsider.com/how-paper-towels-are-made-2017-10

Background for Teachers

Is Bounty really the "quicker picker upper"? Do paper towels really differ? Consumers are constantly exposed to advertisers' claims about their products but rarely test those claims in a systematic way. Although children may not be the purchasers of the products, they are nonetheless consumers of information. In fact, children often have tremendous influence on their parents' purchasing habits, a situation that can lead companies to target their marketing toward children. As students will be future product consumers, it is important that they learn to weigh advertisers' claims and product attributes in a thoughtful manner.

In this lesson, students have the opportunity to test advertisers' claims through laboratory investigations. To do this, they need to develop controlled or fair tests. This means that the tests can't favor one brand over another. Students need to ensure that the only thing that differs between their tests is the paper towel brand; all the other variables, or things that can be manipulated, such as amount of water and time, must be the same, or controlled. In addition, the identity of the paper towel brands being tested must be hidden from students. This is known as a blind test, and it keeps the testers from favoring one brand over another, thus adding to the validity of the fair test.

This lesson focuses on the two main properties of paper towels, absorbency and wet strength, in addition to price, to prompt students to consider how important they think each of these attributes is for making a final decision on which paper towel is best. Of course, there are other ways to clean up spills. Sponges and cloths are alternatives that each have pros and cons. While these products last longer and are less expensive per use than paper towels, they may be less convenient and less sanitary than single-use paper towels. Generally speaking, reusable products such as sponges and cloths create less waste than single-use products, and they may use fewer natural resources for their production, thereby making them more environmentally friendly. However, some paper towel companies are now using high percentages of recycled paper rather than harvesting trees for wood pulp, and they may even use unbleached paper and recyclable or biodegradable packaging materials, all of which adds to the complexity of this issue.

While all these considerations may seem a bit overwhelming, it is important to remember that the goal of this lesson is to introduce students to these issues and allow them to weigh the various factors for themselves, not to derive one "right" answer. For example, students may decide that different products, such as sponges or cloths, are better for certain types of cleanups but that paper towels are best for others. Or they may conclude that a paper towel that performs slightly less well in absorbency or wet strength tests than its competitors is still a better choice because of its low cost or use of recycled materials. Regardless of students' priorities as they weigh their choices, students will recognize the importance of scientific testing in determining product qualities and see that they have the power to use science to inform their decision making.

Additional Resources

Article on "The Best Paper Towel"

- *https://thewirecutter.com/reviews/best-paper-towel*

Article on the pros and cons of paper towels versus cloths

- *www.huffingtonpost.com.au/2016/02/22/paper-towels-versus-cloth_n_9294566.html*

Article on consumer product testing

- Kahn, S. 2005. Savvy consumers through science. *Science and Children* 42 (6): 30–34.

Note: This lesson is based in part on the following:

- Sneider, C. I., and J. Barber. 1990. *Paper towel testing: Teacher's guide.* Berkeley: University of California, Berkeley/ GEMS/Lawrence Hall of Science.

5E Lesson Plan

Engage: Commercial Properties

1. Tell students that they are going to watch some commercials about a very common product: paper towels. Begin by showing students two or three commercials about paper towels such as the following:

Lesson 5: Soaky Doaky

- *www.youtube.com/watch?v=-y30RsG6DJ8*
- *www.youtube.com/watch?v=ljUciFgLArA*
- *www.youtube.com/watch?v=kw9i7vIWxgc*

Then, ask students:

- "What are some of the ways people use paper towels?" (answers may include cleaning up spills, wiping down surfaces, drying hands, wiping faces)

- "What qualities, or *properties*, of paper towels seem to make a paper towel good?" (answers may include absorbency/soaking up liquid, strength/not falling apart when wet, low price, pretty design)

- "How do advertisers, the people who make the commercials, communicate those qualities to you?" (they show people using the paper towels or do tests against other brands)

- "How can we know if the claims advertisers make about their products are true?" (we can test them ourselves)

Explain to students that they are going to have a chance to test paper towels for two properties: wet strength (how strong the paper towel is when wet) and absorbency (the paper towel's ability to soak up water).

Explore and Explain: Paper Towel Testing for Wet Strength

1. Organize students into groups of three or four. Distribute copies of the Paper Towel Testing: Wet Strength data sheet (p. 172). Show students the three paper towel rolls marked A, B, and C. (*Note:* If your school uses paper towels that don't come on a roll, simply place a pile of them alongside the other two other brands. Be sure to make note of what brand each letter stands for!) Ask students, "Why do you think I have replaced the brand names with letters?" (answers should relate to the idea that we want to have a fair test; we don't want to favor some brands over others)

2. Show students the materials, which include measuring cups, plastic basins or tubs, a water supply, and weights (either unit blocks or alphabet blocks). Ask students, "How can we test how strong a wet paper towel is with these materials?" (allow students to propose ideas; typical tests involve wetting the paper towels and seeing how many blocks they can hold without breaking)

3. Then, hold a discussion on conducting a fair test. Ask students, "What if I put a little water on Brand A and a lot of water on Brand B. Would that be a good test?" (no, because they should have the same amount of wetness to be a fair test) Reiterate that to have a fair test, we want to use the same amount of water and do the same test for each of the paper towels.

4. Demonstrate a test for wet strength by inviting two students to assist you. Measure 100 ml of water (or ½ cup, which is 118 ml) in a measuring cup. Have two people hold the paper towel over the basin or tub, and have the third person pour water onto the middle of the paper towel. Then, have the third person demonstrate slowly adding blocks, one at a time, onto the towel (Figure 5.15). Stop after two blocks so that students don't see the paper towel break, but let them know that in their tests, they will continue adding blocks until the paper towel breaks. They will then record the number of blocks the paper towel held before it broke on the data sheet. For example, if the paper towel held five blocks and then broke on the sixth, they would enter 5 on the data sheet because that's the highest number the paper towel could hold.

FIGURE 5.15.
Testing Paper Towel Wet Strength

5. Have student teams repeat the experiment for each of the paper towels and record their answers on their data sheets. *Tip:* Tell students to keep their data sheets away from the table where the testing is being done so that they don't get wet!

6. After students have finished testing, ask:

    - "What question are we trying to answer with this experiment?" (which paper towel is strongest when wet)

    - "Why is this an important property for a paper towel?" (so it doesn't fall apart when you use it)

    - "Which paper towel brand was the strongest when wet?" (answers will vary)

    - "How do you know?" (looking at the data, the paper towel that held the most blocks was the strongest)

Lesson 5: Soaky Doaky

- "Why did [might] we have different results from different teams?" (answers will vary; some teams may have pulled on their towels, the blocks may have been piled up differently, paper towels may be different even though they're on the same roll, teams may have favored a particular towel and helped it "win")

- "What did we do to try to make this a fair test? (used the same amount of water, masked the brand names, used the same procedure for each brand)

- "What is another important property of paper towels?" (absorbency)

Let students know that they'll be testing absorbency next time!

Explore: Paper Towel Testing for Absorbency

1. Have students return to their testing groups. Briefly review the prior test (e.g., what property was tested, how we tested for that property, what made it a fair test). Then, ask students, "What is another important property of a good paper towel?" (absorbency)

2. Show students the materials for this test, which include basins, tubs, or large bowls half filled with water; three clear cups marked A, B, and C; funnels; and either a timer or clock. (*Note:* For young students, you can have them time by counting, "One-one hundred, two-one hundred, three-one hundred ..." to time the tests.) Ask, "Can anyone think of a way that we can test how well paper towels absorb, or soak up, spills?" (allow students to propose different test ideas) Typical tests for absorbency include variations of the *spill test*, creating a spill and seeing how much a paper towel picks up in 10 seconds; the *squeeze test*, dipping the paper towel in a basin of water and then squeezing out the water into measuring cups to see which held the most water; or the *leftover test*, dipping the paper towel into a measuring cup with water in it, taking it out, and then seeing how much water is left in the cup (less water left in the cup means better absorbency).

3. Then, hold a discussion on conducting a fair test for absorbency. Ask students, "What if I created a small spill for Brand A and a big spill for Brand B? Would that be a good test?" (no, because to be fair, they have to have the same size spill) "What if I put Brand A in the water for 10 seconds and Brand B in the water for 20 seconds? Would that be a good test?" (no, because more time might give an advantage; we want a fair test, so they have to have the same amount of time) "What if I put Brand A and Brand B in the water for the same

amount of time, but then squeezed out the water gently for A and used all my strength to squeeze out B? Would that be a good test?" (no, because to make the test fair, you need to squeeze both towels with the same strength)

4. Inform students that today, we are going to focus on the squeeze test, but if their team has time, they can conduct one of the other tests afterward.

5. Distribute copies of the Paper Towel Testing: Absorbency data sheet (p. 173) to students. Demonstrate a test for absorbency by inviting two students to assist you. Place the funnel over the cup, and explain that funnels can help direct the water into the cup. Have one student dip the Brand A paper towel into the water basin, and have the other student time the dipping for 10 seconds using a timer or clock or simply counting to 10. Have the first student pull the paper towel out of the basin and squeeze it out into the funnel over the cup marked A. Encourage the second student to help make sure the cup stays in place and to give the paper towel an additional squeeze to make sure all the water is out. Explain to the class that they will put that cup aside momentarily and repeat the experiment for each of the paper towels, squeezing the water out of each into its own cup. They will then compare the amounts of water in the cups by placing them side by side and recording the results on the data sheet (e.g., high, medium, low).

6. Have student teams repeat the experiment for each of the paper towels and record their answers on their data sheets. *Tip:* Remind students to keep their data sheets away from the table where the testing is being done so that they don't get wet!

7. When students have finished the squeeze test, have students turn their sheets over to the "Our Own Absorbency Test" side. Allow students to try the leftover test, spill test, or a completely original test by writing out the steps of the test, recording data, and answering the questions.

Explain: Results, Ratings, and Reveal!

1. After students have finished testing, ask:

   - "What question are we trying to answer with this experiment?" (which paper towel is the most absorbent)

   - "Why is this an important property for a paper towel?" (because paper towels are supposed to soak up spills)

   - "Which paper towel brand was the most absorbent?" (answers will vary)

- "How do you know?" (looking at the data, the paper towel that held the most water was the most absorbent)

- "Why did [might] we have different results from different teams?" (answers will vary; some teams may not have squeezed as much water out of their paper towels, some may not have kept all the towels in the water for the same amount of time, water may have spilled out instead of getting into the cup, teams may have wanted a paper towel to "win" and didn't squeeze that one as hard)

- "What did we do to try to make this a fair test?" (used the same amount of dipping time and the same amount of squeezing, masked the brand names, used the same procedure for each brand)

2. To analyze the results of the tests, draw a chart like the one in Figure 5.16 on the board.

FIGURE 5.16.

Graphic Organizer for Analyzing Paper Towel Test Results

Paper Towel Brand	Wet Strength Rating (Number of ★s)	Absorbency Rating (Number of ★s)	Total Rating (Number of ★s)
A			
B			
C			

Have volunteers from each team report the rating (number of stars) each paper towel brand received from their team. Record the number of stars from each team in each box. (*Note:* For younger students, draw the stars and count them up to get the total rating; for older students, write the numbers and have students add them). Ask:

- "Which paper towel had the highest overall rating?" (answers will reflect the paper towel brand with the highest total rating)

- "What two properties of paper towels did we test to get this result?" (wet strength and absorbency)

- "Which paper towel was the strongest when wet?" (answers will reflect the paper towel with the highest rating for wet strength)

- "Which paper towel was the most absorbent?" (answers will reflect the paper towel with the highest rating for absorbency)

3. Depending on your results, you may notice that the best overall paper towel may not have been the highest rated for both absorbency and wet strength. For example, your highest overall rated towel may have been the strongest by far, but only second in absorbency. This is an important result to discuss, because it raises the question of whether absorbency and wet strength are *equally important* for paper towels. If you do get this type of result, ask students, "Our best overall paper towel didn't win in both categories that we tested. How can that be?" (it was much better than the others in one category but in the middle for another) "Is there any reason I might want to buy a paper towel that wasn't the overall winner?" (you might have a lot of jobs that require absorbency but not wet strength, such as simply wiping up spills from a counter, or jobs that require wet strength but not absorbency, such as cleaning the sink or a fish tank)

4. Finally, it's time for the big reveal. Unmask the brands and show students the brands that they tested, saying, "Third place goes to …, second place goes to …, and first place goes to …" Allow students a few minutes to share their thoughts on these results.

5. Depending on the age of your students, you can also introduce price as a factor in considering paper towels in a number of ways. For younger students (K–2), you can figure out the cost per sheet of paper towels in advance (by taking the cost of the roll and dividing it by the number of sheets), and add an extra "Price" column to the chart shown in Figure 5.16 (p. 165) where you write in how many cents each sheet costs. You can use actual pennies to make this more concrete, and you can even cover part of a penny with masking tape to represent a price less than 1 cent. Ask students, "Let's look at the cost of each paper towel. Do you think the most expensive one was worth it? Why or why not? Was the least expensive one the lowest rated? How would you weigh this information to decide on a paper towel?" (accept all answers) Older students (grades 3–5) can be asked, "How can we determine whether a paper towel is a good *value*? In other words, how can we tell whether a paper towel performs well for its price?" (allow students to discuss among themselves) If students suggest that they would need to know the cost of the paper towels, tell them the cost of the *entire roll*. Students will likely begin to debate the relative merits of each brand. If no one asks, say, "By the way, this roll had X (number of) sheets, and this roll had Y sheets. Does that matter?" Students will then likely realize that this is important information to make a fair comparison. If

Lesson 5: Soaky Doaky

they don't see this, you can give a hypothetical example such as the following: "Imagine that I have two paper towel brands that performed equally well on absorbency and wet strength tests. One paper towel brand's roll cost $1, and the other brand's roll cost $5. Which is a better value?" (the $1 roll) "Now, imagine that the $1 roll only has one sheet, but the $5 roll has a million sheets! Which is a better value?" (the $5 roll!) "Why?" Students will realize that you shouldn't compare the cost per roll; instead, you need to compare the cost per sheet. Allow students to discuss their thoughts on whether particular brands were worth the cost based on experimental results. (*Note:* If you wanted to be even more precise in grades 3–5, you could have the class look at the size of the sheets to calculate the area or use the total area of the paper towel roll provided on the package, which may differ among brands, for a fair comparison.)

Elaborate: Video on How Paper Towels Are Made and Sequencing Activity

1. Explain to students, "Now that we know a bit about the properties of good paper towels, it's time to think about how paper towels are made. How do you think paper towels are made?" (invite answers and accept all ideas)

2. Show "This Is How Paper Towels Are Made" video (*www.businessinsider. com/how-paper-towels-are-made-2017-10*), which shows the process by which a paper towel brand that uses recycled paper pulp is made.

3. After the video, divide students into small groups and inform them that you have a sequencing task for them. Give each group a packet with the precut cards (p. 175), and ask students to work together with their teams to place the cards in the proper order of how recycled paper towels are made. Teams should draw pictures on the cards to represent each step. As groups finish, check the sequence (see Figure 5.17).

FIGURE 5.17.
Answer Key for Sequencing Activity

4. After all groups have been checked, review with the following questions:

    - "What were the paper towels in the video made of?" (pulp made from recycled office paper)

    - "What are some of the steps this company followed to turn recycled office paper into paper towels?" (you can tape the correct sequence of cards onto the board to help reinforce the sequence)

    - "What can we *infer* from the video about why this company uses recycled office paper to make paper towels?" (the paper towels are made from paper rather than from trees, so no new trees need to be cut down to make them)

    - "What did the company do to make the paper towels absorbent and strong?" (used two layers, or plies, for absorbency and a wet strength agent to make them strong)

Elaborate: Paper Towels Versus Cloths and Sponges Read-Along

1. Distribute a copy of the Paper Towels Versus Cloths and Sponges Read-Along handout (p. 176) to each student. Read aloud to the students, having them follow along. To help students organize the advantages and disadvantages of these methods for cleaning up spills, draw a chart like the one on the handout on the board, and explain the meaning of each box. For example, point to the first box in the upper left-hand corner and ask, "What are some advantages of paper towels?" Elicit some answers from the class, and model writing the answers in the box for students (see Figure 5.18). Allow them to work individually or in pairs to complete the chart on the handout.

FIGURE 5.18.

Answer Key for Advantages/Disadvantages Graphic Organizer on Student Handout

	Advantages ☺	Disadvantages ☹
Paper Towels	Convenient Clean	Makes a lot of waste Some use trees to make
Cloths and Sponges	Absorbent Strong Cheap	Germs if not cleaned Waste when thrown out

Lesson 5: Soaky Doaky

Elaborate: Four Corners Activity: What's Most Important to You?

1. Explain to students that they are now going to have a chance to think about the different features of products used to clean up spills and express their opinions on what they decide is most important to them. The goal of this activity is to help students clarify their values and see that other people can have different but equally important priorities. This activity also helps students form arguments to support their choices. Tape one of the copied corner cards ("Absorbency," "Wet Strength," "Price," and "Environment," [see pp. 177–180]) in each of the four corners of the room.

2. Tell students that they are going to do an activity where they are asked a question that has no right or wrong answer. The question will ask their opinion about something, so it is important that they think about it and then answer honestly.

3. Ask students, "If you were buying a product to clean up spills, which of these four qualities would be most important to you (pointing to the corners/signs): wet strength, absorbency, price, or the environment? Why?"

4. Tell students to think quietly for a moment, reminding them that there is no right or wrong answer. Then, ask students to walk to the corner that reflects their opinion. (If you see that a student is going to be alone in a corner, go to that corner and stand with that student.) *Tip:* You may wish to have your students write their answers first on the back of their read-along sheet if you are concerned that they may follow their friends rather than going to the corner that represents their opinion. This may or may not be necessary depending on whether you have done other activities where students express their opinions on issues.

5. Once students are at their corners, ask them to share the reasons why they chose that corner with the others who are there. (If you are at a corner with a student, stay and discuss the reasons for that choice with him or her; otherwise, circulate and listen.) After three or four minutes, ask students to select a spokesperson who will report the reasons that were given for choosing that corner. Allow students another two or three minutes to choose the spokesperson and finalize their reasons.

6. Ask the spokespeople to report on their groups' reasons, allowing other students to contribute if ideas were missed. When all groups have reported, ask:
   - "Did you hear any reasons that you hadn't thought of? If so, what were they?" (answers will vary)

- "Did any of the reasons you heard persuade you to change your opinion? If so, explain." (answers will vary)

- "If you were the president of a paper towel company, what information might you learn from seeing this activity?" (I would learn what qualities are important to people when buying products that clean up spills; I would improve my paper towels so that they were good at those things; I would target commercials to focus on the qualities that are important to people)

Evaluate: Write a Letter to a Paper Towel Company or Make a Presentation or Video for Your School About Paper Towels

Inform students that they have a choice of final activities for this lesson.

Option I. Write a Letter

1. Students can write and send a letter to a paper towel company of their choice, using the Letter to a Paper Towel Company student handout (pp. 181–182).

2. When students have finished their letters, make copies for students to keep and send the originals to the paper towel companies. The addresses for some of the larger companies are listed below. Addresses for store brands and others not listed can be found on the packages or the companies' websites.

Bounty Paper Towels
The Procter & Gamble Company
1 P&G Plaza
Cincinnati, OH 45202

Brawny Paper Towels
Georgia-Pacific Consumer Products
133 Peachtree St., NE
Atlanta, GA 30303

Cascades Extreme Paper Towels
(the brand in the movie)
404, Marie-Victorin Blvd.
Kingsey Falls, Québec,
Canada J0A 1B0
(or e-mail: info@cascades.com)

Marcal Small Steps Paper Towels
Soundview Paper Company
1 Market St.
Elmwood Park, NJ 07407

Scott Paper Towels
Kimberly-Clark Corporation
Dept. INT
PO Box 2020
Neenah, WI 54957-2020

Seventh Generation Paper Towels
Seventh Generation
60 Lake St.
Burlington, VT 05401

Sparkle Paper Towels
Georgia-Pacific Consumer Products
133 Peachtree St., NE
Atlanta, GA 30303

Viva Paper Towels
Kimberly-Clark Corporation
Dept. INT
PO Box 2020
Neenah, WI 54957-2020

Option II: Make a Presentation or Video for the School (or School Community)

1. Students can use the Presentation or Video for Your School About Paper Towels storyboard handout to organize their project. Provide students with the opportunity to make their presentations or share their videos with other classes or in an assembly.

For Both Projects: Distribute copies of the My Paper Towel Project Checklist (p. 184), which outlines the steps. Evaluate students on completeness and accuracy of their projects.

Going Deeper

- Students can test other common consumer products and share their findings with the school community through a newsletter or webcast.

- Students can observe commercials to identify strategies advertisers use to persuade consumers to buy their products, with particular attention paid to advertising that targets children.

- Students can calculate how much paper toweling is used in their school in a day, a week, and a year.

- Students can interview older adults in the community to learn if consumption of disposable products like paper towels has changed in their lifetime.

- Students can research paper towel consumption in other countries around the world.

- Students can discuss paper towels with their families. Distribute copies of the letter (p. 185) for students to take home.

Name: _____ Date: _____

Paper Towel Testing: Wet Strength

1. Hold a paper towel of Brand A over the basin. Try to hold it by the edges.
2. Pour ½ cup of water in the center of the paper towel.
3. Slowly add blocks, one at a time, onto the paper towel.
4. Record how many blocks the paper towel held until it broke.

Repeat for Brands B and C. Take turns with different jobs.

Paper Towel Brand	How Many Blocks?	Rating (1 star ★, 2 stars ★★, or 3 stars ★★★)
A		
B		
C		

Which paper towel brand was the STRONGEST? _____

Which paper towel brand was the WEAKEST? _____

Draw your test below:

Name: _____ Date: _____

Paper Towel Testing: Absorbency

1. Dip a paper towel of Brand A in the water basin for 10 seconds.
2. Take it out of the basin, and squeeze it out over the cup marked A. Let everyone take a turn squeezing until you are sure it is squeezed out.
3. Repeat for Brands B and C. Take turns with different jobs.
4. Compare the amount of water in the cups. Record your results in the chart.

Paper Towel Brand	How Much Water? (High, Medium, or Low)	Rating (1 star ★, 2 stars ★★, or 3 stars ★★★)
A		
B		
C		

Which paper towel brand was the MOST ABSORBENT? _____

Which paper towel brand was the LEAST ABSORBENT? _____

Draw your test below:

Turn Over

Name: _____   Date: _____

Our Own Absorbency Test

Work with your partners to design a new test for absorbency. Write out the steps of your test below. Then do your test and enter the results in the chart!

The steps for our test are:

1. _____
2. _____
3. _____
4. _____

Paper Towel Brand	How Much Water? (High, Medium, or Low)	Rating (1 star ★, 2 stars ★★, or 3 stars ★★★)
A		
B		
C		

Which paper towel brand was the MOST ABSORBENT? _____

Which paper towel brand was the LEAST ABSORBENT? _____

Draw your test below:

How Recycled Paper Towels Are Made Sequencing Activity

Shredding and mashing paper into pulp	Filtering out staples, plastic, and ink
Squeezing water out of pulp and bleaching	Rolling out paper and adding "wet strength" agent
Pressing patterns on paper	Adding a second layer, or "ply"
Cutting big paper logs into rolls	Wrapping paper towel rolls in plastic wrap

Name: _____ Date: _____

Paper Towels Versus Cloths and Sponges Read-Along

Paper towels are very popular. In the United States, over 13 billion pounds of paper towels are used each year. Most of these towels are used once, thrown into the garbage, and buried in **landfills.** Many paper towels are made from **wood pulp** from trees. Some paper towels are made from **recycled** paper so that fewer trees are cut down. Paper towels are very **clean** and **convenient** because they are used only once and don't need to be washed.

Cloths and sponges are very **strong** and **absorbent.** They may even be stronger and more absorbent than many paper towel brands. They are usually **cheaper** to use than paper towels because they are used several times. They also create less waste than paper towels because they last a long time. Cloths and sponges need to be cleaned because **germs** can grow inside them. Cloths and sponges create waste when they are thrown out.

What are some of the **advantages** and **disadvantages** of paper towels, cloths, and sponges?

Type	Advantages ☺	Disadvantages ☹
Paper Towels		
Cloths and Sponges		

Absorbency

Wet Strength

Price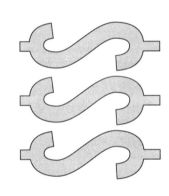

Environment

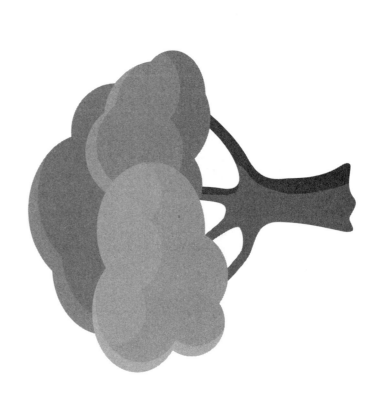

Name: _____ Date: _____

Letter to a Paper Towel Company

Write a letter to a paper towel company to let it know what you've learned about paper towels. Be sure to include the following in your letter:

- Explain how you tested paper towels using the terms **wet strength** and **absorbency**.

- Share the results of your tests to let the company know which brand or brands worked best.

- Give one reason why you will or won't use the company's paper towels.

- Draw a picture to show how you tested paper towels for wet strength and absorbency.

Dear _____,

_____
_____
_____
_____
_____
_____
_____
_____

My Paper Towel Tests Drawing

Sincerely,

_____

Presentation or Video for Your School About Paper Towels

Make a presentation or video to let your school (or school community) know what you've learned about paper towels. In your presentation or video, be sure to do the following:

- Explain how you tested paper towels using the terms **wet strength** and **absorbency**.
- Share the results of your tests to let your audience know which brand or brands worked best.
- Give one reason why you will or won't use paper towels.
- Discuss the advantages and disadvantages of paper towels versus cloths and sponges.
- Use the storyboard below to help you with the steps!

Name: _____ Date: _____

My Paper Towel Project Checklist

Check ✓ the boxes below when you have completed each step!

For all projects:

☐ I have explained how I tested paper towels using the terms **wet strength** and **absorbency**.

☐ I have shared the **results** of our tests to let the company or audience know which brand or brands worked best.

For Letters	For Presentations and Videos
☐ I have given one reason why I will or won't use the company's paper towels.	☐ I have given one reason why I will or won't use paper towels.
☐ I have drawn a picture to show how I tested for wet strength and absorbency.	☐ I have discussed the advantages and disadvantages of using paper towels versus cloths and sponges.

Paper Towel Testing Week

Dear Families,

This week, we conducted tests with paper towels to see which brands work best. After watching some commercials, we tested towels for two main properties: absorbency and wet strength. We also discussed other qualities such as cost and whether the paper towels were made from recycled materials. This study enabled students to learn why materials with certain properties are good for certain purposes. It also helped students design fair scientific tests and begin thinking about the trade-offs that are involved when deciding what brand of a product to buy.

Here's how you can help continue the learning at home:

1. Ask your child about the results of the paper towel tests.

2. Share why you choose a particular brand, or why you choose not to buy paper towels.

3. If you use sponges or cloths instead of paper towels, discuss why.

4. Encourage your child to repeat the tests that he or she did at school with paper towels, sponges, or cloths at home.

5. Watch the video that we watched at school about how recycled paper towels are made (*www.businessinsider.com/how-paper-towels-are-made-2017-10*).

6. In the United States alone, over 13 billion pounds of paper towels are used each year. Track your paper towel use at home for one week and see how many paper towels you use. Then, see if you can reduce your use by 25% the following week.

Thank you!

Lesson Plans

Lesson 6

Bee-ing There for Bees

Are Bees Disappearing?

Suggested Grade Levels

K–2

Driving Questions

- How do bees and plants help each other?
- How do bees help us, and how can we help bees?
- Are bee populations declining, and if so, why?

Lesson Overview

This lesson introduces students to the important role that bees have in the ecosystem as pollinators. Students investigate honeybee behavior through a bee waggle dance simulation game. They then apply their knowledge of bee and flower anatomy to an engineering challenge in which they develop models to demonstrate flower pollination. After considering the importance of bees to human food crops, students are introduced to the controversial issue of whether bee populations are declining, and if so, what the causes are. Students then collaboratively take action by creating a bee hotel that can provide shelter for docile, solitary mason bees, along with informational brochures to share with the community about bees.

Connecting to the *NGSS*

(See full alignment in Table A.6 on p. 507.)

- LS2.A: Interdependent Relationships in Ecosystems

- Plants depend on animals for pollination or to move their seeds around. (2-LS2-2)
- ETS1.B: Developing Possible Solutions
  - Designs can be conveyed through sketches, drawings, or physical models. These representations are useful in communicating ideas for a problem's solutions to other people. (K-2-ETS1-2)

Societal Issues

Contemporary Versus Traditional Farming Methods, Human Impacts on Environment

Nature of Science

- Scientific Knowledge Is Based on Empirical Evidence
  - Scientists look for patterns and order when making observations about the world.
- Scientific Knowledge Is Open to Revision in Light of New Evidence
  - Science knowledge can change when new information is found.
- Science Models, Laws, Mechanisms, and Theories Explain Natural Phenomena
  - Science uses drawings, sketches, and models as a way to communicate ideas.
  - Science searches for cause-and-effect relationships to explain natural events.

CCSS Connections

- English Language Arts
  - RI.K.1. With prompting and support, ask and answer questions about key details in a text.
  - RI.1.1. Ask and answer questions about key details in a text.
  - RI.2.1. Ask and answer such questions as who, what, where, when, why, and how to demonstrate understanding of key details in a text.

- Mathematics
  - MP.4. Model with mathematics. (2-LS2-2)

NCSS Connections

- Theme 8: Science, Technology, and Society
  - Research and evaluate various scientific and technological proposals for addressing real-life issues and problems.
- Theme 10: Civic Ideals and Practices
  - Ask and find answers to questions about how to plan for action with others to improve life in the school, community, and beyond.

C3 Framework

- Dimension 4: Communicating Conclusions
  - D4.1.K-2. Construct an argument with reasons.
  - D4.3.K-2. Present a summary of an argument using print, oral, and digital technologies.
- Dimension 4: Taking Informed Action
  - D4.7.K-2. Identify ways to take action to help address local, regional, and global problems.

Lesson 6: Bee-ing There for Bees

UDL Toolkit

Multiple Means of Engagement	Multiple Means of Representation	Multiple Means of Action and Expression
A collaborative engineering challenge with clear goals and expectations, as well as a whole-class culminating activity, fosters a sense of community and heightens the salience of lesson objectives.	Using a poem, storybook, video, dance, diagram, and engineering design activity to convey information facilitates comprehension through diverse media.	Providing students with the opportunity to engage in the lesson through text, dance, design, and illustration increases entry points for action and expression of ideas.
Collaboratively creating a class bee hotel provides a socially relevant context for learning that can maintain interest and engagement over time.	Using materials to represent pollen that are distinguishable through both tactile and visual cues ensures that most students can participate without additional accommodations.	Providing sentence starters, sentence frames, and graphic organizers helps students organize information and construct and compose writing.

Suggested Schedule and Sequence

- Day 1: **Engage** with "Bees" poem, Bee KWL Chart, and *The Honeybee* read-aloud, and **Explore** with Bee Waggle Dance Game

- Day 2: **Explain** with *The Bee Book* read-aloud (part 1) and Bee KWL Chart

- Day 3: **Elaborate** and **Evaluate** with "Bee" an Engineer!

- Day 4: **Elaborate** with *The Bee Book* read-aloud (part 2) and Build a Mason Bee Hotel

- Day 5: **Evaluate** with Help the Bees! Poster

Materials

For Bee Waggle Dance Game

(per team)

- 1 full-page photo (preprinted) or drawing of a flower
- Tape

Lesson Plans

For "Bee" an Engineer!

(per class)

- Flowers from Bee Waggle Dance Game
- Cake or ice cream sprinkles (try to get 2–3 different shapes such as little rods, round nonpareils, and sugar sprinkles)
- 3–4 small cups
- 3–4 cupcake liners or shallow dishes
- Juice (small amount)
- Craft materials such as craft sticks, pipe cleaners, cotton balls, cotton swabs, yarn, craft paper, pom-poms, googly eyes, Velcro dots, clay, straws, toothpicks
- Scissors
- Glue sticks
- Tape

Note: This activity uses sprinkles to represent pollen that will be transferred between flowers by model bees. You should use different-shaped sprinkles, not just different-colored sprinkles, so that all students, including those with visual impairments or color-blindness, can participate fully. As an alternative, you could use glitter as the pollen for one flower and cornstarch or flour for another. The materials should differ tactually so that the activity doesn't rely only on color differentiation.

For Build a Mason Bee Hotel

(per hotel/per class)

- 2-liter plastic drink bottle or a coffee can
- Utility knife (to cut plastic bottle, if you are using one)
- Butcher paper or brown paper bags
- 4–5 cardboard toilet paper or paper towel tubes
- Tape
- String

- Pencils and markers

(per student)

- Safety glasses or goggles and a nonlatex apron

Student Handouts

- Bees (poem)
- "Bee" an Engineer!
- Bee-come a Mason Bee Expert
- Help the Bees!

Safety Notes

1. All students must wear safety glasses or goggles and nonlatex aprons during the setup, hands-on, and takedown phases of the activity.
2. Use caution when working with glassware or plasticware, which can shatter if dropped and cut skin.
3. Immediately wipe up any spilled water on the floor to avoid a slip-and-fall hazard.
4. Immediately pick up any items dropped on the floor to avoid a slip-and-fall hazard.
5. Use caution when using hand tools to avoid cutting or puncturing skin.
6. Do not eat any food items used in the lab activity.
7. Wash hands with soap and water after completing this activity.

Media

Books

- *The Honeybee,* by Kirsten Hall and Isabelle Arsenault
- *The Bee Book,* by Charlotte Milner

Poem

- "Bees," poem by Aileen Fisher (p. 209)

Videos

- "World's Weirdest: Honey Bee Dance Moves" video from National Geographic
  https://video.nationalgeographic.com/video/weirdest-bees-dance

- "DIY Mason Bee House" video
  www.instructables.com/id/DIY-Mason-Bee-House-great-for-Kids-and-Gardens

Background for Teachers

Bees are small but mighty members of the global environment. Found on every continent except Antarctica, bees play a critical role in pollination. Pollination is the process by which pollen is transferred from the male part of a flower to the female part of another flower of the same species. Pollination is essential for plant reproduction, so we are reliant on it for the foods we eat as well as the air we breathe.

Here's how it works: One of the ways plants reproduce is by making seeds. Seeds are very important because they contain the genetic information needed to make a new plant. Flowers are a tool that plants use to move the genetic information from different plants together to make a seed. The male part of the flower is called the stamen. Its top portion, called the anther, produces pollen. Pollen grains contain sperm. The female part of the flower is called the pistil. It has a sticky top part called the stigma, which receives pollen from the stamen. When a pollen grain from a flower of the same species lands on the stigma, a pollen tube grows down the pistil into the ovary of the flower, which contains an egg cell. The sperm from the pollen grain move down through the pollen tube to fertilize the egg cell. The fertilized egg can then become a seed, while the surrounding ovary becomes fruit (in plants that bear fruit). While some flowers are capable of self-pollination, the preferred way of pollinating (evolutionarily speaking) is by moving pollen between different individual plants so that there is more genetic diversity in the offspring. That's where pollinators come in!

While the pollen of some plants is spread by wind or water, most rely on animals such as bees, birds, bats, beetles, wasps, ants, butterflies, and moths. It is estimated that one-third of the foods we eat are pollinated by animals, including bees. Bees are among the most efficient pollinators, and many flowers have coevolved with bees, changing over time along with them to ensure that their pollen gets carried. Bees eat both nectar and pollen, so flowers produce nectar with sweet scents

Lesson 6: Bee-ing There for Bees

and petals that are attractive to bees. Honeybees have strawlike tongues that can reach into flowers to sip the nectar, which they store in their special nectar stomachs. They carry it back to the hive to make honey, which is stored for eating in the cool months. Pollen attaches to honeybees' fuzzy body and is carried in pollen baskets on honeybees' hind legs. It is fed to baby bees because it is high in protein. Although we don't know what bees are thinking, it is unlikely that bees intentionally pollinate flowers; more likely, pollen drops off of bees as they visit flowers to gather nectar and pollen for food. Evidence to support this is that most bees visit different species of flowers, which wouldn't make sense if pollination were the goal, since pollination can only take place between flowers of the same species. That said, some bee species do prefer to visit a single species of flowers.

Bees are the key pollinators for many crops, including apples, cranberries, melons, broccoli, blueberries, cherries, and almonds. Honey and beeswax are additional contributions that bees make to human lives and sustenance. Of course, bees are also critical in food webs and food chains of many species, since all food webs rely on plants as producers.

While honeybees are often the stars of the bee world, it's important to note that they are not native to North America. Honeybees were brought to North America by European settlers. They have been domesticated for use in agriculture for centuries and are essential to commercial farming. Native bee species, however, existed here long before the honeybees arrived, and many of those native species have struggled to compete. Native bee species were adapted to diverse biomes of grasslands, forests, and shrublands. Modern farms have replaced many of these diverse habitats with single crops, providing fewer food sources for the native bees. Loss of habitat has been a serious threat to all bees, as urban sprawl and fragmenting of natural landscapes make it harder for bees to find food and suitable homes. In addition, bees have suffered from the effects of climate change, which has led to more extreme weather events and temperature fluctuations. The use of pesticides has also decreased bee populations, as have diseases that are in part magnified by the movement of bees around the globe by humans. When bees are moved for agricultural purposes, they may not have the natural defenses to protect them from local parasites or pathogens.

All these factors have had tremendous impact on many bee populations. During the winter of 2006–2007, many beekeepers in the United States noticed a high number of bee deaths that were inconsistent with typical losses. Some beekeepers lost nearly 90% of their hives because of this mysterious situation, which was characterized by the sudden loss of worker bees with very few found near the colony. The term *colony collapse disorder* (CCD) was used to describe this situation. Since then, fewer losses have been reported due to CCD, but other bee losses still continue for many species. The rusty patched bumblebee, for example, was added to the endangered species list in 2017 because of an 87% decrease in its population since the 1990s and the fact that its range has been reduced from 28 states to only 13. Some scientists think that bumblebees, as opposed to honeybees, may be especially susceptible to loss from pesticides. There is ample reason to be concerned about bee declines.

Like all the issues in this book, there is also some controversy "buzzing" around bee decline. Not all species are experiencing declines, and of the ones that are, it is not clear which factor is the most critical. Skeptics of bee declines point to the fact that bee populations have always shown some flux (at least, over the years that people have been keeping track). And some studies have produced conflicting evidence of the impacts of certain types of pesticides on bees. All these questions are excellent teachable moments about the nature of science in that they show that scientists sometimes disagree; scientists must communicate and debate to move understanding forward. These questions also show that scientific knowledge is tentative and subject to change as new studies and new investigative methods are developed. These are just some of the reasons why this issue is so fruitful for science learning.

In this lesson, students learn about bee anatomy and behavior, particularly in relation to foraging and pollination. They develop and test bee models to demonstrate their understanding of the interrelationships between plants and pollinators, as recommended by the *NGSS*. After examining some of the controversies around bee decline, they also build a class mason bee hotel and develop educational bee posters to become advocates for the bees. This lesson promises to keep your students very buzzy!

Additional Resources

Scientific American article on bee decline

- *www.scientificamerican.com/article/u-s-lists-a-bumble-bee-species-as-endangered-for-first-time*

Information about bees from the National Resources Defense Council

- *www.nrdc.org/sites/default/files/bees.pdf*

Information on planting a bee garden

- *https://thehoneybeeconservancy.org/plant-a-bee-garden*

Excellent article on bee hotels and research on native and non-native bees

Lesson 6: Bee-ing There for Bees

- *www.sciencenewsforstudents.org/article/bee-hotels-are-open-business*

Background on mason bees

- *http://erie.cce.cornell.edu/resources/article-22-mason-bee-october-2016*

Article on the importance of protecting pollinators for food security

- *https://news.un.org/en/story/2016/12/547132-countries-urged-prioritize-protection-pollinators-ensure-food-security-un*

Resources on caring for native bees throughout the year

- *https://crownbees.com/raise-bees*

Wonderful book on bees from the NSTA Kids *Next Time You See* series

- Morgan, E. 2019. *Next time you see a bee.* Arlington, VA: NSTA Press.

Audubon Adventures (an environmental education product for kids) on native bees

- *www.audubonadventures.org/docs/AA_Bees_FINAL.pdf*

5E Lesson Plan

Engage: "Bees" Poem, Bee KWL Chart, and *The Honeybee* Read-Aloud

1. Distribute copies of the "Bees" poem handout (p. 209) to students and also project a large image of the poem using a document camera. Use choral reading (teacher and students read aloud together) to read the poem aloud with your students. Ask students:

    - "What is this poem telling us about?" (bees, pollination, pollen, plants that need bees)

    - "What did the poet mean when she wrote, 'There wouldn't be sunflowers, / Wouldn't be peas, / Wouldn't be apples / On apple trees'?" (without bees, these plants wouldn't be alive)

    - "The poem doesn't tell us exactly how bees help plants. We need to make an inference by looking at clues. Are there any clues to tell us how bees might help plants?" (it may have something to do with pollen)

    - "Does the poem give us any other information about bees?" (they're fuzzy, they buzz, and they get pollen on their knees)

2. Show students a Bee KWL Chart like the one on page 210 (you might wish to enlarge it or use a document camera). Ask students what they *think* they Know about bees; accept answers and complete the "K" part of the chart. Let students know they are going to learn a bit more from the book *The Honeybee*, by Kirsten Hall and Isabelle Arsenault. Read the book aloud, and then ask students, "What new information can we add to our Bee KWL Chart?" (answers will vary; add information to the chart) If these topics aren't mentioned, ask:

   - "What two things did the bee collect from flowers?" (nectar and pollen)
   - "Where on the bees did the pollen attach?" (on their legs)
   - "What did the bees do with the nectar?" (turned it into honey that is stored in honey cells)
   - "What do the bees do in the winter?" (the book says they rest; students will learn about hibernation later)
   - "How did the forager bee, the one who found the flowers, communicate where the flowers were to the bees in the hive?" (by dancing)

3. Say, "Let's see what this looks like in action!" Show students the National Geographic video on the bee waggle dance at *https://video.nationalgeographic.com/video/weirdest-bees-dance*. After watching the video, ask students these questions:

   - "What *three* pieces of information does the bee waggle dance communicate?" (the direction of the flowers, the distance from the hive, and the type of pollen)
   - "How does the scout bee show the direction of the flower?" (dancing the straight part of the dance in the direction of the flower)
   - "How does the scout bee show the distance from the hive?" (a short, quick dance means the flowers are close; a long, slower dance means the flowers are farther)
   - "How do the bees learn the type of pollen from the dance?" (they smell it)

Note: To be precise, the waggle dancing scout bee communicates the location of the flowers by indicating the direction of the flowers *in relation to the Sun*. For example, a scout bee might dance straight upward in a hive to show that the flowers can be found by flying in the direction of the Sun at that time. A

Lesson 6: Bee-ing There for Bees

bee that dances downward in a hive would be communicating that the flowers are in a direction away from the Sun at that time. However, for the purposes of this activity with K–2 learners, we simply say that the straight run part of the waggle dance faces the direction of the flowers.

Explore: Bee Waggle Dance Game

Advance Preparation: This game can be done outdoors in an enclosed space (e.g., playground, enclosed garden) or in a classroom. To prepare, take two of the preprinted, full-page flower photos or flower drawings, and tape them in two different spots that would not be easily noticed by students (e.g., if outdoors, tape under a swing or sliding board; if indoors, tape on a bookcase, on the side of a desk, behind the door).

1. Inform students that they are going to have a chance to experience what it's like to communicate like a bee by playing the Bee Waggle Dance Game.

2. Explain to students that you are a forager or scout bee who has just returned to the hive, and they are worker bees. You want to tell them where you've found flowers with a lot of nectar, but since you can't speak, you have to communicate with a dance. You will do one dance by yourself first, so they can watch, and then do it again with them following. The dance is shown in Figure 5.19. In addition to demonstrating the bee waggle dance, displaying the diagram by drawing it on the board, projecting it with a document reader, or having it available as a handout provides additional cues for students.

3. When you demonstrate the waggle dance, remind students that the straight run shows the direction of the flowers, and the speed of your waggle shows the distance (short, quick waggles = close flower; slow, long waggles = far flower). After you and the students have danced, let them know that they

FIGURE 5.19.

Bee Waggle Dance Diagram

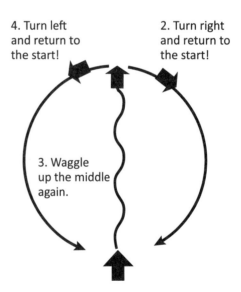

Lesson Plans

need to find a flower that is hidden by calmly flying to it. Allow students to look for and find the flower.

4. After students find the first flower, do another demonstration with them so that they can locate the second flower.

5. Explain to students that they will have a chance to work in hive groups of six to eight students. Each group will be given a flower and tape. They should take turns being the scout. The worker bees cover their eyes while the scout hides the flower. Then the scout returns to the hive and does the waggle dance, and the worker bees look for the flower. The child who finds it becomes the scout for the next round unless he or she has already had a chance, in which case this child passes the flower to a student who hasn't been the scout. Continue until all students in the hive groups have been scouts. After all students have had a chance to be the scout bee, ask:

- "Why do you think bees have developed this special language for communicating about flowers?" (because bees need to find nectar and pollen to survive)

- "I wonder, do the flowers need the bees, too?" (answers will vary; students may mention pollination but likely have little prior knowledge and may have misconceptions) Tell students, "We will learn more about this part of the bee-flower partnership next time!"

Misconception Alert

Children often think that bees visit flowers for the purpose of pollinating them. Bees actually gather nectar and pollen for food and may accidentally drop pollen on other flowers of the same species, which can result in pollination. Another misconception students may have is that bees gather *honey* from flowers. This is easy to understand since nectar and honey are both described as sweet liquids. These concepts will be elaborated upon as you continue through this lesson, but you should "bee" ready to address these very common misconceptions!

Explain: *The Bee Book* Read-Aloud (Part 1) and Bee KWL Chart

The Bee Book, by Charlotte Milner, is an informational picture book that addresses various topics such as bee species, anatomy, behavior, pollination, and the decline

Lesson 6: Bee-ing There for Bees

of bees in two or three pages each. Each new topic is introduced with a leading questions (e.g., "Where does honey come from?"). This reading session will cover bee basics, including anatomy, honey making, pollination, and the importance of bees to humans. A later class reading will introduce bee decline.

1. Show students the cover of the book. Open the book to the table of contents, and show students some of the topics that will be covered. Tell them, "Today, we will be reading pages from the first half of the book, and later in the week, we'll continue with pages in the second half." Read pages 4–9 and 14–21, with emphasis on page 6 ("What Is a Honeybee?"), page 8 ("The Bee's Knees"), and pages 18–21 ("What Is Pollination?" and "Why Do We Need Pollination?").

2. After reading, return to the Bee KWL Chart and ask students, "What new information and questions can we add now?" Allow students to contribute information that they've learned (under "L") and information that they Want to Know (under "W"). Depending on the information that they have contributed, you may wish to supplement with the following questions:

   - "How do the honeybees use pollen?" (they feed it to baby bees)

   - "How do the honeybees use nectar?" (they eat it and use it to make honey, which they store and eat during the fall and winter when there are no available flowers)

   - "What special body parts do bees have that help them survive?" (they have five eyes that help them detect flowers, antennae that help them smell and taste nectar, a long strawlike tongue that helps them suck up nectar, pollen baskets on their back legs to carry pollen, a honey stomach to carry nectar back to the nest, and tiny holes in their body called spiracles to help them breathe; females have a stinger but die after stinging) (*Note:* Students may wonder why a honeybee would sting if it dies after stinging; after all, the sting doesn't really protect the bee! Evolutionary biologists, who are scientists that study how different species adapt and change over time, believe this feature evolved in honeybees to protect the hive. Bees die after stinging because they have barbed stingers that are attached to their digestive tracts. When a bee stings, its barbed stinger gets stuck in the skin of the mammal it has stung, and the bee's digestive tract gets pulled out. The bees who can sting are all female, but they can't reproduce. Only the queen [which happens to have a smooth stinger and doesn't die after stinging] can reproduce. So when the female worker bees sting, they are essentially sacrificing themselves for the good of the hive [and for their relatives inside the hive, which share many of their genes]. The sting sends a powerful message to an intruding mammal to stay away from the nearby hive. Students also sometimes wonder whether bees *know* they are going to die if they sting. Scientists obviously don't know what a bee knows, but they hypothesize that the *instinct* to sting, as well as the stinger anatomy, have evolved over a very long time to help the nest [and the species] survive.)

- "How do bees help plants?" (bees pollinate flowers, or move pollen from flower to flower, as they visit different flowers; flowers need pollen to make seeds, which is how plants reproduce) (*Note:* You can review this process on pages 18–19 in the book.)

- "Why is this important for people?" (one-third of the food we eat comes from crops that are pollinated by bees and other pollinating animals)

- "Are bees *trying* to pollinate the flowers? In other words, do bees pollinate flowers on purpose?" (no, bees are collecting nectar and pollen for food, but pollen can accidentally fall off and pollinate flowers)

Inform students that next time, they are going to "bee" engineers!

Elaborate and Evaluate: "Bee" an Engineer!

Advance Preparation: Set up craft items in a central location or divided among table groups. Lay the paper flowers that were created for the Bee Waggle Dance Game on a table that will become a "flower patch." Put a cupcake liner or shallow dish filled with sprinkles and a small cup of juice near each flower's center. The sprinkles represent pollen, and the juice represents nectar. Use a different shaped sprinkle for each flower (for example, use rod-shaped sprinkles in one, round nonpareils in another, sugar sprinkles in a third). Students will work individually on this challenge but can be at table groups to share materials and ideas.

1. Distribute copies of the "Bee" an Engineer! handout (p. 211) to students. Explain that they are going to have a chance to apply their knowledge of bees to an engineering challenge. Read the "Bee" an Engineer! handout to students.

Lesson 6: Bee-ing There for Bees

> **A Word About Models**
>
> Models help scientists and engineers understand, visualize, or develop solutions to problems. Models are essentially external representations of what we are thinking or planning in our minds. We often think that models need to be three-dimensional physical replicas, such as model cars or animal figurines made of modeling clay. However, diagrams, drawings, dioramas, storyboards, dramatizations, formulas, and even computer simulations are also types of models. In this activity, the word *model* refers to the physical objects that students are developing to represent bees and flowers. But it's important to note that students' storyboard diagrams are also models, since they represent students' thinking about the form and function of bee and flower anatomy and the interrelationship between the two through pollination.

2. Then, show students the flower patch that has been set up, pointing out the different sprinkles and the juice cups. Ask:

   - "Why do bees visit flowers?" (to collect nectar and pollen for food)

   - "What does the juice represent in our model?" (nectar)

   - "What do the sprinkles represent in our model?" (pollen)

   - "What special body parts do bees have to collect nectar and pollen?" (strawlike tongues for sipping up nectar, fuzzy bodies for collecting pollen, and pollen baskets on their back legs for carrying pollen)

   - "Do bees pollinate flowers on purpose?" (no; when bees visit flowers to get food, pollen from one flower accidentally drops off their fuzzy bodies onto another flower of the same kind)

   - "In our models, how will we know if bees pollinated flowers?" (if the different sprinkles get mixed up after bees visit the flowers)

   Point out that we are using different kinds of sprinkles so that we can easily notice whether the pollen has traveled between flowers. In the real world, flowers can only be pollinated by pollen from the same kind (species) of flower, so their pollen wouldn't look different.

   Safety Alert: Be sure to remind students that they are not to eat the sprinkles or drink the juice. Those materials are there to help us to model pollination.

3. Show students the craft materials that are available to them. Explain to students that they must include all the parts that a bee needs to fly from flower to flower to collect food and to accidentally transfer pollen between flowers. But, the models *do not* have to be precise in terms of other bee body structures (such as three body parts and five eyes) because our focus is on pollination. Also, be sure to clarify that *accidentally* transferring pollen means that students cannot pull pollen off their bee and place it on other flowers. The models must be able to transfer pollen between flowers on their own.

4. Give students their time limit and have them start working. As students are building their models, circulate around the room to check for understanding. Allow students to test their models at their tables with small amounts of sprinkles. As students experience success with their models (sprinkles are sticking to the bee and moving from one small pile to another), have them explain how the various parts are suited for their purpose (e.g., the straw is like a bee's tongue to suck up nectar; the cotton, pom-poms, or yarn is fuzzy like a bee's body to pick up pollen; the pipe cleaner legs are fuzzy to hold pollen).

5. When students have completed their bee models and tested them on a small scale at their tables, explain that their bees are going to come and feed at the flower patch. Have students come up to the table by forming two lines on either side. Encourage students to buzz and slurp as they walk by the patch, visiting flowers by flying down, sipping nectar, and catching pollen on their bodies. When all the bees have visited the patch, examine the flowers with your students. (The sprinkle pollen will now be widely distributed among the flowers, with different-colored glitter.) Ask students:

    - "Did our bees pollinate the flowers?" (yes)

    - "How do you know?" (the sprinkles have moved between flowers)

    - "What are some of the body parts on your models that helped collect nectar and pollen?" (Have several students describe their models; provide a sentence frame such as "The _____ is for _____." Ask students to explain how the parts are like those on a real bee. Also ask why they used certain materials for certain parts. For example, "Why did you use pom-poms on the bees' body?" "Why did you choose pipe cleaners for the legs instead of toothpicks or straws?")

Lesson 6: Bee-ing There for Bees

6. Have students complete the storyboards on their handouts. Remind them to label all the parts in their pictures. They can use the same sentence frame as above or a similar one ("_____ for _____") for their labels. (For example, a student might write, "Fuzzy legs for catching pollen" or "Straw tongue for sipping nectar.")

7. Finally, ask students to think about the strengths and weaknesses of their models. Ask:

   - "How are our model bees like real bees?" (carry pollen, travel from flower to flower)

   - "How are our model bees different from real bees?" (they're not alive, they don't really fly, they don't transfer real pollen, they don't eat)

   - "How are our model flowers like real flowers?" (they have pollen and nectar, they are in a patch)

   - "How are our model flowers different from real flowers?" (they're not alive, they don't have real pollen or nectar, they don't grow or make seeds or fruits)

8. Explain to students that scientists and engineers use models to understand how things work and to test and share ideas. But every model has strengths and weaknesses, because models aren't exactly the same as the thing that they're modeling. Collect student handouts and assess the storyboards and student verbal explanations with the rubric on page 214.

Elaborate: *The Bee Book* Read-Aloud (Part 2) and Build a Mason Bee Hotel

This read-aloud covers pages 34–43 of *The Bee Book*, by Charlotte Milner. The first few pages of this reading session discuss the issue of bee decline. Students are introduced to the idea that many types of bees have been in decline as a result of different factors, including climate change, loss of habitat, diseases, and pesticides. The book then reiterates the importance of bees and ways that people can help, including making a bee hotel. The book does not, however, discuss bee decline as a controversial issue. Because there are controversies about which species may or may not be in decline, whether the declines of some species are parts of natural cycles, and whether pesticides have contributed to bee loss as much as bee advocates claim, it is important to raise these issues with students so that they have a more nuanced understanding. This discussion also introduces important aspects of the tentativeness of scientific knowledge. A recommended discussion after the reading would proceed as follows:

- Ask, "What is happening with bees?" (they are declining; there are fewer of them)

- "Are *all* types of bees declining?" (answers will vary)

- Point out the following sentence on the top of page 36: "There are lots of reasons why *many types* [italics added] of bees are in decline." Ask, "What does that sentence imply, or tell us in an indirect way?" (that not *all* bees are in decline)

- Explain to students that people became very concerned about bees in 2006, when a very large number of bees in the United States and Canada died. Some scientists thought that it was just part of a natural cycle (in which population numbers go down sometimes and up other times), while others thought it could be due to diseases or pesticides. This shows that sometimes scientists disagree, and that's a good thing, because they keep researching and communicating to learn from each other. Since 2006, there haven't been as many big die-offs of bees, but some types of bees are definitely declining because of the factors discussed in the book.

- Ask, "What are some of the factors that are causing some bee types to decline?" (climate change, loss of habitat, disease, and pesticides)

- Point out that some companies that make pesticides say that pesticides don't cause big die-offs of bees if they're used properly. But some people think that the companies just say that so they don't lose business. Both sides have scientific research to back their claims! This shows that sometimes, scientific studies can have different results. More studies are needed.

- Then, ask, "What do you think about the bee issue?" (allow students to share thoughts and ideas; accept all answers)

- "Even if not all bees are declining, why is it important to protect bees?" (they are needed to pollinate plants that we eat and other wildlife eats, too)

- "What are some ways we can help bees?" (build bee hotels, plant bee-friendly gardens, don't use pesticides or kill bees, learn more about them, spread the word about bees)

Lesson 6: Bee-ing There for Bees

- Say, "Let's start by building our very own class bee hotel!"

Note: Even bee hotels have some controversy buzzing around them. One study suggested that bee hotels attract both native and non-native bees, and non-native bees often outcompete native ones. Bee hotels can also attract bee predators, such as wasps (a docile kind that are not the stinging variety). They also require a little maintenance each year, as the tubes need to be cleaned or replaced in preparation for spring. Keeping bee hotels small and simple, like the one in this lesson, should keep some native mason bees very buzzy!

Safety Alert: Be sure to check with your school nurse about bee-sting allergies and how to deal with them before doing this activity. Although mason bees are very docile (males cannot sting, and females will typically sting only if they are squeezed or pinched, and even then it is like a mosquito bite), sensible precautions should be taken. Students should observe occupied bee hotels from a safe distance and should never handle the hotel or the bees. Students with known bee allergies should be kept away from occupied hotels and can observe using binoculars or a bee-cam video.

Advance Preparation: Cut the top off the plastic bottle using a utility knife to make the bottle approximately the height of a pencil. If you use a coffee can instead, clean it out and ensure there are no sharp edges by covering with tape. Cut paper (either butcher paper or brown paper bags) into squares that are about the height of your container. Prepare enough squares to provide three pieces per student. Cut cardboard paper towel tubes to about the height of your container, or tape toilet paper tubes together to make them long enough. This is a whole-class activity, but students should work at tables in groups of four or five to share materials. Place tape, paper squares, pencils, one or two cardboard tubes, and one or two markers on each table.

1. Begin by explaining to students that some of the most important pollinators are native bees, bees that were in North America before honeybees were brought from Europe. One type of native bee that's an excellent pollinator is called the mason bee. Mason bees don't make honey and don't live in hives. They are solitary bees. But they are very docile, friendly bees that are some of the best pollinators around! Their populations have been decreasing because of all the factors mentioned in *The Bee Book*, especially habitat loss. Ask, "What is habitat loss?" (when people build houses, towns, and cities, the bees' natural environment of fields, wildflowers, and forests is removed) "By building a

Lesson Plans

bee hotel, we can give them a habitat for nesting and raising their young bees. But we need to know more about mason bees first!"

2. Distribute copies of the Bee-come a Mason Bee Expert handout (p. 212) to students. Display the handout using a document camera or projector so that you can guide students as you are reading. You might read the handout aloud to students, have student volunteers read facts, or use choral reading (teacher and students). Ask students:

    - "Are there any facts about mason bees that surprised you?" (answers will vary)

    - "Do you still have any questions about mason bees?" (answers will vary, and questions can be added to the "W" section of the Bee KWL Chart for later research)

3. Show and describe the materials to the class. Explain that there are many different ways to build a bee hotel. *The Bee Book* shows one method (on pp. 42–43) that uses bamboo stems as the tubes. Tell the class, "We are going to use paper instead, as it is less expensive and will allow us all to participate." Show students the instructional video on building a bee hotel at *www.instructables.com/id/DIY-Mason-Bee-House-great-for-Kids-and-Gardens*.

Then, have students build a bee hotel following the steps outlined below:

1. Have students each take a square of paper and a pencil. Instruct them to place the pencil on one edge of the paper, then roll the paper up around the pencil as tightly as possible. Once they have finished rolling a tube, they should tape it together before removing the pencil. Encourage students to help each other with this part, as it is sometimes hard for students to tape while they are holding the tube.

2. Instruct students to remove the pencil from the tube by shaking it out. Then, they should use a small piece of tape to seal up one end of the tube.

3. Students can repeat this process two more times. Ask one student from each table to use a marker instead of a pencil for one of the tubes. This will provide a few larger tubes for the bees.

4. Have members of each team work together to place their paper tubes inside a cardboard tube, with the *open ends all facing the same way*, until the cardboard tube is stuffed with their paper tubes. The cardboard tubes will help give strength to the hotel. Students can print their names on their team's cardboard tube to personalize their excellent effort!

Lesson 6: Bee-ing There for Bees

5. When each team has completed filling a cardboard tube, have one student from each team bring the tube to you. You can then place the tubes in the containers with the open ends facing outward. If teams have extra paper tubes, these can be used to fill in the gaps between the cardboard tubes. Your class now has a bee hotel!

6. Tie the string around the hotel so that you can hang it up outside. Choose a spot that faces south or southeast. The bees love morning sun! The hotel should be hung up away from busy traffic areas and should be high enough that it is not accessible to students' reach. Try to secure the hotel so that it won't shake in the wind. Fences, trees, and poles are all good places for hanging the hotel. If you have a school garden, this can be a great choice as long as you can place the hotel at a height where students can observe but not bother the bees. Rooftop gardens in urban areas work as well.

7. Have students observe the hotel regularly to see if bees have moved in.

Evaluate: Help the Bees! Poster

1. Review the tips for helping bees on pages 42–45 of *The Bee Book*, writing them on the board or chart. They include the following:

    - Plant a bee-friendly garden.
    - Stay calm around bees.
    - Study bees to learn more about them.
    - Spread the word about the importance of bees.
    - Allow wildflowers and weeds to grow in your garden.
    - Don't use pesticides in your garden.
    - Build a bee hotel.

    In addition to reading *The Bee Book* and providing the list of tips, you can display the following great posters as references for students:

    - Poster of bee-pollinated foods: *www.fix.com/blog/how-bees-impact-our-food*
    - Poster on creating a bee-friendly garden: *www.fix.com/blog/creating-a-bee-friendly-garden*

You can also research pollinator-friendly plants specific to your geographic region:

- *www.pollinator.org/guides*

Tell students that they are going to help protect bees by creating posters to educate the school and community about them. Distribute copies of the Help the Bees! handout (p. 213). Have students fill in the blanks, including their favorite bee-pollinated food (e.g., honey, strawberries, almonds, oranges). Explain that they should complete the sentence with their favorite bee-helping tip. Then, they should draw a picture of the tip. (For example, if a student chooses "Plant a bee-friendly garden," he or she would draw a picture of someone planting a garden with some of the plants included in *The Bee Book* or the pollinator guide website.)

Going Deeper

- Students can develop a map showing the ranges of different types of bees around the world.

- Students can take photos of bumblebees that visit home and school gardens and upload them to the Bumblebee Watch citizen science project at *www.bumblebeewatch.org*.

- Students can research the history of honey and its uses in different cultures. Rock art of honey harvesting dating back to approximately 8000 BC has been found!

- Students can write a fictional "bee-ography" of their life as a bee.

Bees

by Aileen Fisher

There wouldn't be sunflowers,

Wouldn't be peas,

Wouldn't be apples

On apple trees.

If it weren't for fuzzy old,

Buzzy old bees,

Dusting pollen

From off their knees

On apple blossoms

On apple trees

And clover and daisies

And such as these.

REPRINTED WITH PERMISSION OF THE BOULDER LIBRARY FOUNDATION.

Bee KWL Chart

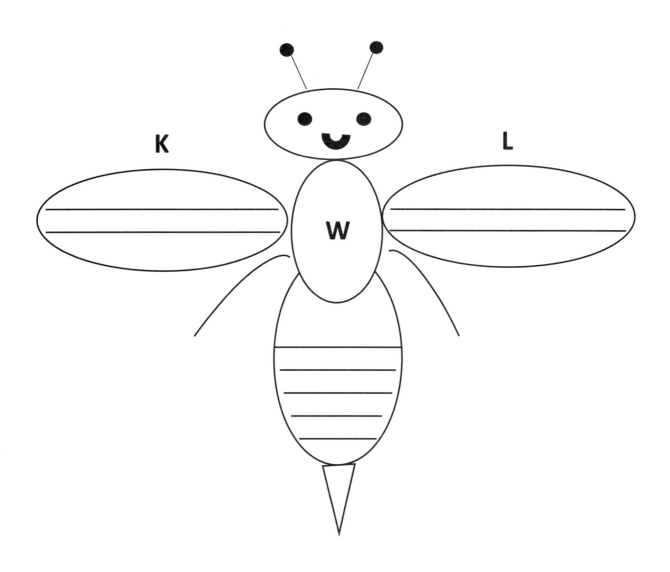

Name: _____ Date: _____

"Bee" an Engineer!

Your Challenge: Design and build a model bee that can accidentally pollinate flowers while collecting food.

Your model bee should mimic a real bee. It must …

1. Have special body parts for collecting nectar and pollen.
2. Move from flower to flower to collect nectar and pollen.
3. Demonstrate how bees can accidentally pollinate flowers by moving pollen from one flower to another while collecting food. (Our "flowers" will use juice to represent nectar and sprinkles to represent pollen.)

Draw and **label** your model in the storyboard below.

1. My bee and its special body parts	2. My bee moving from flower to flower to collect nectar and pollen	3. My bee accidentally moving pollen from one flower to another while collecting food

IT'S STILL DEBATABLE! USING SOCIOSCIENTIFIC ISSUES TO DEVELOP SCIENTIFIC LITERACY, K–5

Bee-come a Mason Bee Expert

Mason bees are amazing pollinators! Here are some important facts about mason bees:

- Mason bees got their name from the female bee's habit of using mud to plug up holes in wood or hollow plant stems to protect their eggs. Masons are people who build structures out of bricks or stone.

- Mason bees are native to North America. That means that they were in North America before honeybees and other non-native species were brought by European settlers.

- There are about 150 different species of mason bees in the United States.

- Mason bees are solitary bees, which means they live alone, not in hives. But mason bees will live near each other.

- Because mason bees live alone, the female bees must build their own nests, find their own food, and lay their own eggs. It's a lot of work!

- Mason bees are very docile, probably because they are not protecting a hive. The males do not sting, and the females will typically sting only if they are squeezed or pinched. The sting is similar to a mosquito bite.

- Mason bees are very productive! They are able to work in different types of weather (including cool, rainy days) and are not very choosy about the flowers they pollinate.

- Mason bees are a little smaller than honeybees, so they can get into flowers that are too small for honeybees.

- Mason bees pollinate **many** different plants, including blueberries, raspberries, oranges, apples, pears, almonds, roses, and wildflowers.

- Mason bees don't make honey. They make special food for their baby bees (larvae) out of pollen, nectar, and their own saliva.

- Baby mason bees, like other bees, begin as eggs, then become larvae, pupae, and finally adults. This is called metamorphosis.

- Adult mason bees live only about four to six weeks once they emerge in the spring.

Name: _____ Date: _____

Help the Bees!

We need bees because they _____ many plants. This helps plants to make seeds and fruits. Without bees, we wouldn't have many foods, including _____. But many types of bees are disappearing! You can help bees by _____.

Rubric for "Bee" an Engineer!

Criteria	3 pts.	2 pts.	1 pt.	Not Done
Student describes (verbally or through labeled drawings or both) the parts of the model that mimic a real bee.				
Student describes (verbally or through labeled drawings or both) the parts of the flower model that mimic a real flower.				
Student describes (verbally or through labeled drawings or both) how pollen is transferred between flowers.				
Student developed a physical model of a bee that successfully participated in pollinating the flower patch.	■	■		

Total: ___ / 10 pts.

Lesson 7

Weather or Not

Should We Rebuild in Twisterville?

Suggested Grade Levels

K–3

Driving Questions

- What are tornadoes?
- How do weather scientists predict tornadoes?
- What can people do to reduce the impacts of tornadoes?

Lesson Overview

Students investigate tornadoes as an example of severe weather. After being introduced to tornadoes through media and a water-filled tornado model, students create and test an anemometer, an instrument to measure wind speed, then build and test wind-resistant houses. Next, they learn about tornado forecasting and steps that can be taken to prepare and respond. Finally, students are introduced to the controversial question of whether a town in Tornado Alley called Twisterville should rebuild after a destructive tornado.

Connecting to the *NGSS*

(See full alignment in Table A.7 on pp. 508–509.)

- ESS3.B: Natural Hazards
  - Some kinds of severe weather are more likely than others in a given region. Weather scientists forecast severe weather so that the communities can prepare for and respond to these events. (K-ESS3-2)

- ESS3.B: Natural Hazards
  - A variety of natural hazards result from natural processes. Humans cannot eliminate natural hazards but can take steps to reduce their impacts. (3-ESS3-1)
- ETS1.A: Defining and Delimiting Engineering Problems
  - A situation that people want to change or create can be approached as a problem to be solved through engineering. Such problems may have many acceptable solutions. (K-2-ETS1-1)
- ETS1.A: Defining and Delimiting Engineering Problems
  - Possible solutions to a problem are limited by available materials and resources (constraints). The success of a designed solution is determined by considering the desired features of a solution (criteria). Different proposals for solutions can be compared on the basis of how well each one meets the specified criteria for success or how well each takes the constraints into account. (3-5-ETS1-1)

Societal Issues

Risk Assessment, Economic Costs of Natural Hazards

Nature of Science

- Scientific Knowledge Is Based on Empirical Evidence
  - Scientists look for patterns and order when making observations about the world. (K–2)
  - Science findings are based on recognizing patterns. (3–5)
  - Science uses tools and technologies to make accurate measurements and observations. (3–5)
- Science Models, Laws, Mechanisms, and Theories Explain Natural Phenomena
  - Science uses drawings, sketches, and models as a way to communicate ideas. (K–2)
  - Science searches for cause-and-effect relationships to explain natural events (K–2)

- Science explanations describe the mechanisms for natural events. (3–5)

CCSS Connections

- English Language Arts
    - RI.K.1. With prompting and support, ask and answer questions about key details in a text.
    - SL.K.3. Ask and answer questions in order to seek help, get information, or clarify something that is not understood.
    - W.3.1. Write opinion pieces on topics or texts, supporting a point of view with reasons.
    - W.3.7. Conduct short research projects that build knowledge about a topic.
- Mathematics
    - MP.4. Model with mathematics. (K-ESS3-2)
    - K.CC. Counting and cardinality. (K-ESS3-2)
    - MP.2. Reason abstractly and quantitatively. (3-ESS3-1)
    - MP.4. Model with mathematics. (3-ESS3-1)

NCSS Connections

- Theme 2: Time, Continuity, and Change
    - Use sources to learn about the past in order to inform decisions about actions on issues of importance today.
- Theme 10: Civic Ideals and Practices
    - Evaluate positions about an issue based on the evidence and arguments provided, and describe the pros, cons, and consequences of holding a specific position.

C3 Framework

- Dimension 3: Developing Claims and Using Evidence (3–5)

- D3.4.3-5. Use evidence to develop claims in response to compelling questions.
- Dimension 4: Communicating Conclusions
    - D4.1.K-2. Construct an argument with reasons. (K–2)
    - D4.1.3-5. Construct arguments using claims and evidence from multiple sources. (3–5)

UDL Toolkit

Multiple Means of Engagement	Multiple Means of Representation	Multiple Means of Action and Expression
Using cooperative learning groups with team roles (in the Tornado Testing Challenge) and a yes/no argument line fosters collaboration and a sense of community to sustain effort and persistence.	Using picture books, hands-on models, wind-measuring tools, and class discussions provides multiple options for perception.	Having students show what they know using drawing, writing, speaking, designing, and building provides multiple media for communication.
Providing different goals for the Tornado Testing Challenge helps sustain interest by varying demands and resources to optimize the challenge.	Using KLEW chart and yes/no argument line activates prior knowledge and guides information processing and visualization.	Providing graphic organizers and templates for data collection and assessments facilitates managing information and resources.

Suggested Schedule and Sequence

- Day 1: **Engage** with *Otis and the Tornado* read-aloud and KLEW chart, and **Explore** with Tornado in a Bottle Exploration
- Day 2: **Explain** with *Tornadoes* read-aloud (part 1) and How Good Was Our Model? T-chart, and **Explore** and **Explain** with Measuring the Wind
- Day 3: **Elaborate** and **Evaluate** with Tornado Testing Challenge
- Day 4: **Elaborate** with *Tornadoes* read-aloud (part 2), and **Evaluate** with My Tornado Safety Tip
- Day 5: **Elaborate** and **Evaluate** with Should We Rebuild in Twisterville? Town Hall Meeting yes/no argument line

Materials

For Tornado in a Bottle Exploration

(per team of 2–3)

- 2 plastic 1- or 2-liter bottles
- 1 Tornado Tube coupler (available at *www.amazon.com* or *www.teachersource.com*) or a metal washer and duct tape
- Water

(per student)

- Indirectly vented chemical splash goggles and a nonlatex apron

For Measuring the Wind

(per student)

- Ping-Pong ball
- Ruler, dowel, or stick (for anemometer handle)
- String
- Tape
- Copy of anemometer template (p. 241) on card stock
- Scissors
- Safety glasses or goggles
- Clipboard (optional)

(per class)

- Compass or app such as Smart Compass or Compass 360
- 4 index cards with one direction (East, West, North, South) written on each
- Bubble solution and wands
- Electric fan or blow dryer (optional)

For Tornado Testing Challenge

(per team of 2–3)

- 30 index cards
- Tape
- Paper plate (for the foundation)
- Scissors
- Ruler (for young students, mark the 15 cm point with masking tape to create a standard unit)
- Pencils

(per class)

- Electric fan or blow dryer

For Should We Rebuild in Twisterville?

(per class)

- 1 copy of "Yes" and "No" signs (pp. 152–153)
- Masking tape

Student Handouts

- Tornado in a Bottle Exploration
- Measuring the Wind
- Tornado Testing Challenge Sheet (2 levels available)
- My Tornado Safety Tip
- Should We Rebuild in Twisterville?

Safety Notes

1. All students must wear safety goggles and nonlatex apron during the setup, hands-on, and takedown phases of the activity.

2. Use caution when working with plasticware, which can shatter if dropped and cut skin.

3. Immediately pick up any items dropped on the floor to avoid a slip-and-fall hazard.

4. Immediately wipe up any spilled water on the floor to avoid a slip-and-fall hazard.

5. Use caution when using hand tools or sharp materials to avoid cutting or puncturing skin.

6. Use caution when working near fan; warn students never to put their fingers or other items into or near the fan, which can cut fingers or create projectiles.

7. Use caution when working near blow dryer, which can burn skin on medium and high settings. Be sure to keep it on the low or cool setting.

8. Use only GFI protected circuits when using electrical equipment, and keep the fan and blow dryer away from water sources to prevent shock.

9. Wash hands with soap and water after completing this activity.

Media

Books

- *Otis and the Tornado,* by Loren Long
- *Tornadoes,* by Gail Gibson
- *Green City,* by Allan Drummond (optional)

Video

- "Engineers Test Storm Resistant Homes," from the Weather Channel *www.youtube.com/watch?v=oF_zBqY9aGU*

Background for Teachers

"Toto, I've a feeling we're not in Kansas anymore." When Dorothy says these words in *The Wizard of Oz*, it's clear that the tornado has delivered her to a fantastical place far from home. Unfortunately, the reality of tornadoes can be much more devastating. Contrary to popular belief, tornadoes are not limited to small geographic areas. Tornadoes have been documented in every state in the United States and on every continent except for Antarctica. The United States sees over 1,200 tornadoes a year on average, leading the world in tornado strikes (see Figure 5.20, p. 222).

FIGURE 5.20.

Average Annual Number of Tornadoes per State, 1991–2010

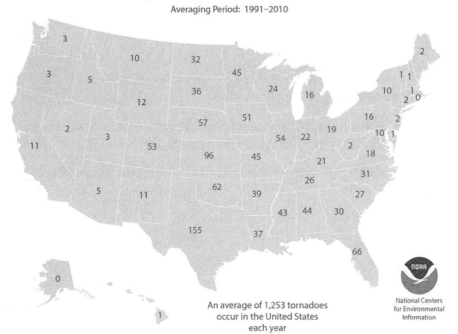

Source: Courtesy of the National Oceanic and Atmospheric Administration (NOAA), *www.ncdc.noaa.gov/climate-information/extreme-events/us-tornado-climatology*.

Within the United States, several areas are disproportionately affected by tornadoes. These include portions of the south-central plains states of Texas, Oklahoma, Kansas, eastern Colorado, Missouri, and Iowa, colloquially referred to as Tornado Alley, as well as the state of Florida, which has among the highest number of tornadoes per year in the country. In addition, high numbers of tornadoes occur in the region commonly referred to as Dixie Alley, which includes portions of the Gulf Coast and the southeastern states of Arkansas, Louisiana, Alabama, Mississippi, Georgia, Tennessee, Kentucky, and western parts of the Carolinas, while Hoosier Alley in Indiana and Carolina Alley in the eastern part of the Carolinas are also quite active. So why are certain areas more susceptible to this type of extreme weather?

Tornadoes are violently swirling columns of air that extend from clouds (usually cumulonimbus clouds) to Earth's surface. Tornadoes form when moist air near Earth's surface is warmed by the Sun and rises, causing an updraft. As the air cools above, the moisture condenses and precipitation (such as rain or hail) falls, bringing a downdraft of cool air. This by itself would create a typical thunderstorm. However, in some instances, wind shear from air moving quickly in different directions or at different speeds at different levels above Earth's surface helps form a vortex, or spinning motion, which can cause a spinning effect within the thunderstorm, forming a supercell storm. The supercell essentially feeds on itself, with the vortex, or eye, of the tornado bringing more warm, moist air up as the cool air and precipitation come down, all while swirling violently. While this can occur anywhere, certain places on the globe, such as the south-central portion of the United States, are points where cold, polar air commonly meets warm, subtropical air. The collision of these fronts creates a boundary where thunderstorms regularly occur. Additionally, such places in the middle latitudes of the globe (between 30 and 50 degrees north or south) are prone to experiencing a lot of wind shear and thus are susceptible to tornadoes.

Tornadoes are measured by the Enhanced Fujita (EF) Scale, which rates tornadoes from EF0 to EF5. This scale is a variation of the original Fujita Scale, named after meteorologist Ted Fujita, who devised it in 1971. Both the original and the enhanced scales estimate the wind speeds of tornadoes based on the amount and type of damage they do. This is because it is impossible to accurately measure the speeds inside most tornadoes as they are occurring (because the tornado damages the equipment). When it was found that the original scale's wind speeds were likely inaccurate at the higher levels, the enhanced scale was adopted in 2007. Tornado wind speeds on the EF Scale range from 65 mph to over 200 mph. Some tornadoes have been estimated at over 300 mph!

Thanks to a large amount of data obtained through tornado tracking and modeling of that data, predicting the occurrence of tornadoes has improved in the past several decades. Meteorologists are better able to identify the conditions that lead to tornado formation earlier than ever before, so warnings can be communicated earlier to the public. However, unlike hurricanes, which allow for a long lead time before they strike, tornadoes can occur with little warning. Often, by the time they are spotted, they have already caused damage. Fortunately, most tornadoes don't result in deaths. Better forecasting, improved preparation for severe weather, more stringent building codes, and faster response systems all are factors in reducing fatalities. In building tornado-resistant homes, costs and livability are often weighed against the likelihood of a severe tornado strike. Innovations that help make houses more tornado resistant include reinforced roofs that have stronger clips to keep the roof stable, sometimes with four slopes rather than two; anchors that hold houses onto their foundations; and even a dome-shaped design rather than the traditional box shape. Many homes in tornado-prone areas have safe rooms, which are cement or steel reinforced and are built to withstand strong winds that might destroy the rest of the house.

The *NGSS* for kindergarten (K-ESS3-2) expects students to ask questions about local weather and connect this to the idea that certain weather phenomena are more likely to occur in certain places. The same standard explores how weather forecasting can use knowledge of weather patterns to help people prepare for severe weather. The related third-grade standard (3-ESS3-1) challenges students to develop arguments about models that are tested to minimize damage from hazardous weather. In this lesson, students are introduced to tornadoes through a picture book, after which they make observations using water-filled bottles that function as Tornadoes in a Bottle when students create a swirling motion, or vortex, in the bottles. After learning more about the science of tornadoes, students evaluate the accuracy of this model to reinforce the idea that models can never be completely accurate, as they are representations of the real phenomena. Students then explore and measure wind by making a Ping-Pong ball anemometer. This activity helps students begin to think about the importance of having a standard scale for weather phenomena in order for scientists to communicate. Students then build and test tornado-resistant homes with material constraints (index cards and tape). They also learn many of the steps recommended by the Federal Emergency Management Agency (FEMA) to prepare for and respond to tornadoes. Finally, students engage in a class activity about whether their town should rebuild after a destructive tornado. By reading quotes gleaned from articles about post-tornado communities, students realize that fear of weather hazards is often countered by preparation, knowledge, and resilience.

Additional Resources

High-quality short video about the science of tornadoes

- *https://video.nationalgeographic.com/video/101-videos/tornadoes-101*

Explanation of how tornadoes are rated

- *https://weather.com/storms/tornado/news/enhanced-fujita-scale-20130206*

Information on tornado climatology

- *www.ncdc.noaa.gov/climate-information/extreme-events/us-tornado-climatology*

Types of tornadoes

- *www.nssl.noaa.gov/education/svrwx101/tornadoes/types*

Fact sheet for kids about tornadoes

- *www.fema.gov/media-library-data/a4ec63524f9fd1fa5d72be63bd6b29cf/FEMA_FS_tornado_508_8-15-13.pdf*

The Young Meteorologist Program's "Severe Weather Preparedness Adventure," an interactive game

- *http://youngmeteorologist.org*

FEMA mobile app with real-time weather alerts and emergency tips

- *www.fema.gov/mobile-app*

Facts on tornadoes for kids

- *www.ready.gov/kids/know-the-facts/tornado*

NOAA/National Weather Service Storm Prediction Center

- *www.spc.noaa.gov*

5E Lesson Plan

Engage: *Otis and the Tornado* Read-Aloud and KLEW Chart

1. *Otis and the Tornado*, by Loren Long, tells the story of a tractor named Otis, who helps his farm friends, as well as an unfriendly bull, through a tornado. Before reading, ask students to pay careful attention to the descriptions the author provides of the oncoming storm. After reading, ask:

    - "What were some of the clues that warned Otis about the oncoming storm?" (skies began to swirl and turn dark, the rain and wind came and then suddenly stopped, the sky turned green, the birds stopped chirping, the animals became restless)

    - "Where did Otis and the animals go for safety? Why?" (they went down to the lowest point in the farm, below the bank of Mud Creek, because this would give them cover from the winds and the lightning)

    - "Where did the farmer go when the storm was approaching?" (a storm cellar below ground)

    - "What sound did the tornado make?" (it roared like a freight train)

- "What kind of damage did the tornado leave?" (it took the roof off the barn, pulled down trees, turned over a truck, knocked down the wind vane)

- "How do the illustrations in this book communicate the seriousness of the tornado?" (dark colors, swirling skies, Otis's worried looks)

- "Why do you think the bull's attitude changed after the tornado?" (he was grateful to Otis for rescuing him)

2. Draw a KLEW chart like the one in Figure 3.2 (see p. 17). Ask students, "What do you Know about tornadoes?" Write their comments in the "Know" column. Ask, "What do you Wonder about tornadoes?" Write student questions in the "Wonderings" column. Alternatively, you can have students write what they Know and Wonder about on sticky notes and have them post these on the chart. They can then move their Wonderings to the "Learnings" column as they address those questions throughout the study.

3. Inform students that they will be revisiting this chart during their study of tornadoes to help organize their learning.

Explore: Tornado in a Bottle Exploration

Advance Preparation: To create a model Tornado in a Bottle, first fill a bottle approximately two-thirds full with water. Then, connect it to an empty bottle using the Tornado Tube coupler. Alternatively, place the metal washer on top of the bottle with water, place the empty bottle upside down on the bottom bottle, aligning the openings of the bottles with the washer in between, and tape the connection together tightly with duct tape. Depending on your time limitations, you can either prepare all the Tornadoes in a Bottle in advance or have teams create their own during class. Either way, keep one unmade so that you can show the components to students.

1. Show students a complete Tornado in a Bottle, as well as the component parts. Be sure to show students that the coupler is simply a small tube that holds the bottles together and has a hole in it (or that the metal washer is simply creating a small opening between the bottles). Explain that the Tornado in a Bottle is a model that has some features that are similar to real tornadoes and some that are quite different. Tell students that their challenge is to work with their teammates to explore their Tornadoes in a Bottle, making observations and developing questions (Wonderings), which they will record on their Tornado in a Bottle Exploration sheets (p. 240).

Lesson 7: Weather or Not

2. Divide students into groups of two or three. Distribute Tornado in a Bottle Exploration sheets and either premade Tornadoes in a Bottle or materials for teams to make them. Show students that the first part of the sheet asks them to simply turn their bottles over and draw their observations. (Ask students to first predict what they think will happen when they do this; most will think that the water from the top bottle will pass into the bottom bottle. Accept all predictions.) Tell them to give their bottles a big swirl (demonstrate this motion using an empty bottle so that you don't give away how the Tornado in a Bottle works!), then draw their observations in the second box. *Tip:* Tornadoes in a Bottle can sometimes leak. Have students keep papers off to the side of their desks while they are testing. If a tube leaks, decouple the bottles and reconnect them. Tell students to try not to squeeze the bottles as they are joining them.

FIGURE 5.21.
Tornado in a Bottle Exploration

3. Give students time to explore freely with the bottles (see Figure 5.21). A few minutes before the time is up, remind them to record their observations and wonderings on their sheets.

4. Using a Tornado in a Bottle to demonstrate during this discussion, ask students:

   - "What happened when you turned your Tornado in a Bottle upside down?" (most of the water stays in the top bottle, usually a surprising result) "Why do you think this happened?" (guide students through questioning to think about what might be keeping the water from coming into the lower bottle: air!)

   - "What happened when you gave the Tornado in a Bottle a swirl?" (a "tornado" formed, and the water moved from the top to the bottom bottle)

- "How do you think this works?" (guide discussion leading to the idea that the swirl created a small opening that allowed the air from the bottom bottle to be drawn up into the top bottle, creating space for the water to come down as it is pulled by gravity)

- Explain to students that the swirling of the water is referred to as a vortex. Ask students, "Where have you observed a vortex before?" (bathtubs, sinks, toilets)

5. Invite students to share their other observations and wonderings. (Students may observe that swirling several times or with greater force creates a larger tornado; that they can form a vortex by swirling in either a clockwise or counterclockwise direction; and that they can make a "rope" tornado by swirling the bottles while the tornado is moving.) Add any student Wonderings about tornadoes to the KLEW chart, as well as any Learnings about vortices (the plural of *vortex*).

6. Explain to students that they will be learning more about real tornadoes, but that you'd like them to begin thinking about the ways that this model tornado might be similar to and different from real tornadoes. Invite students to share some of these initial ideas. (Students may note that the model has a vortex like a tornado, but that it is small and made of water instead of air, and of course, it's in a bottle!) This comparison will be analyzed more fully in the next lesson.

Explain: *Tornadoes* Read-Aloud (Part 1) and How Good Was Our Model? T-Chart

1. Position the KLEW chart so that it is visible to students. Read *Tornadoes*, by Gail Gibson, up through page 21. This portion of the book describes how tornadoes form and how they are measured. Ask students to turn to a partner and explain to each other, in their own words, how a tornado forms. Allow students to discuss for two to three minutes. Ask volunteers to share what they have learned.

2. Invite students to add Learnings and Wonderings to the KLEW chart. They can also add Evidence from the book or from their Tornado in a Bottle experience that provides support for their Learnings.

3. Create a T-chart like the one in Figure 5.22 for students.

FIGURE 5.22.

Tornado in a Bottle T-Chart

How was our Tornado in a Bottle **like** a real tornado?

How was our Tornado in a Bottle **not like** a real tornado?

Like a real tornado	Not like a real tornado

4. Explain to students that every model has strengths and weaknesses. Because models are only estimations of the real phenomenon, they will have ways that they are like and not like the real thing. Invite students to contribute items to the T-chart. A sample completed chart might look like Figure 5.23.

FIGURE 5.23.

Sample Completed Tornado in a Bottle T-Chart

How was our Tornado in a Bottle **like** a real tornado?

How was our Tornado in a Bottle **not like** a real tornado?

Like a real tornado	Not like a real tornado
Has a vortex (swirling motion)	Made of water instead of air
Has a funnel shape	Inside a bottle
Air is pulled up through the center (or eye)	Small
	Human made
	Doesn't cause damage

5. Ask students to imagine an improved model for a tornado. Elicit ideas on what that would entail. (Students may mention wind tunnels or other physical models.) Explain to students that scientists also use computer models and

mathematical models (using data that they have collected from actual tornadoes) to help simulate what happens during a tornado.

6. Review the EF Scale and how tornadoes are measured. Remind students that the scale ranges from EF0 to EF5, with the numbers representing the *estimated* wind speeds based on the damage done by the tornado (Figure 5.24). Ask students, "Why do you think it was important for scientists to have scales like the Fujita (F Scale) and Enhanced Fujita Scale (EF Scale) to measure tornadoes?" (a scale helps scientists communicate about tornadoes with a common or standard language; using a scale is more precise than saying "a big tornado" or "a little tornado"; comparing the locations and ratings of tornadoes that have occurred helps them study and predict tornadoes)

FIGURE 5.24.

Enhanced Fujita (EF) Tornado Scale

> EF0 (weak): 65–85 mph, light damage.
> EF1 (weak): 86–110 mph, moderate damage.
> EF2 (strong): 111–135 mph, considerable damage.*
> EF3 (strong): 136–165 mph, severe damage.*
> EF4 (violent): 166–200 mph, devastating damage.*
> EF5 (violent): Over 200 mph (rare), incredible damage.
>
> *Note that in *Tornadoes*, the author lists these estimated wind speed categories 1 mph higher than the categories established by the National Weather Service, which are listed above.

7. Explain to students that scientists are able to measure wind speed under more normal conditions (not in a tornado). Measuring the wind can help *predict* storms like tornadoes and hurricanes. It can also help in everyday life. Ask students, "Can you think of any reasons why measuring the wind can be helpful?" (helping airplanes take off and land, helping sailors on sailboats and other sailing ships, helping us know how to dress in the morning, warning us to tie down or bring in yard furniture and other items) Instruments that measure wind speed are called anemometers. Inform students that they are going to build their own anemometers and go outside to measure the wind.

Lesson 7: Weather or Not

Explore and Explain: Measuring the Wind

1. Distribute one copy of the anemometer template (p. 241) on card stock to each child. Have children cut out the wedge-shaped scale.

2. Have students attach the wedges to rulers using tape, taking care to match the corner of the straight edge of the wedge with the edge of the ruler. The top number on the scale should sit at the middle of the ruler. (See Figure 5.25.)

3. Tell students to tape the string to the edge of the ruler so that it hangs straight down to the 0 mark. Then, students can tape the bottom of the string to the Ping-Pong ball.

4. Distribute copies of the Measuring the Wind handout (p. 242). Distribute clipboards (if available) and pencils.

5. Go outside to an open space. Using a compass or compass app, identify the four directions: East, West, North, and South. Mark them with index cards placed on the ground so that students have a visual reminder. (You can set a small rock on each to weigh them down or stand them vertically in the grass.)

6. Have students take turns blowing bubbles and recording the directions that the bubbles move from and to on their data sheets. This tells them the direction that the wind is moving.

7. Then, have students measure the wind with their anemometers. Explain that to use the anemometer, you hold the ruler with the wedge and ball facing away from you (Figure 5.26). Turn into the wind, and watch the ball fly up! The wind can be measured by simply noting where the string crosses on the scale. Students should record their measurement on the handout.

8. Have students continue to measure the wind for several minutes, recording the strongest gust (short, strong wind).

9. When students return to the classroom, discuss their findings. Ask:

FIGURE 5.25.

A Completed Anemometer

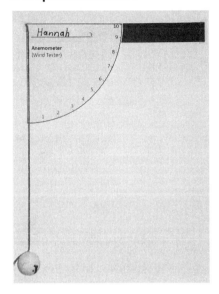

FIGURE 5.26.

Students Learn How to Read the Anemometer

- "Why might we have some different measurements?" (students measured wind at different times, faced in different directions as the wind changed directions, tilted the anemometer)

- "Why do meteorologists measure the wind?" (to help predict weather including storms, to help people prepare for wind and storms)

10. If time allows, have students use their anemometers with an electric fan or blow dryer (with the teacher holding it and set on low or cool air) to see how their anemometers work in high winds. They can also try blowing on each other's anemometers to see how strong a wind they can create.

Safety Alert: Be sure to warn students about the danger of putting their fingers on or near the fan. Also, be sure that the blow dryer is on the low or cool setting. Keep both of these electrical items away from water and sinks.

Elaborate and Evaluate: Tornado Testing Challenge

In this activity, teams of students design and build houses out of index cards and tape to try to develop tornado-resistant houses. The houses will be tested with an electric fan or blow dryer.

1. Ask students, "What do you think engineers and architects need to think about when they design houses in places that have a lot of tornadoes?" (accept all answers)

2. Explain to students that they are going to work in teams to design and build houses that can withstand a tornado (from the fan or blow dryer). Distribute the Tornado Testing Challenge handout (pp. 243–245). Note that two levels of this challenge are available. Choose the handout for either Level 1 or 2, depending on which one you feel is more appropriate for your students.

3. Divide students into teams of two or three. Distribute the index cards, tape, rulers, paper plates, and scissors.

4. For Level 1, explain to students that their challenge is to build a house that can stay together in the wind for five seconds, using only index cards, tape, and a paper plate (as the foundation).

5. For Level 2, explain to students that the *criteria* for their house are as follows:

- Your house must be at least 15 cm wide and 15 cm tall.

- It must have a roof, a door, and two windows.

Lesson 7: Weather or Not

- It must be attached to the foundation (the paper plate).
- It needs to stay together for 5 seconds in the wind!

6. For Level 2, explain the *constraints* of the challenge:
   - Your house can only be made out of these materials:
     - 30 index cards
     - Tape
     - 1 paper plate (for the foundation)
   - You can also use scissors, a ruler, and pencils for planning, measuring, and building.

7. Give students an amount of time (about 20 minutes) to complete the challenge. (This is an additional constraint!)

8. As students are working, circulate around the teams to ensure that students understand the criteria and constraints of the challenge.

9. When the time is up, have students come to the testing center (a table with the fan or blow dryer) to test their houses. Have students place their houses on the table about 2 feet from the fan or blow dryer. Test the houses to see whether they are able to withstand the tornado test for 5 seconds. You can either have students count together to 5 or use a clock or stopwatch.

10. Ask students to explain on the data sheet why they think their houses did or did not stay together in the tornado. Allow teams to revise their houses as necessary for a second test. If houses survived the first test, you can make a harder challenge by either moving the wind source closer to the houses or increasing the speed of the fan. You can also increase the amount of time that the house needs to withstand the wind.

11. Have teams report on their results. For each team, ask: "What did you do to make your house withstand the tornado? What revisions did you make to make it even stronger?"

12. When all teams have presented, ask, "What are the features of a tornado-resistant house?" (strong connections between walls, roof, and foundation; flatter roof; dome-shaped or rounded houses; reinforced walls)

13. Show students the Weather Channel video "Engineers Test Storm Resistant Homes" at *www.youtube.com/watch?v=oF_zBqY9aGU*. The video shows how

houses are tested in a wind laboratory (very much like the test the students just did). Note that although the video refers to "hurricane-force winds," the tests are for all storms, including tornadoes. Discuss with students the similarities (e.g., using fans, building different houses, testing different features) and differences (e.g., size) between the test in the video and their test.

14. Revisit the KLEW chart; have students add Learnings, Evidence, and Wonderings.

15. Assess student work with the Rubric for Tornado Testing Challenge (p. 248).

Elaborate: *Tornadoes* Read-Aloud (Part 2)

1. Read aloud the rest of the *Tornadoes* book, starting on page 22. This portion of the book discusses the idea that Tornado Alley and Florida are both areas of the United States that experience many tornadoes. It also discusses a few historical tornadoes and their destruction, leading into a section on the importance of tornado forecasting to warn people and the steps we can take to stay safe in a tornado. Use the following prompts to guide reading:

    - After reading pages 22–23, ask, "Why do you think certain parts of the country experience more tornadoes than others?" (they have the ingredients for tornadoes, including a lot of wind and frequent thunderstorms)

    - After pages 24–27, ask, "Why do you think the author tells us about these historical tornadoes?" (to show the damage that was done and make the point that better predicting, or forecasting, has helped keep people safer)

    - After pages 28–29, ask, "How do meteorologists predict and communicate about tornadoes?" (they monitor radar and computer data about storms, and then broadcast information on television and radio) *Note:* Although not mentioned in the book, the National Weather Service also has an app to warn of storms called NWS NOW.

    - After reading to the end of the book, ask, "What can you do to stay safe when a tornado approaches?" (if inside, go to a basement or safe room away from windows, crouch down low and cover your head, cover yourself with a mattress or heavy blankets to protect you from falling debris; if outside, find a low spot such as a ditch, lie flat on your stomach, and cover your head with your hands)

Lesson 7: Weather or Not

2. Finally, ask students, "Why is weather forecasting about severe weather such as tornadoes important?" (it helps people prepare for and respond to severe weather)

3. Revisit the KLEW chart and have students add Learnings, Evidence, and Wonderings from this section of the book.

> **Misconception Alert**
>
> Students often believe that tornadoes cannot occur in cities. This misconception is very understandable since most tornadoes do occur in more rural areas in the United States. But this isn't because tornadoes can't occur in cities; it is simply because cities and suburbs make up a much smaller percentage of U.S. land than rural areas. This is particularly true in areas like Tornado Alley where there are great expanses of open land. Statistically speaking, tornadoes are more likely to be in rural places because there is more rural land! However, tornadoes can (and do) hit cities. St. Louis, Missouri, Birmingham, Alabama, and Topeka, Kansas, are just a few of the large cities that have experienced major tornadoes. This fact bolsters the need for tornado safety awareness for all students.

Evaluate: My Tornado Safety Tip

1. Distribute copies of the My Tornado Safety Tip handout (p. 246). Students are first asked to complete the sentence "Tornado forecasts keep us safe by _____." The following are some sample answers:

    - Giving people time to go to storm cellars or safe rooms
    - Giving people time to get blankets, mattresses, and other materials to protect them from falling debris
    - Giving people time to cover their heads
    - Helping people know that they live in tornado-prone areas so that they can make their houses and schools more tornado resistant

2. Then, students should draw a picture of a way to stay safe from tornadoes. These safety tips can cover any of the suggestions from the *Tornadoes* book or the other teacher resources listed in this lesson, such as going to a storm

cellar or safe room, putting a mattress or heavy blanket on top of you, covering your head, going under stairs, staying away from windows, or building houses with storm ties and reinforced walls.

3. Assess student work with the Rubric for My Tornado Safety Tip (p. 248).

Elaborate and Evaluate: Should We Rebuild in Twisterville? Town Hall Meeting Yes/No Argument Line

Advance Preparation: Place a piece of masking tape on the floor to divide a portion of the room into two sides. Post the "Yes" sign (with the thumbs-up picture) on one side of the room and the "No" sign (with the thumbs-down picture) on the other. Write the following sentence frame on the board: I think we should/should not rebuild (claim) because _____ (evidence) according to _____ (source). You might wish to use the CES template in Figure 3.9 (p. 22).

1. Ask students to imagine the following scenario:

   You are a resident of a town called Twisterville in Tornado Alley. Twisterville experienced an EF3 tornado last week. It did tremendous damage to most of the houses and destroyed many of the public buildings such as schools and stores. Amazingly, no one was badly hurt because people had enough warning from weather forecasters and were able to go to storm cellars and safe rooms in homes and other buildings. No animals were hurt, either. Although the townspeople have food and water that were brought in for them, they are very shaken up by all the damage. Many people lost everything. There is also a huge mess to clean up, and it will be very expensive to rebuild the houses and other buildings. They have experienced smaller tornadoes but nothing like this one.

2. Ask students to think about how they might feel in this situation. Allow students to turn and talk to a partner for one to two minutes, and then ask for volunteers to report. (comments will likely include sadness about losing possessions, fear about another tornado, desire to go somewhere else, and worries about how to pay to repair homes)

3. Explain that when homes and towns are damaged badly, people need to decide whether they want to rebuild there or move somewhere else. In Twisterville, because so many of the homes and buildings were damaged, the mayor has asked the residents to come to a Town Hall Meeting to decide whether they should rebuild the town.

Lesson 7: Weather or Not

4. Tell students that the Town Hall Meeting will progress this way: You (the mayor) will ask, "Should we rebuild in Twisterville?" You will then ask the residents (the students) to go to either the "Yes" side of the line or the "No" side of the line. When they are all on one side of the line or the other, we will have a town meeting. Remind them that there is no wrong answer; you are just looking for their opinions. *Tip:* If you haven't done this type of activity before, you may wish to have students jot down "Yes" or "No" on paper (or on sticky notes and post them on their shirts) before they go to their sides so they will be less likely to follow friends. The more students do this kind of activity, the more comfortable they will be with following their real opinions.

5. Have students go to their chosen sides, and then ask students to share *why* they chose the way they did. Show students the sentence frame, and explain that you would like them to share reasons using that frame. For example, students could say, "I think that we should not rebuild (claim) because there are a lot of tornadoes in Tornado Alley (evidence) according to our book (source)." Or, they might say, "I think that we should rebuild (claim) because we can use storm ties when we build our house to make them safer (evidence) according to our video (source)". (Students will likely have more arguments for leaving than for staying; allow all arguments.) Reference the KLEW chart to remind students of what they learned and where they learned it. You can also project the Sources of Evidence template (Figure 3.14 on p. 27) to remind students that they can use information they learned from the book, the video, their building experiment, other students' building experiments (or the experiment in the video), their own experiences with tornadoes, or others' experiences.

6. After students have had the opportunity to share their thoughts, ask if anyone wants to change sides. If someone does, ask him or her which arguments were convincing.

7. Distribute copies of the Should We Rebuild in Twisterville? handout (p. 247). Explain to students that these are some of the comments from neighbors who couldn't come to the meeting. Have students read the comments with a partner, or read aloud as a whole class. Then, have students answer the question at the bottom, which asks them to list two reasons that neighbors gave for rebuilding in Twisterville. (*Note:* The purpose of this handout is to help students understand the reasons why people who have been through devastating losses like this stay where they are. These statements are paraphrased from real articles and interviews with people who have experienced severe tornadoes. Children often see the arguments for leaving more easily than for staying, especially in places that experience frequent tornadoes. The sheet

is meant to offer authentic voices to assist students in seeing the arguments for staying. From a literacy standpoint, it helps them to practice restating an argument in their own words.)

8. Inform students that because they've heard some additional arguments from their neighbors (others' experiences), you'd like to see where they now stand on the question of whether to rebuild in Twisterville. Remind them that there is no wrong answer; as the mayor, you are just trying to get information. Ask students to decide and go to their chosen side of the line. Invite students to share their ideas again using the CES sentence frame.

9. Then, ask students the following questions:

    - "How should we make our final decision whether to rebuild?" (invite ideas such as voting, delaying a decision, determining that it's best for people to decide for themselves without trying to get agreement)

    - "Is there any other information you'd like to have before you make a decision?" (invite answers, which may include finding out how much it will cost to rebuild, how much other places would cost, how often tornadoes really do come through this area, whether they can afford to build tornado-resistant homes and schools)

    - "How does weather forecasting affect your decision, if at all?" (weather forecasts help people be safer even in areas where tornadoes are frequent)

10. Evaluate informally based on students' ability to support their claim with evidence and a source.

Optional: *Green City* Read-Aloud

Read *Green City*, by Allan Drummond, which tells the story of how a real town in Tornado Alley named Greensburg recovered from a devastating tornado and rebuilt in a sustainable way. (*Note:* I did not make this a foundational book for this lesson because much of the book focuses on sustainability rather than tornadoes specifically. Nevertheless, it is a great extension for this lesson.)

Going Deeper

- Students can develop a public service announcement for their school on tornado safety.

Lesson 7: Weather or Not

- Students can interview architects and engineers about building storm-safe structures.

- Students can repeat the Tornado Testing Challenge with additional materials, such as clay, rubber bands, or craft sticks, to see how tornado resistant they can make them.

- Students can interview the school building manager or custodian to learn about storm safety features that are used in the school building. They can advocate for improved features if not present.

Name: _____ Date: _____

Tornado in a Bottle Exploration

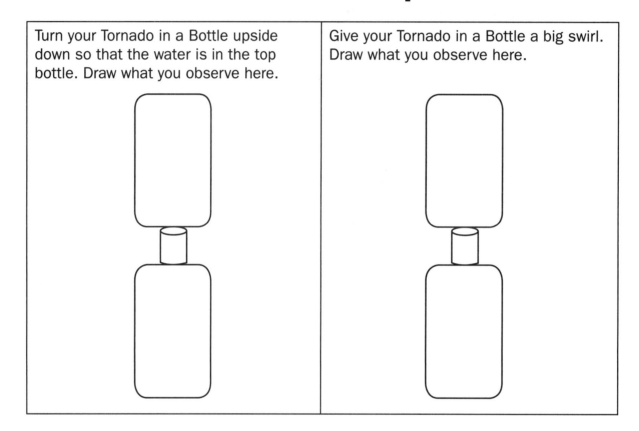

| Turn your Tornado in a Bottle upside down so that the water is in the top bottle. Draw what you observe here. | Give your Tornado in a Bottle a big swirl. Draw what you observe here. |

Spend some time exploring your Tornado in a Bottle. Write your observations and wonderings here:

I observed …

I wonder …

Anemometer Template

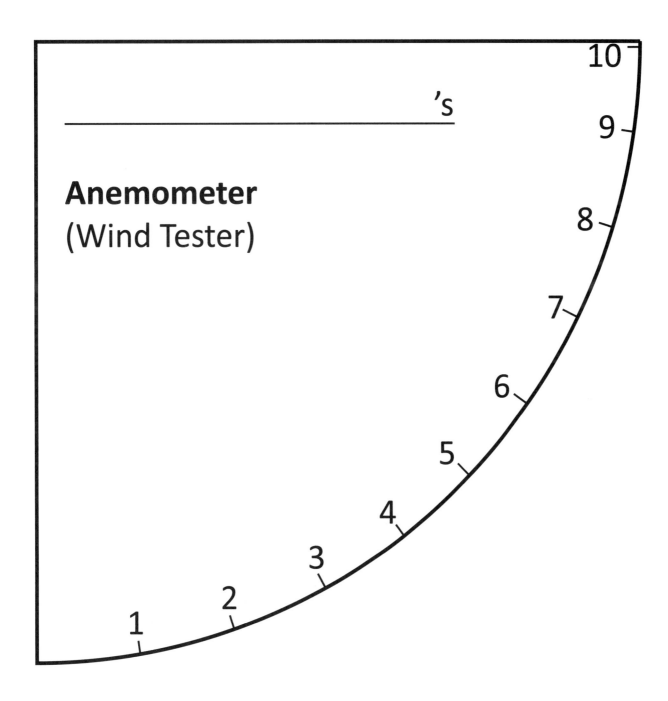

Meteorologist: _____   Date: _____

Measuring the Wind

1. With your teacher, identify the directions (East, West, North, and South).

2. Blow some bubbles and watch the direction that they move as they are carried by the wind. Draw yourself, the directions, and the bubbles in the box below.

The wind is blowing from the _____ to the _____.

3. Hold your anemometer so that it is facing *into* the wind (facing the direction from which the wind is blowing) and measure the wind.

    The wind measures _____ on my anemometer.

4. Keep measuring the wind for several minutes for *gusts* (short, quick winds).

    The highest gust measured _____ on my anemometer.

5. Meteorologists measure the wind because _____
_____.

Engineer: _____

Tornado Testing Challenge (Level 1)

Our House	Our Tornado Test Results
	☹ 😐 🙂

Why do you think your house did or did not stay together in the tornado?

_____

Revise your house and see if you can make it stay together in a longer, stronger tornado!

Our FINAL Tornado Test Results (circle one)   ☹   😐   🙂

Engineer: _____   Date: _____

Tornado Testing Challenge (Level 2)

Your challenge: To design and build a house that can withstand a tornado!

The criteria:

- Your house must be at least 15 cm wide and 15 cm tall.
- It must have a roof, a door, and two windows.
- It must be attached to the foundation (the paper plate).
- It needs to stay together for 5 seconds in the wind!

The constraints:

- Your house can only be made out of these materials:
  - 30 index cards
  - Tape
  - 1 paper plate (for the foundation)
- You can also use scissors, a ruler, and pencils for planning, measuring, and building.

Draw and label your team's house below. Use the word bank for help!

Word Bank
Roof
Foundation
Window
Door
Wall

1. Did your house stay together in the Tornado Test?   Yes   No

2. Why do you think your house did or did not stay together in the Tornado Test?

_____

_____

_____

3. If your house <u>did not</u> stay together, revise it and test again! If your house <u>did</u> stay together, revise it to see if you can make it withstand a longer, stronger tornado!

Our revised house

4. Did your <u>revised</u> house stay together in the Tornado Test?   Yes   No

5. What are some important features of a tornado-resistant house?

_____

_____

_____

Meteorologist: _____

My Tornado Safety Tip

Tornado forecasts keep us safe by _____
_____
_____

My Tornado Safety Tip: _____

Name: _____ Date: _____

Should We Rebuild in Twisterville?

To help you decide whether you want to rebuild your home and our town, read these statements from some of your Twisterville neighbors:

Joanne: "I was born here and tornadoes are a way of life. But tornadoes don't touch down often, and they probably won't touch down again here."

Beau: "My family has lived here for five generations. We would never think of living anywhere else!"

Kira: "The cost of living is very low here. I couldn't afford to have nearly as nice a home in other places."

Wynton: "There are problems no matter where you live: hurricanes, earthquakes, snowstorms, wildfires. At least we only really have to worry during our tornado season between March and June."

Javier: "We're going to put in a safe room with strong, steel-reinforced walls. We know there's no such thing as a tornado-proof house, but this will keep us safe."

Ashley: "My job is here. I can't just pick up and go!"

Chris: "The people here are the best! Neighbors all pitch in to help neighbors. We all watch out for each other. I love living here and can't imagine a better place!"

List two (2) reasons that your neighbors gave for rebuilding in Twisterville.

1. _____

   _____

2. _____

   _____

Rubric for Tornado Testing Challenge

Criteria	3 pts.	2 pts.	1 pt.	Not done
Student describes (through labeled drawings) the construction of his or her team's house.				
Student describes (verbally or through labeled drawings or both) the reasons why he or she thinks the house did or did not stay together during the test.				
Student describes (verbally or through labeled drawings or both) the features of a tornado-resistant house.	▓			
Student demonstrates understanding of criteria and constraints in this engineering challenge.	▓			

Total: ___ / 10 pts.

Rubric for My Tornado Safety Tip

Criteria	2 pts.	1 pt.	Not done
Student describes in writing how tornado forecasts make people safe.			
Student draws one tip about tornado safety.			

Total: ___ / 4 pts.

Lesson 8: *Egg*streme Sports

Lesson 8

Eggstreme Sports

Is Football Too Dangerous for Kids?

Suggested Grade Levels

3–5

Driving Questions

- What role do forces play in sports?
- Why is it important to protect your head during certain sports?
- How can we design and test solutions to the problem of concussions in children's sports?

Lesson Overview

Students learn about the importance of protecting their heads in contact sports to avoid concussions. They then design, create, and test different helmets using an egg as a model for the human head. Finally, they read about concussion statistics for children in different sports and develop an informational pamphlet about sports safety to share with others.

Connecting to the *NGSS*

(See full alignment in Table A.8 on p. 510.)

- PS2.A: Forces and Motion
  - Each force acts on one particular object and has both strength and a direction. An object at rest typically has multiple forces acting on it, but they add to give zero net force on the object. Forces that do not sum

to zero can cause changes in the object's speed or direction of motion. (3-PS2-1)

- PS2.B: Types of Interactions
  - Objects in contact exert forces on each other. (3-PS2-1)
- LS1.A: Structure and Function
  - Plants and animals have both internal and external structures that serve various functions in growth, survival, behavior, and reproduction. (4-LS1-1)
- ETS1.C: Optimizing the Design Solution
  - Different solutions need to be tested in order to determine which of them best solves the problem, given the criteria and the constraints. (3-5-ETS1-3)

Societal Issues

Government and School Regulation, Personal Autonomy

Nature of Science

- Scientific Knowledge Is Based on Empirical Evidence
  - Science findings are based on recognizing patterns.
- Scientific Knowledge Assumes an Order and Consistency in Natural Systems
  - Science assumes consistent patterns in natural systems.
- Scientific Knowledge Is Open to Revision in Light of New Evidence
  - Science explanations can change based on new evidence.

CCSS Connections

- English Language Arts
  - RI.3.1. Ask and answer questions to demonstrate understanding of a text, referring explicitly to the text as the basis for the answers.

- RI.4.9. Integrate information from two texts on the same topic in order to write or speak about the subject knowledgeably.
- RI.5.9. Integrate information from several texts on the same topic in order to write or speak about the subject knowledgeably.
- Mathematics
  - MP.2. Reason abstractly and quantitatively. (3-5-ETS1-3)
  - MP.4. Model with mathematics. (3-5-ETS1-3)
  - MP.5. Use appropriate tools strategically. (3-5-ETS1-3)

NCSS Connections

- Theme 6: Power, Authority, and Governance
  - Examine issues involving the rights and responsibilities of individuals and groups in relation to the broader society.
- Theme 8: Science, Technology, and Society
  - Research a scientific topic or type of technology developed in a particular time or place, and determine its impact on people's lives.
- Theme 10: Civic Ideals and Practices
  - Evaluate positions about an issue based on the evidence and arguments provided, and describe the pros, cons, and consequences of holding a specific position.

C3 Framework

- Dimension 3: Developing Claims and Using Evidence
  - D3.3.3-5. Identify evidence that draws information from multiple sources in response to compelling questions.
  - D3.4.3-5. Use evidence to develop claims in response to compelling questions.
- Dimension 4: Communicating Conclusions
  - D4.1.3-5. Construct arguments using claims and evidence from multiple sources.

UDL Toolkit

Multiple Means of Engagement	Multiple Means of Representation	Multiple Means of Action and Expression
Focusing on a subject that is directly related to school sports and children's health and involves educating the community optimizes relevance and authenticity.	Having helmets available for students to examine, in addition to pictorial and written expressions of helmets, presents information using various means.	Students expressing their understanding in written, pictorial, and oral forms allows for multiple means of expression.
Providing self-assessment rubrics, such as the one included in the pamphlet project, optimizes motivation and autonomy.	Having budgets for materials with prices and play money for helmet design provides students with tangible, varied means of understanding cost.	Providing students with a four square writing organizer, pamphlet guide, and letter template aids in organization, construction, and composition.

Suggested Schedule and Sequence

- Day 1: **Engage** with Helmet Helpers Sorting Cards and data sheet, Brain Basics article, and egg model introduction

- Day 2: **Explore** with *Eggs*treme Sports helmet design and testing

- Day 3: **Explain** with "Football Helmets: The Last Line of Defense?" video, helmet design revisions and testing, egg model T-chart, and "The New Football Helmet Test That Could Save Kids From Concussions" video

- Day 4: **Elaborate** with "The Science of Football" video, Is Football Too Dangerous for Kids? handout, and yes/no argument line activity

- Day 5: **Evaluate** with SAC: Stop Athletics Concussions letter, pamphlet, and rubric

- Additional Assessment (optional): Football Safety Letter to school administration

Lesson 8: *Egg*streme Sports

Materials

For Helmet Helpers

(per pair)

- Helmet Helpers Sorting Cards (p. 268), precut and in a sandwich bag

(per class)

- An assortment of sports helmets, such as bicycle, football, and baseball helmets (can be borrowed from school or students)

For *Egg*streme Sports Helmet Design and Testing

(per team of 3–4)

- Ruler
- 2 raw eggs
- 1 copy of Team Spending Money for Helmet Materials (p. 275), can be precut and placed in a sandwich bag
- Scissors

(per class)

- Craft materials for helmets, such as egg carton sections, cotton balls, nonlatex rubber bands, pipe cleaners, nonlatex balloons, paper and plastic cups, bubble wrap squares, packing peanuts, string, fabric squares, paper towels, and sandwich bags
- Tape
- Large plastic garbage bag or tarp to cover testing area floor

(per student)

- Indirectly vented chemical splash goggles and a nonlatex apron

For Yes/No Argument Line Activity

- 1 copy of "Yes" and "No" signs (pp. 152–153)

For SAC: Stop Athletic Concussions Pamphlet

(per student)

- 1 sheet of paper (8.5 × 11 inch office or construction paper works well)
- Markers, crayons, pencils

Student Handouts

- Helmet Helpers
- Brain Basics
- *Eggs*treme Sports Challenge Data Sheet
- Revised *Eggs*treme Sports Helmet Planner (Day 2)
- Is Football Too Dangerous for Kids?
- SAC: Stop Athletics Concussions
- Football Safety Letter (optional)
- Four Square Writing Template for Football Safety Letter (optional)

Safety Notes

1. All students must wear safety goggles and nonlatex aprons during the setup, hands-on, and takedown phases of the activity.
2. Immediately pick up any items dropped on the floor to avoid a slip-and-fall hazard.
3. Immediately wipe up any spilled liquids on the floor to avoid a slip-and-fall hazard.
4. Use caution when using hand tools or sharp materials to avoid cutting or puncturing skin.
5. Wash hands with soap and water after completing this activity.

Media

Videos

- "Football Helmets: The Last Line of Defense?"
  www.youtube.com/watch?v=meWQ2K2p0c8

Lesson 8: *Eggs*treme Sports

- "The New Football Helmet Test That Could Save Kids From Concussions"
  www.youtube.com/watch?v=Aa53zQlnpe4
- "Newton's First Law of Motion—Science of NFL Football" (watch until 1:54; stop at discussion of inertia)
  www.youtube.com/watch?v=08BFCZJDn9w

Background for Teachers

Youth sports, both in and outside of school, are extremely popular in the United States. Among the most popular sports are basketball, baseball, soccer, football, hockey, track and field, and lacrosse. Student participation in sports has many positive advantages, including engaging students in physical activity (particularly important in an age of growing screen time), encouraging social-emotional growth, learning about sportsmanship, gaining self-confidence, and promoting lifelong healthy habits. However, youth sports have recently come under scrutiny in part because of a stream of professional athletes, particularly football players, raising concerns about concussions.

The word *concussion* comes from the Latin *concutere*, which means "to shake violently." A concussion is a brain injury caused by a blow to the head or strong shaking of the head. The human brain has to withstand a great deal of movement in the normal course of each day. Fortunately, the brain is covered with a membrane containing cerebral spinal fluid, which help cushion the brain from hitting the skull. The skull, or cranium, provides protection for the brain, as it is hard and bony and forms a cavity within which the brain, membranes, and fluid sit. But if the head is hit hard enough, the membranes surrounding the brain can swell, putting pressure on the brain. This causes a concussion. Signs of concussion include dizziness, nausea, loss of consciousness, confusion, and loss of memory, among many others. There is evidence that repeated concussions can cause permanent brain damage. Since children's brains are still forming, concussions can be particularly worrisome.

Helmets can help prevent head injuries such as concussions, but it is important to note that no helmet can give complete protection. Helmets work in two ways: First, the hard outer shell of a helmet helps distribute or spread out the forces that are applied from a collision (such as being hit by another athlete or a ball) or a fall to the ground. And second, the crushable foam padding inside a helmet helps cushion the blow, giving the head a slightly longer stopping time before impact. It is helpful here to understand Newton's first law, which essentially states that an object in motion will stay in motion until an unbalanced force acts on it and changes its direction. In football, we could say that a running back will continue to run in a line to the goal unless an unbalanced (stronger) force acts on him, causing the player to change direction or speed (and possibly stop completely). When objects collide, such as when football players tackle another player or when a baseball hits a batter's helmet, the force is transferred. A helmet absorbs the impact of the force, decreases movement of the brain within the skull, and distributes the impact over the greater surface area of the helmet to decrease the risk of a skull fracture.

In this lesson, students learn about brain anatomy to reinforce the idea that the brain is an internal structure that has a specific role in the body's functioning, an *NGSS* DCI. Students then create and test helmets undergoing impact, using eggs as a model for the brain. This activity serves as an application of the DCI that (1) objects in contact exert forces on each other (3-PS2-1); and (2) an object at rest typically has multiple forces acting on it, but they add to give zero net force on the object. Forces that do not sum to zero can cause changes in the object's speed or direction of motion. When the egg is at rest in your hand, gravity is exerting a pulling force downward, but your hand is exerting an equal force pushing it up. These are balanced forces. But when you release the egg, the force of gravity is now unbalanced (because there's no opposing force from your hand), and the egg quickly drops to the floor. When the egg hits the floor, the floor exerts a force upward that stops (and cracks) the egg. The egg rests on the floor because the downward pulling force of gravity is now balanced with an upward pushing force from the floor. By creating helmets, students try to distribute the impact forces and slow the impact down enough to reduce the force. Students also assess the strengths and weaknesses of the egg model, recognizing that every model scientists and engineers use has limitations, since it is only an approximation of the real phenomenon. Students also watch a video of a new test that has been developed to ensure helmet safety. This test is based on increased understanding of concussions; in this way, students learn that as science knowledge changes, scientists and engineers work together to improve both product and safety test design.

It should be noted that while this lesson focuses on forces and motion, it could easily be extended to conservation of energy, as the energy of the moving egg is transferred to the floor and the surrounding air when it hits. That is why you hear the crack of the egg (or in football, the crunch of the players colliding). This transfer of energy slows the egg (or football players) to a stop. The helmet works to slow the egg before it makes an impact with the floor. I tried to limit the concepts covered in this lesson to keep it manageable, but if your students are proficient in forces and motion, you can extend the lesson to include energy transfer and conservation of energy by having students ask questions and predict outcomes about the changes in energy (expressed as a change of speed) when the egg hits the floor or football players collide (4-PS3-3).

Lesson 8: *Eggstreme Sports*

Additional Resources

Centers for Disease Control and Prevention (CDC) information site on concussions in sports for parents and schools

- *www.cdc.gov/HeadsUp*

"Rocket Blades," CDC's mobile game app for children on brain and concussion safety

- *http://apple.co/2m5OY0g*

Pamphlet on CDC's "Rocket Blades" game app

- *www.cdc.gov/headsup/pdfs/resources/RocketBlades_HandoutforKids-a.pdf*

Book on the physics of football

- Gay, T. 2005. *The physics of football: Discover the science of bone-crunching hits, soaring field goals, and awe-inspiring passes.* New York: HarperCollins.

Series of videos on the "Science of NFL Football"

- *www.nbclearn.com/science-of-nfl-football*

Information on a study linking brain diseases to tackle football for children under 12

- *www.nytimes.com/2017/09/19/sports/football/tackle-football-brain-youth.html*

Infographic on the importance of playing only flag football until age 14

- *https://concussionfoundation.org/programs/flag-football/why-age-14*

Book about the brain and nervous system

- Simon, S. 2006. *The brain: All about our nervous system and more!* New York: HarperCollins.

5E Lesson Plan

Engage: Helmet Helpers Sorting Cards and Data Sheet, Brain Basics Article, and Egg Model Introduction

1. Provide students with an opportunity to examine different sports helmets (for example, football, bicycle, baseball batter's helmet). Distribute the premade packets of Helmet Helpers Sorting Cards (p. 268) to pairs of students, and distribute a copy of the Helmet Helpers data sheet (p. 269) to each student. Ask students to work together to try to match each helmet card to its

sport card. Once students believe they have matched a helmet to a sport, have them record this on the data sheet by drawing a line between the helmet and the sport, and then work together to answer the questions at the bottom of the sheet. The answers are as follows:

1. Why do athletes wear helmets? (to protect their heads/brains)

2. How are the helmets alike? How are they different? (students will notice different shapes and face coverings, but all are built to fit on the head; if you have actual helmets available, allow students to compare the different materials as well)

3. How do you think helmets work? (accept all answers)

The answers to the Helmet Helpers matching are as follows:

- 1 = e. Baseball
- 2 = c. Hockey
- 3 = a. Football
- 4 = d. Horseback riding
- 5 = b. Bike riding (cycling)

Ask students, "Why is protecting our heads important?" Briefly discuss the role of the brain in controlling our bodies' functions.

2. Distribute copies of the Brain Basics article (p. 270), and have students work in pairs to read and complete it. Ask:

- "What is the function of the skull?" (protect the brain)

- "What is the function of the fluid surrounding the brain?" (cushion the brain so that it doesn't hit the skull)

- "What are some of the body functions that can be affected by a concussion?" (e.g., thinking, speaking, hearing)

- "How can you protect your brain during sports?" (wearing a helmet, avoiding hard hits to the head)

- "How does a helmet work?" (students typically discuss helmets as an apparatus that either cushions the head or spreads out the forces during an accident)

Lesson 8: *Eggs*treme Sports

3. Next, introduce the egg model by asking students to imagine that an egg represents a human head. Ask, "How is the egg similar to a human head?" (the egg can get "injured" from a collision or fall) "What parts of the egg are similar to parts of the human head?" (the shell is similar to the skull, the egg white is similar to the brain fluid, and the yolk is similar to the brain) "How is the egg different?" Draw or project Figure 5.27 to show students the comparison. Inform students that the next activity will involve designing, building, and testing "helmets" to protect our egg "heads" from concussions.

FIGURE 5.27.

Comparison of an Egg and a Human Head

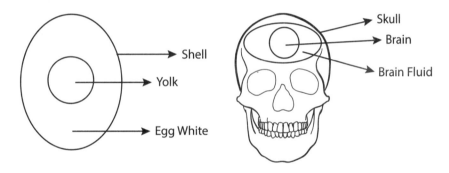

Explore: *Eggs*treme Sports Helmet Design and Testing

Advance Preparation: Set up a table with an array of craft materials that allow for open-ended exploration, such as plastic sandwich bags, paper towels, cotton balls, sections from an egg carton, paper and plastic cups, string, and tape. Also set up a testing area that consists of a cut-open plastic garbage bag or tarp and a ruler.

1. Distribute copies of the *Eggs*treme Sports Challenge Data Sheet (p. 271) to students. Form teams of three or four students who will develop safety helmets for eggs. Give each team one raw egg. Teams can draw faces on their eggs to personalize them. *Tip:* It is best to provide the eggs to teams in small containers, such as paper cups or bowls cushioned with paper towels or egg carton sections, to avoid unplanned breakage.

2. Explain to students that there are certain *criteria* that must be met for their helmets to be successful:

    - The helmet must protect the egg from breaking.

- The helmet must allow for visibility (so the egg can "see") and breathability (so the egg can "breathe").
- The helmet must be removable (not permanently attached to the egg) so that students can check the egg after testing.

Also, there are certain cost and time *constraints* that limit their helmet designs:

- The helmet must cost less than $10 to make.
- The helmet must be completed in _____ minutes.

Having a budget makes the activity more realistic and provides an important engineering design constraint, and having "dollars" to spend makes the budget more concrete. Place a price list near the materials, and distribute a copy of the Team Spending Money for Helmet Materials sheet (p. 275) or the precut "dollars" in a sandwich bag to each team. Figure 5.28 is a sample materials list with costs.

FIGURE 5.28.
Sample Materials Price List

Helmet Material	Price
Egg Carton Section	$5
Cotton Ball	$1
Pipe Cleaner	$1
Rubber Band	$2
Paper Cup	$3
Balloon	$2
Packing Peanut	$2
Bubble Wrap Square	$1
Fabric Square	$1
String	$1
Sandwich Bag	$2
Tape	Free

Allow students to plan their designs and their budgets. When teams are ready to shop, have them bring their team spending money to the materials area and pay you for the materials.

Lesson 8: *Eggs*treme Sports

3. After each team has developed a helmet (Figure 5.29), bring the class together to test each team's helmet by dropping the egg from a height of 5 cm. An "eggsaminer" from each group can check the group's egg. Surviving eggs can be tested in ongoing rounds of increasing heights. If eggs survive drops from heights greater than 20 cm, students can extend the data table. Allow teams to discuss their designs, and have students complete the questions at the bottom of the sheet:

FIGURE 5.29.

Samples of Student Helmets and Test Eggs

- Did your egg survive all the drops? (answers will vary)
- What do you think are the best features of your helmet? (answers will vary)
- What changes would you make if you had a chance to modify your helmet? (answers will vary)
- How do helmets protect our heads (and our eggs)? (distributing the forces and slowing the egg down)

Explain: "Football Helmets: The Last Line of Defense?" Video, Helmet Design Revisions and Testing, Egg Model T-Chart, and "The New Football Helmet Test That Could Save Kids From Concussions" Video

1. Begin by discussing the following questions:
   - "What problem have we tried to solve?" (head injuries from sports)
   - "What criterion did we use to determine a successful design?" (egg didn't crack at a specific height)
   - "What were our design constraints?" (cost and time)
   - "What were some of the characteristics of successful helmets?" (e.g., cushioning to absorb force and slow egg down, a hard outside that "spread out" forces)

2. Inform students that they are going to have a chance to revise their helmets, but it would be helpful if they had a little more information on helmet design. Watch the video "Football Helmets: The Last Line of Defense?" at *www.youtube.com/watch?v=meWQ2K2p0c8*.

Lesson Plans

3. Ask students, "Was any of the information in the video new to you? Can any of it inform your new designs?" (answers will vary, but students will likely mention the history of helmets, as early helmets were made of leather; helmets began to be regulated under certain standards after many athletes died from head injuries; forces on the brain during a football collision are much stronger than the forces on a fighter pilot in a tight turn or a person on a roller coaster; new plastic and foam materials have contributed to better helmets; the idea to use Kevlar in helmets failed because it didn't slow down the impact) It is important to note that the changes over time in helmets reflect new knowledge in science and engineering. Also important to note is that not every idea worked well!

4. Allow student teams to revise and retest their helmet designs using the Revised *Eggs*treme Sports Helmet Planner (Day 2) handout (p. 273). Have teams repeat the tests with a second egg, then have students answer the questions at the bottom of the page:

    - Did the egg in your revised helmet survive all the drops? (answers will vary)

    - Which helmet design was more successful: your original or your revised version? How do you know? (answers will vary; the egg didn't crack and met all other criteria)

    - Why do you think that version was able to perform better in your test? (answers will vary)

    - Why do you think engineers test different designs when they're trying to solve a problem (such as preventing concussions)? (engineers try to *optimize* their solutions to problems by testing different solutions to find the best one)

5. Remind students that in this activity, we modeled the effect of helmets for protecting the brain by using eggs and everyday craft materials. Ask, "Why didn't we just use real helmets and real heads?" (people could get hurt!) Explain that all models have strengths and limitations, because they're not exactly the real phenomena. Ask, "How was our experiment an accurate model of the effects of helmets for protecting the brain? How is the egg a good model for our experiment?" (it resembles a fragile head) "How was our experiment not an accurate model? How is the egg not a good model?" (people don't crack easily, football accidents aren't exactly like being dropped, real helmets use other materials) Write student responses in a T-chart. Your T-chart might look something like the one in Figure 5.30.

Lesson 8: *Egg*streme Sports

FIGURE 5.30.

T-Chart to Evaluate Our Egg Model

Ways the Model Is Accurate	Ways the Model Is Not Accurate
Heads and eggs both have hard coverings.	People don't "crack" easily.
Heads and eggs both have some liquid inside.	Real sports collisions aren't the same as dropping.
Heads and eggs can both get damaged from forces.	Real helmets are made of different materials.
Hitting the floor and hitting (or being hit) by another athlete both involve forces.	
Our helmets and real helmets "spread out" force and slow objects (eggs or heads) down.	

6. Invite students to suggest ways of developing a better model for testing helmets for student sports. For example, students may suggest having collisions between egg "heads" rather than individual drops (to simulate contact sports like football), using melons to represent human heads (closer in size to the real thing), or testing rolling and spinning rather than just dropping or colliding (more realistic than a single impact).

7. Show students the video "The New Football Helmet Test That Could Save Kids From Concussions" at *www.youtube.com/watch?v=Aa53zQlnpe4*. This video shows a new side impact test, as opposed to the conventional drop test (which was very much like our egg model testing). The new test takes into account rotational, or turning, forces on children's necks. Take a few minutes to allow students to discuss the two tests and compare the tests with our test. Point out that as we continue to learn more about concussions, engineers and scientists work together to make improvements to both helmet and safety test design.

Elaborate: "The Science of Football" Video, Is Football Too Dangerous for Kids? Article, and Yes/No Argument Line Activity

1. Explain to students that certain sports have become controversial lately because of concerns about injuries. Many people are especially concerned about youth sports (sports for kids) because of the possibility of concussions. To understand that debate, it's important to understand why certain sports, such as football, are more prone to cause head injuries. This has to do with

Lesson Plans

forces and motion. Tell them, "We're going to watch part of a video on the science of football." Watch "Newton's First Law of Motion—Science of NFL Football," available at *www.youtube.com/watch?v=08BFCZJDn9w*, up to the discussion of inertia at 1:54. This video explains Newton's first law and its relationship to football.

2. After the video, ask students to turn and talk with a partner about these three questions (which should be written on the board):

   - "What is an 'unbalanced force'?" (a force that's stronger than the other forces acting on an object; when forces are balanced, there is no motion)

   - "How would you express Newton's first law of motion in your own words?" (answers will vary but should be something like this: objects at rest or standing still will continue to stand still until a force that's stronger than the ones acting on it collide with it; objects that are moving will continue to move at a particular speed and in a particular direction until they collide with a stronger force)

   - "What two things can change about the motion of an object, such as a football player carrying a ball, when an unbalanced force collides with it?" (speed and direction)

3. Ask students if they have any questions about the video. Let students know that they are now going to learn a bit more about the controversy surrounding kids and football.

4. Distribute copies of the Is Football Too Dangerous for Kids? handout (p. 276). In pairs, have students read the article and answer the questions in writing. While students are working, prepare for the yes/no argument line activity (below). When students are done, review the answers, but try to avoid having students debate at this point. Simply focus on having students state their opinions, support them with evidence, and identify additional information they might need. You can make note of the additional information so that you can locate resources for their final project.

5. In preparation for the yes/no argument line activity, post the "Yes" sign (with the thumbs-up picture) on one side of the room and the "No" sign (with the thumbs-down picture) on the other. Write the following sentence frames on the board:

   "I think that _____ because _____."

"I disagree with you because _____."

"According to the article/our experiment/my personal experience/others' experiences, _____."

"I think you should also consider _____."

You can add other sentence frames or provide students with copies of Sentence Frames for Arguments (Figure 3.13, p. 26).

6. Explain to students that you'd like to hear their opinions on football safety, and they're going to share their opinions in an interesting way. Remind students that there is no wrong answer. There are good reasons on both sides of the issue; the important thing is that they share their ideas in a thoughtful and respectful manner.

7. Show students the posters on either side of the classroom and ask them to imagine that there is a line connecting both signs. Explain that in a moment, you are going to ask them whether football is too dangerous for kids; they are then going to get up and move to a spot along the line that shows their opinion. If they believe strongly that football is too dangerous, they should move close to the "Yes" sign; if they feel strongly that it is not too dangerous, they should move close to the "No" sign. If they feel somewhere in between, they should move somewhere along the imaginary line connecting the two signs.

8. Then, ask students, "Is football too dangerous for kids?" Allow students to move to their chosen sides of the room. Encourage students to stand anywhere along the spectrum from yes to no, with undecided students standing somewhere in the middle.

9. When students are on their chosen sides, show them the sentence frames and ask them to share their opinions in that format. Remind them that they need to take turns and listen hard to other people's opinions. Have students share their opinions, making sure that students justify their opinions (claims) using evidence from the article, their experiment, their personal experiences, or others' experiences. Pay careful attention to whether you see anyone shifting along the line as they are persuaded by others' arguments. If you notice that, you can say, "I notice that some people may be rethinking their positions on the issue, and that's fine. Would anyone like to share their thoughts on that?" Ask, "What arguments *persuaded* you to change your mind?"

10. When everyone who wants to share has had an opportunity, ask, "Is there any other information you might want to have to make an informed decision?"

Write student suggestions on the board so that additional research can be conducted for the final activity. If you are aware of the school or district policy on football (or better yet, if you have your students find out about it), you can ask an additional question that relates to it, such as, "Our school district doesn't allow tackle football to be played until age 12. Do you agree?" You can also explore broader questions such as "Right now, football helmet design requirements are set by a national organization called the National Operating Committee on Standards for Athletic Equipment (NOCSAE) to make sure helmets are safe. Do you think it's a good idea to have a national standard for regulating helmet safety?" This opens up some discussion of the role of government and federal versus local control. Again, as you engage in the yes/no activity, remind students of the sentence frames to help them articulate their opinions and provide supporting evidence.

Evaluate: SAC: Stop Athletics Concussions Letter, Pamphlet, and Rubric

1. Distribute copies of the SAC: Stop Athletics Concussions letter (p. 277), which asks students to develop a trifold pamphlet that can educate members of the school community about sports helmet safety. Discuss the letter, then have students create their pamphlets.

2. Distribute copies of the SAC: Stop Athletics Concussions Pamphlet Scoring Rubric (p. 278), which can help students self-assess their efforts. You can use the rubric after students complete their pamphlets to add your assessment of their work.

3. Then, make photocopies of the pamphlets for students to distribute around the school community.

Going Deeper

- Students can write an advocacy letter to a school administrator. Distribute copies of the Football Safety Letter handout (pp. 280–281) and read it aloud. The handout explains the requirements of the letter. Distribute copies of the Four Square Writing Template for Football Safety Letter (p. 279), which will serve as a planner. Students can circle "agree" or "disagree," as well as "school's" or "school district's" policy. Elicit an example of evidence and source from students, which might be something like this: "In our Brain Basics reading (source), we learned that concussions can cause problems such as dizziness and headaches (evidence)." Remind students that they should use the same sentence frames they used for the class discussions to help with their sentences. When the letters are finished, allow students to send them to school administrators.

- If you have students do both the pamphlet and the letter, ask students to compare and contrast the purposes of the two types of writing (the pamphlet is primarily educational, while the letter is advocating for a position).

Helmet Helpers Sorting Cards

1.		(a) Football
2.		(b) Bike Riding (Cycling)
3.		(c) Hockey
4.		(d) Horseback Riding
5.		(e) Baseball

Name: _____   Date: _____

Helmet Helpers

Can you match the helmet to the sport? Draw a line to connect the helmet to its sport. Then answer the questions below.

1.		(a) Football
2.		(b) Bike Riding (Cycling)
3.		(c) Hockey
4.		(d) Horseback Riding
5.		(e) Baseball

1. Why do athletes wear helmets? _____

2. How are the helmets alike? How are they different?

   _____

   _____

3. How do you think helmets work?

   _____

   _____

Name: _____ Date: _____

Brain Basics

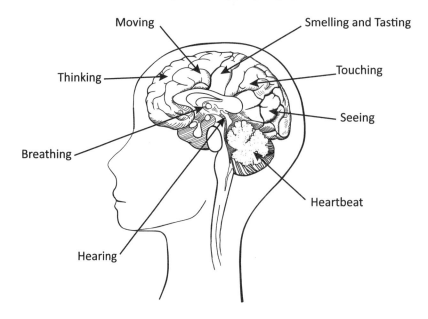

The human brain is located inside the hard, bony skull. The brain is surrounded by a fluid-filled sac that protects it from hitting against the skull during normal movements. If the head is hit or shaken hard enough, the brain can hit the skull and swell up. This is called a **concussion.** Concussions are serious because the brain controls many functions of the body, with different regions of the brain controlling different functions. A concussion can cause problems with several different body functions. Helmets can help protect the skull and brain during sports in which falls or blows to the head are common. However, concussions can still happen, so it is important to tell your coach, parent, or another adult if you hit your head during sports.

Questions to Think About:

1. **What is the purpose of the skull?**

2. **What is the purpose of the fluid surrounding the brain?**

3. **What are some of the body functions that can be affected by a concussion?**

4. **How can you protect your brain during sports?**

5. **What should you do if you hit or fall on your head during sports?**

Name: _____ Date: _____

Eggstreme Sports Challenge Data Sheet

Draw a diagram of your helmet in the box below. Be sure to label all the materials.

Enter your budget in the chart below. Remember that the total cannot exceed $10.

Material	Cost per Item	Number of Items	Total Cost (Cost of Item × Number)
Example: Cotton Balls	$1	2	$2

TOTAL: _____

Name: _____   Date: _____

Helmet Testing Data

Drop your egg "head" with the helmet side down from the heights listed below. Measure from the bottom of the helmet. Record your results in the "Results" column. Be specific, and include drawings if they help you communicate the results.

Trial	Height (cm)	Results
1	5 cm	
2	10 cm	
3	15 cm	
4	20 cm	

1. Did your egg survive all the drops? _____

2. What do you think are the best features of your helmet?

   _____

   _____

3. What changes would you make if you had a chance to modify your helmet?

   _____

   _____

4. How do helmets protect our heads (and our eggs)?

   _____

   _____

Name: _____ Date: _____

Revised *Eggs*treme Sports Helmet Planner (Day 2)

Draw a diagram of your revised helmet in the box below. Be sure to label all the materials.

[ ]

Enter your budget in the chart below. Remember that the total cannot exceed $10.

Material	Cost per Item	Number of Items	Total Cost (Cost of Item × Number)
Example: Cotton Balls	$1	2	$2

TOTAL: _____

IT'S STILL DEBATABLE! USING SOCIOSCIENTIFIC ISSUES TO DEVELOP SCIENTIFIC LITERACY, K–5

Name: _____ Date: _____

Helmet Testing Data (Revised Design)

Drop your egg "head" with the helmet side down from the heights listed below. Measure from the bottom of the helmet. Record your results in the "Results" column. Be specific, and include drawings if they help you communicate the results.

Trial	Height (cm)	Results
1	5 cm	
2	10 cm	
3	15 cm	
4	20 cm	

1. Did the egg in your revised helmet survive all the drops? _____
   _____
   _____

2. Which helmet design was more successful: your original or your revised version? How do you know?
   _____
   _____

3. Why do you think that version was able to perform better in your test?
   _____
   _____

4. Why do you think engineers test different designs when they're trying to solve a problem (such as preventing concussions)?
   _____
   _____

Team Spending Money for Helmet Materials

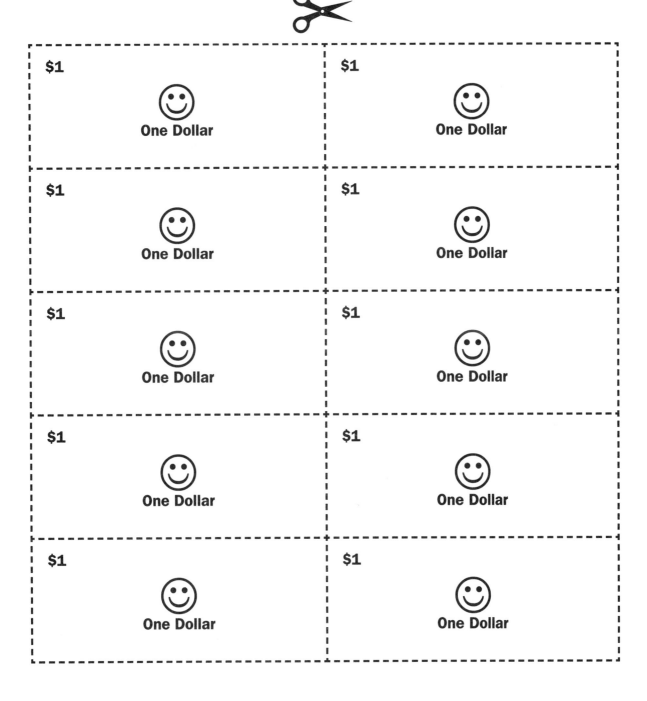

Name: _____ Date: _____

Is Football Too Dangerous for Kids?

More students are participating in sports than ever before. Soccer, basketball, baseball, football, volleyball, and track and field are among the most popular. Football has received a lot of attention lately because of reports of professional players struggling with the effects of concussions from repeated blows to the head. Some of these effects include loss of memory, dizziness, confusion, and difficulty walking and speaking.

The question of whether football is too dangerous for children is controversial. People who feel that football is too dangerous for children point to statistics that indicate a higher rate of concussions for football than any other sport. They also cite the fact that repeated blows to children's heads can cause lifelong damage as children's brains are still growing. Finally, they note that even with helmets and padding, concussions are still common in football.

Advocates of youth football point to the fact that while football might have the highest rate of concussions, concussions can happen in all sports, especially soccer, hockey, gymnastics, and lacrosse. They also state that with proper instruction and equipment, including helmets and padding, concussions can be reduced tremendously. Finally, they argue that the benefits of playing football, including learning sportsmanship, practicing teamwork, and gaining speed, strength, and quick thinking, outweigh the relatively low risk of getting seriously hurt.

1. Do you think football is too dangerous for kids? Why or why not?

   _____

   _____

2. What evidence do you have to support your opinion?

   _____

   _____

3. What other information would be helpful for you to know?

   _____

   _____

SAC: Stop Athletics Concussions

Dear Scientists and Engineers,

As president of the organization Stop Athletics Concussions (SAC), I am writing to ask you to help us get the word out about the importance of helmets for sports safety. Please support us by developing pamphlets that can be distributed to parents, teachers, school administrators, students, and coaches!

Your pamphlet should be neat and attractive and explain the following:

1. Why is it important to wear helmets during certain sports?

2. What is a concussion?

3. How do forces cause concussions?

4. How do helmets work? (Be sure to use the word *force* in your explanation.)

5. What evidence do you have that helmets work? (Describe your experiment and include any other data that you have researched.)

Your pamphlet should include written explanations and diagrams. It can be created by taking a piece of paper and folding it into three parts to get six pages (front and back) like this:

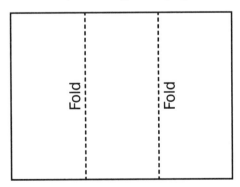

Thank you in advance. I appreciate your assistance with this very important project!

Sincerely,

Brainy Smith, President

SAC

SAC: Stop Athletics Concussions

Pamphlet Scoring Rubric

Dear _____:
  Scientist/Engineer Name

Your pamphlet will be assessed using this 4-3-2-1 rubric:

- Includes clear explanations of how forces cause concussions

                                                  4   3   2   1

- Includes clear explanations of how helmets work (using the word *force* in the explanation)

                                                        3   2   1

- Includes evidence that helmets work (using your experiment data or other data you have researched)

                                                                2   1

- Is neat and attractive

                                                                         1

  Total: ___/ 10 pts.

Four Square Writing Template for Football Safety Letter

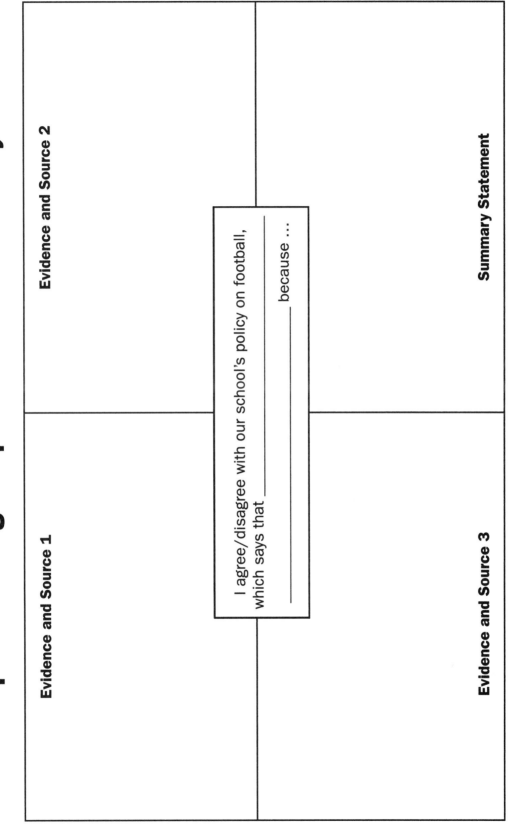

Evidence and Source 1	Evidence and Source 2
I agree/disagree with our school's policy on football, which says that _____ because …	
Evidence and Source 3	Summary Statement

IT'S STILL DEBATABLE! USING SOCIOSCIENTIFIC ISSUES TO DEVELOP SCIENTIFIC LITERACY, K–5

Football Safety Letter

Write a letter to someone in your school or district's administration (such as your principal, athletics director, or superintendent) about your thoughts on the school's or district's football policy. The policy may include such things as the age when children can play football, whether tackling is allowed, and the equipment that is required. You may agree or disagree with the policy (or there may be parts of the policy that you agree with and parts that you disagree with). Use the **Four Square Writing Template** to plan your letter. In your letter, be sure to do the following:

- State your claim. Begin with a statement of whether you agree or disagree with the policy. Include a brief description of the policy.

- Back up your claim with evidence. Evidence might include results from your helmet tests, results from other people's helmet tests, information from your readings or videos, personal experience, or experience of others.

- Include the source of the evidence, such as the name of the reading or video, stating, "In my helmet test (or in someone else's test) …"

- Include a summary statement that supports your claim. Then, send your letter!

Date: _____

Dear: _____,

_____
_____
_____
_____
_____
_____
_____
_____
_____
_____
_____
_____
_____
_____
_____
_____

Sincerely,

Lesson 9

Marsh Madness

What's Your Plan for Bullfrog Pond?

Suggested Grade Levels

3–5

Driving Questions

- What are wetlands, and what important functions do they serve?
- What organisms live in a wetland, and how do changes in the environment affect those organisms?
- How does development affect wetlands, and what steps can mitigate or lessen those impacts?

Lesson Overview

Students investigate the ecological relationships between living and nonliving components of a wetland and collaboratively determine the type, extent, and consequences of nearby development.

Connecting to the *NGSS*

(See full alignment in Table A.9 on p. 511.)

- LS2.C: Ecosystem Dynamics, Functioning, and Resilience
  - When the environment changes in ways that affect a place's physical characteristics, temperature, or availability of resources, some organisms survive and reproduce, others move to new locations, yet others move into the transformed environment, and some die. (secondary to 3-LS4-4)

- LS4.D: Biodiversity and Humans
  - Populations live in a variety of habitats, and change in those habitats affects the organisms living there. (3-LS4-4)

Societal Issues

Property Rights, Land Use, Environmental Sustainability

Nature of Science

- Science Models, Laws, Mechanisms, and Theories Explain Natural Phenomena
  - Science explanations describe the mechanisms for natural events.

CCSS Connections

- English Language Arts
  - SL.3.1. Engage effectively in a range of collaborative discussions (one-on-one, in groups, and teacher-led) with diverse partners on grade 3 topics and texts, building on others' ideas and expressing their own clearly.
  - RI.4.9. Integrate information from two texts on the same topic in order to write or speak about the subject knowledgeably.
- Mathematics
  - MP.2. Reason abstractly and quantitatively. (3-LS4-4)
  - MP.4. Model with mathematics. (3-LS4-4)

NCSS Connections

- Theme 10: Civic Ideals and Practices
  - Evaluate positions about an issue based on the evidence and arguments provided, and describe the pros, cons, and consequences of holding a specific position.

C3 Framework

- Dimension 3: Gathering and Evaluating Sources
  - D3.2.3-5. Use distinctions among fact and opinion to determine the credibility of multiple sources.
- Dimension 3: Developing Claims and Using Evidence
  - D3.4.3-5. Use evidence to develop claims in response to compelling questions.
- Dimension 4: Communicating Conclusions
  - D4.1.3-5. Construct arguments using claims and evidence from multiple sources.

UDL Toolkit

Multiple Means of Engagement	Multiple Means of Representation	Multiple Means of Action and Expression
An interactive checkpoint lab that consists of several repetitions of action (raining on the wetland) under different circumstances creates a safe space for students to engage with materials while promoting autonomy.	Technology integration through use of an iPad app and a low-tech option (human food web) that uses physical engagement to convey information support comprehension through diverse media.	Students express their learning through checkpoint lab, iPad research, town hall meeting, and flip-book creation to allow for multiple opportunities through varied media.
Using small-group interaction (wetland lab, Bullfrog Pond activity), whole-group lessons (town hall meeting), and individual work (iPad and flip-book) fosters collaboration and a sense of community while providing flexible groupings.	Making explicit connections between science learning and literacy (e.g., during the discussion of media literacy) helps students comprehend and organize information, as well as generalize and transfer concepts across disciplines.	Providing a word bank, brainstorming matrix, and flip-book frame helps students organize information and construct and compose written and oral communication.

Lesson 9: Marsh Madness

Suggested Schedule and Sequence

- Day 1: **Engage** with *About Habitats: Wetlands* read-aloud, and **Explore** and **Explain** with Wetland in a Pan Checkpoint Lab
- Day 2: **Explore** with iBiome-Wetland or "Living" Wetland Food Web
- Day 3: **Explain** with *Here Is the Wetland*, and **Elaborate** with Bullfrog Pond
- Day 4: **Evaluate** with Town Hall Meeting and rubric
- Day 5: **Evaluate** with Wetlands Flip-Book

Materials

For Wetland in a Pan Checkpoint Lab

(per team of 3–4)

- Paint roller tray
- 2 large sponges
- Clay
- Florist moss
- Measuring cup
- Water
- Soil
- Stickers, toothpicks, other craft materials to represent plants and animals
- Scissors
- Tape

(per student)

- Indirectly vented chemical splash goggles and a nonlatex apron

For Explore with iBiome-Wetland

- Computers, iPads, or other devices with internet access
- iBiome-Wetland app

For "Living" Wetland Food Web

(per student)

- 1 piece of construction paper
- Crayons or markers
- String or yarn to hang placard around neck

(per class)

- 1 or 2 copies of "Living" Wetland Food Web Roles sheet (p. 307), cut into strips and placed in a bag
- *Wetland Food Chains* and other resource books
- Ball of string or yarn

For Bullfrog Pond Activity

(per team of 4–5)

- 1 copy of Bullfrog Pond Map (p. 312) on 11 × 17 inch paper
- 1 copy of Development Cutout Pieces (p. 313) on 8.5 × 11 inch paper
- Tape
- Scissors
- 1 copy of team signs (pp. 308–310)

Student Handouts

- Wetland in a Pan Checkpoint Lab
- iBiome-Wetland Exploration
- How Should We Develop Bullfrog Town? Brainstorming Matrix
- Wetlands Flip-Book template

Safety Notes

1. All students must wear safety goggles and nonlatex aprons during the setup, hands-on, and takedown phases of the activity.

Lesson 9: Marsh Madness

2. Immediately pick up any items dropped on the floor to avoid a slip-and-fall hazard.

3. Immediately wipe up spilled water on the floor to avoid a slip-and-fall hazard.

4. Use caution when using hand tools or sharp materials to avoid cutting or puncturing skin.

5. Wash hands with soap and water after completing this activity.

Media

Books

- *About Habitats: Wetlands,* by Cathryn Sill
- *Here Is the Wetland*, by Madeleine Dunphy
- *Wetland Food Chains,* by Bobbie Kalman and Kylie Burns (optional)

Apps

- iBiome-Wetland: School Edition by Springbay Studio
  https://itunes.apple.com/ca/app/ibiome-wetland-school-edition/id1069411327?mt=8

- A free but limited (freshwater marsh only) demo version of this app for PCs
  www.brainpop.com/games/ibiomewetlandschooledition

Background for Teachers

Wetlands are incredibly productive and diverse habitats that provide a wonderful context for learning about ecosystems and human impacts. Wetlands are areas where water collects over soil all or part of the time; the water may come from a number of sources, including rainfall, tides, underground springs, and overflow from lakes and rivers. Some common types of wetlands include marshes, swamps, bogs, wet meadows, fens, estuaries, and vernal pools. Wetlands are found all around the globe, except for in Antarctica, and they can collect fresh water or salt water (or a mix of both). Wetlands are essentially the link between land and water, serving as a transition zone where water and nutrients cycle through. Wetlands play an integral role in aquatic ecosystems as they act as both filters and sponges; they filter out pollutants and sediment that run off from the land into bodies of water such as rivers, streams, and lakes, and they absorb overflow from those bodies of water, which helps minimize flooding and erosion. Wetlands also serve as safe, protective nurseries for wildlife and critical habitat for both resident and migratory species. Many foods that we eat are also native to wetlands, including cranberries, rice, and a variety of fish and shellfish species.

Unfortunately, many wetlands have been destroyed, in part because for many years, people didn't recognize their importance in the ecosystem. Wetlands were commonly drained or filled so that development could take their place. Some wetlands, such as the Meadowlands near New York City, were considered wasteland and were polluted and neglected. However, many recent efforts to balance land development with the need to protect wetlands have improved the outlook for these amazing habitats. And many concerted efforts to clean up polluted wetlands such as the Meadowlands have resulted in improved environmental quality of the wetlands. Today, wetlands are protected under the Clean Water Act, which requires that developers take steps to avoid impacts to wetlands, and when it is considered unavoidable, they must try to mitigate the effects of the development. However, competing interests and different ideas about what constitutes mitigation can sometimes make this a contentious issue.

In this activity, students learn about and model wetland dynamics through readings and a checkpoint lab. They then investigate wetland food webs and the potential impacts to the food webs from intrusions such as pollution, development, and draining. Finally, students role-play different constituencies as they try to determine a development plan for a wetland area.

Additional Resources

"Wetlands Protection and Restoration" factsheet series

- *www.epa.gov/wetlands*

"Marsh Market," a student activity about food webs

- *www.fws.gov/uploadedFiles/Region_1/NWRS/Zone_2/Inland_Northwest_Complex/Turnbull/Documents/EE/Wetland_Ecology/Marsh%20Market.pdf*

5E Lesson Plan

Engage: *About Habitats: Wetlands* Read-Aloud

1. Read the book *About Habitats: Wetlands,* by Cathryn Sill, as a whole class read-aloud to begin a study of the wetland environment. This book describes various types of wetlands (e.g., bogs, marshes, swamps) and their importance to their plant and animal inhabitants as well as to people. Informally assess students by asking the following questions:

Lesson 9: Marsh Madness

- "What is a wetland?" (a place covered by shallow water)

- "What are some of the different types of wetlands?" (bog, marsh, prairie pothole, swamp, vernal pool, riparian wetland)

- "What are some of the living and nonliving parts of the wetland environment?" (the Sun, water, plants, fish, invertebrates)

- "What are some of the important functions of a wetland?" (provides food, shelter, and protection for animals and their young; provides food for people; prevents flooding and erosion: acts as a filter for water)

FIGURE 5.31.
Wetland in a Pan Setup

Explore and Explain: Wetland in a Pan Checkpoint Lab

1. Explain to students that they are going to create a model of a wetland to explore how wetlands work (Figure 5.31). Divide students into groups of three or four, and distribute copies of the Wetland in a Pan Checkpoint Lab handout (pp. 303–305) to each student. Read through the instructions on the sheet, which explain how students will construct their wetland. *Tip:* You might want to have a sample wetland made in advance to display, but explain to teams that they will be able to personalize their wetlands with different flora and fauna.

2. Explain to students that this is a checkpoint lab, which means that they need to have you check their work at each of the lab's three checkpoints, A, B, and C. The instructions guide students through a series of investigations in which they simulate rain on the pan, with and without the wetland in place, to see the impact of the wetland on the pond and surrounding environment. In Part A of the lab, students

build the model and predict the impact of rain. In Part B, students test the impact of rain on the pond and surrounding land with and without the wetland (sponge) in place (Figure 5.32). In Part C, students put real soil over the clay and repeat the rain simulation with and without the wetland in place. Students will observe that when the wetland (sponge) is present, water is absorbed, thereby reducing flooding. They will also notice that when the wetland is present, it filters out soils before the rainwater enters the pond. This helps keep the pond water clear. Finally, when the wetland is present, the soil stays in place, thereby reducing erosion.

FIGURE 5.32.

"Raining" on the Wetland

3. As teams work, check in with them to make sure that they are following the directions and answering the questions. Following are the answers to some of discussion questions included in the checkpoint lab:

Part B Questions

1. Measure 1 cup of water in a measuring cup. "Rain" 1 cup of water over the land. What happens? (a lot of water is absorbed by the sponge, some goes into the pond)

2. What do you think would happen if the wetland (sponge) is removed and it rains again? (answers will vary)

3. Remove the wetland, and rain 1 cup of water over the land. What happens? (water runs right into pond and pond level gets high)

4. What would happen if it kept raining for a long time? (there could be a flood)

5. What might happen if houses were built in place of the wetland? (houses and higher land could get flooded)

Part C Questions

1. Squeeze out the wetland (sponge) and put it back in place. Then spread a thin layer of soil over the clay.

2. Rain 1 cup of water over the soil. Observe what happens to the soil, the wetland, and the pond, and describe below. (the wetland filters out the soil and allows clean water into the pond; the soil stays in place)

3. Now remove the wetland again and spread another thin layer of soil over the clay. Predict what you think will happen when it rains with NO wetland. (answers will vary)

4. Let it rain! Rain 1 cup of water over the land. Observe what happens to the soil and the pond, and describe below. (the soil runs into the pond, making it very muddy)

Lab Conclusion Questions

- Why are wetlands important to people and the environment? List three reasons. (prevent flooding, prevent erosion, keep pond water clean, home to wildlife)

- What are two impacts that people have on wetlands? (e.g., build homes on wetlands, drain or fill wetlands, pollute wetlands)

4. To complete the lab, have the class discuss the following questions:

- "How might muddy water from the loss of wetlands affect fish, plants, and other animals?" (reduce light and oxygen, make it harder for fish and other animals to breathe)

- "How might the loss of wetlands affect you?" (more floods, more erosion, less food and wildlife)

Explore: iBiome-Wetland or "Living" Wetland Food Web

Two activity options are presented here: (A) iBiome-Wetland, an app that lets students research and build virtual food webs, or (B) "Living" Wetland Food Web, in which students research and represent various wetland elements to become part of a "living" food web. These activities present high- and low-tech options that meet similar objectives. If you have time to extend the lesson plan, you can certainly do both options!

FIGURE 5.33.
iBiome-Wetland Screenshot

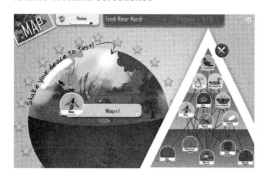

Option A. iBiome-Wetland (High-Tech)

In this activity, students work individually or in pairs to investigate and manipulate wetland habitats using the interactive app iBiome-Wetland at *https://itunes.apple.com/ca/app/ibiome-wetland-school-edition/id1069411327?mt=8* (Figure 5.33). On this app, students build wetland environments by reading about various organisms, adding living and nonliving elements to their experimental dome, and testing the impacts of introduced species on the wetland food web. On the paid version of the app, four wetland types are available (freshwater marsh, saltwater marsh, and two mangrove swamps). On the free version, only the freshwater marsh is available; it has most functions but is missing some species as noted below. For those using PCs, there is a free but limited (freshwater marsh only) demo version of this app at *www.brainpop.com/games/ibiomewetlandschooledition*. This lesson plan focuses on the freshwater marsh.

1. Project the iBiome-Wetland app for the class. Select the Freshwater Marsh. Begin by viewing the introduction and then navigating through the next several screens (reading aloud or asking for student volunteers to read). Demonstrate a few tasks, including researching new species (by clicking on the bouncing + sign), adding new species to the dome, and creating "Crazy Webs" by clicking and dragging species into their appropriate spots in a food web.

2. Distribute copies of the iBiome-Wetland Exploration handout (p. 306) and assign students partners as needed. Allow students to begin; circulate the room to ensure that students understand how to navigate the app. Also ensure that students undersand the vocabulary, including *producers* and *consumers*.

3. If students complete the assignment on the handout, they can work on building more food webs (or visit other types of wetlands if using the paid app). Once everyone has finished, go over the answers to the questions:

   1. Name three (3) PRODUCERS in your wetland. (water lily, willow, pickerelweed, algae, cabomba)

   2. Name three (3) CONSUMERS in your wetland. (mosquito, viceroy butterfly, snail, wasp, hoverfly, dragonfly, red-winged blackbird, blue-winged teal [the last three species are available only on the paid version])

3. Choose one (1) of the species above and write an interesting fact you learned from your research. (answers will vary)

4. To informally assess this activity and help students organize their conceptual thinking, discuss the following:

    - "If a wetland becomes polluted, how would it affect the living things?" (water would not be healthy for plants or animals, oxygen and light could be reduced and have an impact on plants and animals)

    - "If a wetland is drained for buildings, how would it affect the living things?" (animals and plants would lose their homes, flooding would increase from lack of absorption from wetland, more erosion would occur if plant roots aren't there to keep soil in place)

    - "If people build dams that prevent water from entering a wetland, how would it affect the living things?" (without water, plants and animals would die)

Option B. "Living" Wetland Food Web (Low-Tech)

Advance Preparation: Copy the "Living" Wetland Food Web Roles sheet (p. 307) and cut into strips. Explain to students that they will be creating a "living" classroom wetland food web. To do this, they will research and represent various species within a wetland ecosystem.

1. Have each student randomly choose one of the slips of paper with various wetland species on them. Students should create their own placards using the paper with their designated species, construction paper, crayons or markers, and string or yarn (to hang the placards around students' necks). It's fine to have duplicates, so more than one student can represent a particular species, each with his or her own card. *Tip:* You can use color coding to help students remember different organisms' roles in the food web. For example, give students representing producers (plants) green paper, primary consumers (herbivores) blue paper, secondary consumers (carnivores or omnivores) red paper, and decomposers brown paper. You can be the Sun with yellow paper!

2. Students then research their organisms using various sources, including *Wetland Food Chains,* by Bobbie Kalman and Kylie Burns, to familiarize themselves with what their organisms eat and what eats their organisms (see Table 5.2 on p. 294 for an assortment of excellent books on wetlands).

TABLE 5.2.

Picture Books About Wetland Environments

Literature	Description
*Dunphy, M. 2007. *Here is the wetland.* Berkeley, CA: Web of Life Children's Books.	A cumulative, lyrical prose book that introduces young readers to the complexity of the wetland habitat and food chain. Final pages provide critical information on wetland threats and resources for further information.
Fredericks, A. D. 2005. *Near one cattail.* Nevada City, CA: Dawn Publications.	Told in rhyming stanzas, a young girl takes readers on a journey through a wetland. Beginning with a cattail and continuing with frogs, turtles, muskrats and many other organisms, the book explores the wide array of living creatures that live in a wetland.
Heinz, B. J. 2000. *Butternut Hollow Pond.* Minneapolis: Millbrook Press.	Beginning with daybreak and continuing until nighttime, readers experience a day in the food chain at Butternut Hollow Pond. While not every creature gets a meal, readers are treated to vivid descriptions of the struggles and beauty of life in a freshwater pond ecosystem.
Kalman, B., and K. Burns. 2006. *Wetland food chains.* New York: Crabtree Publishing Company.	Using vivid photographs and instructive drawn illustrations, this book provides readers with multiple examples of wetland food chains and the threats to their survival. An easy-to-use glossary and index make this an invaluable classroom resource.
Ridley, K. 2013. *The secret pool,* Thomaston, ME: Tilbury House Publishers.	Vernal pools, springtime wetlands that form on forest floors around the world, are one of the forest's best-kept secrets. With lovely illustrations and clear informational text, children will delight in learning about the animals that make their homes in and around these unique forest wetlands.
Sidman, J. 2005. *Song of the water boatman.* Boston: Houghton Mifflin Company.	Told through poetry, this Caldecott Honor Book lets readers listen to the peepers sing the first song of springtime, discover the wonders of aquatic plants, and watch the burrowing of the hibernating painted turtle. Informational text and hand-colored woodcut artwork accompany each fanciful wetland poem.
*Sill, C. 2013. *About habitats: Wetlands.* Atlanta: Peachtree.	Beautiful color illustrations walk students through the various types of wetlands, their inhabitants, and the importance of wetlands to the environment. An informative Afterword references earlier pages providing captivating additional facts.
Yezerski, T. F. 2011. *Meadowlands: A wetlands survival story.* New York: Farrar, Straus, Giroux.	Reminiscent of Lynne Cherry's *A River Ran Wild*, this book tells the history New Jersey's Meadowlands by showing readers the tragedy of unbridled development, the resilience of life, and the signs of recovery in this critical urban wetland.

*NSTA Recommends®

Source: Reprinted from Kahn and Hartman (2018, p. 38, Table 1), with permission from NSTA.

3. Have students form a large circle, with you at the center. Inform the students that you are the Sun, and you have energy in your hands (the ball of string or yarn). Wrap the end of the string around one hand, and ask students, "Who should get my energy?" (a plant, or producer) Choose a student representing a plant, and give that student the ball of string as you continue to hold the end in your other hand. Ask the student to wrap the end once around his or her hand, then pass the ball to a student representing an organism that eats the plant, connecting the producer to the consumer. This student should wrap the string around his or her hand, then pass the ball to a student representing either an organism that eats his or her organism or an organism that is a food source for his or her organism. Allow students to help each other with information on who eats whom. Remember that many of the plants and animals should be connected to several others; if a student receives the ball of string more than once, he or she should pass it to a student that hasn't yet received it. Continue in this manner until everyone is connected to the chain at least once.

4. Ask students about various threats to the wetland, such as pollution, draining the wetland in order to build on it, or building a dam that prevents water from entering the wetland. For example, ask:

- "If the wetland becomes polluted, or if it is drained for building, what living things would be affected?" (less light passing through the water or less water in the wetlands would kill the aquatic plants, then the fish that eat the plants, and then birds that eat the fish, and so on)

- Have producers drop their strings, and then ask, "Who should drop their strings next?" Have successive groups of students drop their strings as the food web becomes affected. Decomposers can pick up the strings that are dropped.

- If time allows, you can have students create a new web and ask, "What will happen if the wetland grasses and other plants are cut down and removed for houses or other buildings?" Again, have the producers drop their strings, followed by consumers that are herbivores and then consumers that are carnivores or omnivores. Decomposers can pick up the strings.

Explain: *Here Is the Wetland* Read-Aloud

1. Begin with a whole-class read-aloud of *Here Is the Wetland*, by Madeleine Dunphy, which uses cumulative, lyrical prose to illustrate the wetland food

ecosystem. Using think-pair-share, discuss and describe the following: What is a wetland? Why are wetlands important? What is an example of a wetland food chain? What are some of the threats to wetlands? Use these questions as a review and to prime students for the next activity.

2. The fact that this is the second or third book on wetlands that the students have read provides an excellent opportunity to discuss evaluating media sources. (For a fuller discussion of this topic and a template, see pp. 27–28) Say to students:

- "You know, I noticed that the author of this book, Madeleine Dunphy, includes a short biography that says [reading from the inside back cover of the book], 'Madeleine Dunphy is an educational consultant who teaches children about endangered environments. She has written several books for children.' Why do you think the author included this?" (to show that she is an expert in this subject)

- "When an author has expertise in a subject, it helps make them *credible*, or believable and *trustworthy*. Can you think of any other ways we might know if an author is credible or trustworthy?" (answers will vary; may include that the author has a lot of education, is respected in his or her field, is recommended by other experts)

- Share with students that both this book and *Wetlands: About Habitats*, by Cathryn Sill, whose biography indicates that she is a former elementary teacher and author of a popular science series, received "NSTA Recommends" designations from the National Science Teaching Association, a big organization that reviews science books to make sure they are of good quality. This is another way of knowing that a book and its author are trustworthy and credible.

- Ask, "Did the authors of our two books pretty much agree on the information they presented?" (yes) "This is another way of evaluating whether media are trustworthy."

- "How about bias? Do the authors seem to present the information in a fair way, or did it seem biased or designed to promote only one particular viewpoint?" (answers may vary; students may perceive that the information is fair and credible but that the authors are advocates for wetlands, so they may not have presented any negative aspects of wetlands) "Perhaps we can look at other books and websites over the next few days to see if other authors agree with these two authors and if they present similar information." You can show students some of the books from Table 5.2 (p. 294), which all emphasize the importance of wetlands, as well as websites on wetlands that support their importance.

- "One way to keep track of these types of questions is to use a rubric or chart like this one." Show students the Evaluating Media Sources template (Figure 3.15 on p. 28), and review the discussion you just had about *Here Is the Wetland*, highlighting evidence of Credibility, Accuracy, Reasonableness, and Support (CARS). "We will continue to think about ways of ensuring that the sources we use are credible and trustworthy!"

Elaborate: Bullfrog Pond

Advance Preparation: In this land use activity, teams of four or five students will represent various constituencies in Bullfrog Town who are charged with proposing development plans for a nearby wetland called Bullfrog Pond. To prepare, make one copy per team of the Bullfrog Pond Map, Development Cutout Pieces, and How Should We Develop Bullfrog Town? Brainstorming Matrix (p. 311). Prepare slips of paper (one per team) that will be used for teams to randomly pick their constituency groups. Write "Homeowners," "Farmers," or "Business Owners" on each piece of paper. Write the question "How should we develop Bullfrog Town?" on the board.

1. Show students maps of Bullfrog Pond and explain that the surrounding area, a freshwater marsh, was being considered for development of a town. Ask, "How should we develop Bullfrog Town?"

2. Divide students into teams of four or five, and have one student from each team randomly pick a slip of paper with a constituency on it. Distribute the team signs (pp. 308–310) for teams to display at their tables. Explain to the students that they are going to represent their group's interests in the pond's development. This means that they are role-playing that group, or constituency, during this activity, so they should try to imagine what a homeowner, farmer, or business owner might think about developing a town around the pond. In the next class, they will participate in a town hall meeting to decide what development plan should be followed for Bullfrog Pond, but today, they are going to collaborate to develop their plans.

3. Distribute one copy of the How Should We Develop Bullfrog Town? Brainstorming Matrix handout to each team. Project it using a document camera or

projector so that you can write on it in front of the room. Have teams discuss the advantages and disadvantages of wetland development *to their constituency group* and write these on the row that corresponds to their group (for example, the homeowners would put the advantages and disadvantages of building homes in the wetland area, while the farmers would put the advantages and disadvantages of having farms in the wetland area, and so on). Do one example for the class. Say, "For example, homeowners, what is one advantage of having homes near Bullfrog Pond?" (nice views, fishing, wildlife) "What is one disadvantage?" (pollution, flooding, mosquitoes) Allow teams to try to think of two or three advantages and disadvantages of their group's development in the wetland area.

4. Have teams share their answers, and as they do, write them on the matrix on the board. Invite students to share any other ideas that may have been missed. Have one student from each team complete the matrix with the other teams' responses, so that each team has a completed matrix. A possible completed matrix might look like the one in Figure 5.34

FIGURE 5.34.

Sample Completed How Should We Develop Bullfrog Town? Brainstorming Matrix

	Advantages ☺	Disadvantages ☹
Homes	• Pretty views • Lots of wildlife • Quiet and peaceful • Sports and recreation	• Would bring pollution • Could have flooding • Mosquitoes • Overfishing
Businesses	• Would bring jobs and money • Would bring people • Would have nice views • Restaurants could serve local fish and other wildlife	• Would produce garbage • Flooding • Need to build more roads • Could be hard to build on soggy, squishy land
Farms	• Natural water supply and grazing area for farm animals • Could grow food for town • Brings jobs and money • Could sell at farmers' market	• Fertilizers could harm wetland • Animal wastes could get into pond • Wetland predator animals could hurt farm animals

FIGURE 5.35.

Sample Farmers' Team Development Map

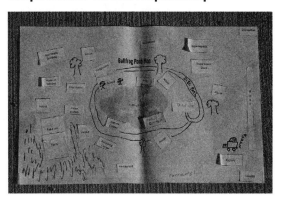

5. Provide each group with a copy of the Bullfrog Pond Map and the Development Cutout Pieces to represent aspects of the proposed built environment, such as houses, farms, and stores. Tell teams that they need to cut out the pieces and decide where they should go on the map. (Figure 5.35 shows a sample map.) They should make sticky rings of tape (rolling a piece of tape with the sticky side on the outside) and use these to place pieces on the map so that they can be moved around easily. The blank extra shapes can be used to create any other built environment places they'd like. The rules are as follows:

- All pieces must be used (so that no land use type, such as businesses, homes, or farms, is excluded).
- All team members must come to a consensus on the plan.

6. As students are working, listen to team discussions. If you find students are having difficulty reaching a consensus, ask some guiding questions like the following:

- "What is most important to you? Can you compromise on anything?"
- "What are the trade-offs of getting what you want? Are there any disadvantages?"

At the end of the class, explain that next time, teams will present their plans at the Bullfrog Pond Town Hall Meeting!

Evaluate: Town Hall Meeting and Rubric

1. Welcome students to the Bullfrog Pond Town Hall Meeting. The purpose of this meeting is to present various plans to the community and ultimately decide on a plan for developing Bullfrog Town in a wetland area around Bullfrog Pond. Inform teams that they will be asked to present a three-minute opening statement while showing and describing their plans to the class. After the opening statements, teams will have an opportunity to question other teams about their plans. Write the criteria for the opening statement

(which parallels the assessment) on the board. The opening statements must include the following:

- Why did you decide on that layout? (rationale)

- What good and bad effects might your plan have on the environment? (impacts)

- What did you do to protect the wetland? (mitigation)

Give teams several minutes to prepare their opening statements, and then begin presentations. After a team presents, have the team tape its plan onto the board so that all plans are visible to the class.

2. After all teams have presented their opening statements, allow time (5 to 10 minutes) for students to ask each other questions and politely challenge proposed plans. Emphasize that the questions should not be aimed at selling their own plans, but rather they should question the other teams' plans and ask for clarification. Provide students with questioning frames such as "Can you please explain why you _____?" "Did you consider _____?" or "Can you please clarify _____?" (*Note:* You can also provide each team with a copy of Sentence Frames for Arguments on p. 26.)

3. To promote students' use of evidence in their arguments challenging other teams' plans, you can write the following frame and example on the board, or you might wish to use the CES template in Figure 3.9 (p. 22):

Claim + Evidence + Source

Example: We think fishing shouldn't be allowed in the wetland (claim) because we saw that removing fish affects the food web (evidence) on the wetland app (source).

4. Allow teams to present and respond to claims, ensuring that different team members have the opportunity to respond to and present claims.

5. Once questioning is complete, draw a large-scale map of Bullfrog Pond on the board (or project the included map using a document camera), and ask, and ask students to identify what they see as strengths of various plans. Then, ask each team to place one or two of their most important elements (their non-negotiables) on the new class wetland area. After all teams have placed their most important elements, have the class try to come to a consensus on a plan. Allow students to debate and discuss for a few minutes if needed.

6. To close the activity, ask these questions:

- "What are some of the challenges in deciding how to develop an area of land?" (answers will vary; may include that different groups have different priorities and that any development has impacts on the land)

- "Were there any groups that had common interests? Anything that you agreed on?" (students may notice commonalities such as that "green businesses" can be a positive for business owners as well as for homeowners and farmers and that setting aside wildlife areas with no development can be a positive choice for all groups)

- "What do you think would be the impact of your development on another ecosystem downstream from Bullfrog Pond?" (our development will likely have impacts on others even when we do our best to mitigate the effects in our area)

7. Assess each team using the rubric on page 318.

Evaluate: Wetlands Flip-Book

1. Distribute copies of the Wetlands Flip-Book template (pp. 314–317), and ask students to create a flip-book using books, websites, and other media. Allow students to write and draw answers to the flip-book questions.

2. Assess each student's work using the rubric on page 319.

Going Deeper

- Students can write advocacy letters to town councils regarding proposed wetland development.

- Classes can visit local wetlands with naturalists and historians.

- Students can conduct interviews with local land developers, business leaders, farmers, or their own families about their perspectives on the environmental impacts of land development.

- Students can conduct research on historical land use issues that have arisen in their own neighborhoods through research at local libraries, historical societies, or via online newspaper websites, as well as through interviews with older local residents.

- Students can use the CARS rubric to evaluate additional books about wetlands.

- Students can extend the Bullfrog Town development activity to include other constituencies such as environmentalists, a local tourism company, and sport and fishing enthusiasts.

Reference

Kahn S., and S. L. Hartman. 2018. Debate, dialogue, and democracy through science! *Science and Children* 56 (2): 36–44.

Name: _____ Date: _____

Wetland in a Pan Checkpoint Lab

Part A. Prepare the Wetland

1. Spread a thin layer of modeling clay over the sloping portion of the paint tray to represent land. Leave the bottom part of the pan empty to represent a pond. You can form streams in the clay that lead into the pond by using your finger or a pencil.

2. Place sponges across the pan along the bottom edge of the clay. This represents a wetland between dry land and the pond.

3. Add moss, as well as plants and animals, on top of your wetland (sponges). The sponges are similar to the masses of thick, tangled plant roots.

4. Predict what you think will happen if we "rain" water over the land (clay):

   _____
   _____
   _____
   _____
   _____
   _____
   _____
   _____
   _____

Modeling clay with sponges

Moss, plants, and animals added on top

Checkpoint A ☐

Name: _____   Date: _____

Part B. It's Raining!

1. Measure 1 cup of water in a measuring cup. "Rain" 1 cup of water over the land. What happens?

   _____

   _____

   Adding rain

2. What do you think would happen if the wetland (sponge) is removed and it rains again?

   _____

3. Remove the wetland, and rain 1 cup of water over the land. What happens?

   _____

4. What would happen if it kept raining for a long time?

   _____

5. What might happen if houses were built in place of the wetland?

   _____

6. Draw your experiment here:

Checkpoint B ☐

Name: _____ Date: _____

Part C. Soil Toil

1. Squeeze out the wetland (sponge) and put it back in place. Then, spread a thin layer of soil over the clay.

2. Rain 1 cup of water over the soil. Observe what happens to the soil, the wetland, and the pond, and describe below.

   _____
   _____

Soil added

3. Now, remove the wetland again and spread another thin layer of soil over the clay. Predict what you think will happen when it rains with NO wetland.

   _____
   _____

4. Let it rain! Rain 1 cup of water over the land. Observe what happens to the soil and the pond, and describe below.

   _____
   _____

Lab Conclusion Questions

Why are wetlands important to people and the environment? List three reasons.

1. _____
2. _____
3. _____

What are two impacts that people can have on wetlands?

1. _____
2. _____

Checkpoint C ☐

IT'S STILL DEBATABLE! USING SOCIOSCIENTIFIC ISSUES TO DEVELOP SCIENTIFIC LITERACY, K–5

Name: _____ Date: _____

iBiome-Wetland Exploration

Word Bank		
Algae	Hoverfly	Snail
Blue-Winged Teal	Mosquito	Viceroy Butterfly
Cabomba	Pickerelweed	Wasp
Dragonfly	Red-Winged Blackbird	Water Lily
		Willow

1. Name three (3) PRODUCERS in your wetland:
   _____ _____ _____

2. Name three (3) CONSUMERS in your wetland:
   _____ _____ _____

3. Choose one (1) of the species above and write an interesting fact you learned from your research.

   _____

   _____

4. Draw your final wetland food web below or on the back of this sheet.

"Living" Wetland Food Web Roles

Sawgrass	Snail
Cattails	Snake
Duckweed	Earthworm
Muskrat	Bacteria
Beaver	Algae
Rabbit	Alligator
Hawk	Goose
Duck	Otter
Fox	Turtle
Fish	Mosquito
Great Blue Heron	Dragonfly

Homeowners

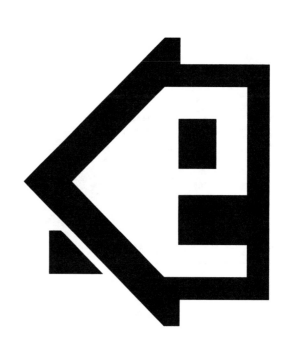

Farmers

Business Owners

How Should We Develop Bullfrog Town?

Brainstorming Matrix

Our team represents the _____

	Advantages :)	Disadvantages :(
Homes		
Businesses		
Farms		

IT'S STILL DEBATABLE! USING SOCIOSCIENTIFIC ISSUES TO DEVELOP SCIENTIFIC LITERACY, K–5

Bullfrog Pond Map

Bullfrog Pond

Development Cutout Pieces

Road

- Department Store
- Supermarket
- Apartment Building
- School
- House
- House
- House
- House
- House
- House
- Playground
- Restaurant
- Restaurant
- Boat House
- Nature Center
- House of Worship
- Recycling Center
- Gas Station
- Police Station
- Fire House
- Feed Lot
- Farm
- Factory
- Landfill

IT'S STILL DEBATABLE! USING SOCIOSCIENTIFIC ISSUES TO DEVELOP SCIENTIFIC LITERACY, K–5

Wetlands Flip-Book

By _____

------- Fold Here -------

How can we protect wetlands?

Why are wetlands important?

What are wetlands?

---------- Fold Here ----------

Rubric for Town Hall Meeting

Each team should make a three-minute clear, organized, and evidence-based opening statement that describes the following: (1) the rationale for the team's map layout, (2) the impacts of the plan on the environment, and (3) how the team has attempted to mitigate negative impacts on the wetland environment. The team will then respond to questions from town members.

	Early (1 pt.)	**Emerging (2 pts.)**	**Sophisticated (3 pts.)**	**Points and Comments**
Use of Evidence	Students use opinion without evidence to back their claims.	Students use tenuous or incomplete evidence to back claims.	Students demonstrate complete and accurate use of evidence to back claims.	
Source and Quality of Evidence	Students are unable to identify sources of evidence.	Students demonstrate some effort in identifying and evaluating the sources of evidence.	Students thoughtfully identify and evaluate the sources of evidence.	
Science Content Understanding	Students demonstrate minimal understanding of science content.	Students demonstrate a moderate degree of understanding of science content.	Students demonstrate strong understanding of science content and consistently apply it to their arguments.	
Clarity and Organization of Presentation	Presentation is unclear and disorganized.	Presentation is somewhat clear and organized.	Presentation is clear, organized, and compelling.	
Response to Questions	Students are unable to respond to questions.	Students respond in inappropriate manner or with inaccurate information.	Students respond appropriately, thoughtfully, and accurately.	

Total: ___ / 15 pts.

Lesson 9: Marsh Madness

Rubric for Wetlands Flip-Book

Criteria	3 pts.	2 pts.	1 pt.
Student is able to communicate the key characteristic of wetlands (e.g., places that collect water all or part of the time, such as marshes, fens, bogs, estuaries, swamps).			
Student is able to identify one or more reasons for the importance of wetlands (e.g., filter water, reduce erosion and flooding, nurseries for juvenile wildlife, habitat for adult wildlife, importance of food web as food sources for humans and other animals).			
Student is able to identify one or more ways that humans have negatively affected wetlands and suggest one way that impacts can be reduced or reversed (e.g., filling, draining, polluting; reducing development near wetlands, reducing water pollution, restoring wetlands).			
Flip-book is accurate and complete.	▓	▓	

Total: ___ / 10 pts.

Lesson 10

Finders Keepers?

Who Owns the Dinosaur Bones?

Suggested Grade Levels

3–5

Driving Question

- What can fossils tell us about the past?

Lesson Overview

Students learn how fossils provide clues about Earth's past. After being introduced to the concept of fossils, they examine various objects to determine whether they are fossils. They then engage in a fossil bed excavation and an online simulation in which they replicate the work of paleontologists. Finally, they are introduced to a fossil controversy, based on the real dispute about ownership of the famous *Tyrannosaurus rex* dinosaur skeleton nicknamed Sue. Students then determine and discuss their views on who should own fossil finds.

Connecting to the *NGSS*

(See full alignment in Table A.10 on p. 512.)

- LS4.A: Evidence of Common Ancestry and Diversity
  - Some kinds of plants and animals that once lived on Earth are no longer found anywhere. (3-LS4-1)
  - Fossils provide evidence about the types of organisms that lived long ago and also about the nature of their environments. (3-LS4-1)

Lesson 10: Finders Keepers?

Societal Issues

Property Rights, Public Versus Private Goods, Ownership of and Access to Scientific Discoveries

Nature of Science

- Scientific Investigations Use a Variety of Methods
    - Science investigations use a variety of methods, tools, and techniques.
- Science Is a Human Endeavor
    - Most scientists and engineers work in teams.
- Scientific Knowledge Is Based on Empirical Evidence
    - Science findings are based on recognizing patterns.

CCSS Connections

- English Language Arts
    - RI.3.7. Use information gained from illustrations (e.g., maps, photographs) and the words in a text to demonstrate understanding of the text (e.g., where, when, why, and how key events occur).
    - RI.4.9. Integrate information from two texts on the same topic in order to write or speak about the subject knowledgeably.
    - RI.5.3. Explain the relationships or interactions between two or more individuals, events, ideas, or concepts in a historical, scientific, or technical text based on specific information in the text.
    - SL.3.1. Engage effectively in a range of collaborative discussions (one-on-one, in groups, and teacher-led) with diverse partners on grade 3 topics and texts, building on others' ideas and expressing their own clearly.
- Mathematics
    - MP.2. Reason abstractly and quantitatively. (3-LS4-1)
    - MP.4. Model with mathematics. (3-LS4-1)
    - MP.5. Use appropriate tools strategically. (3-LS4-1)

NCSS Connections

- Theme 7: Production, Distribution, and Consumption
  - Examine and evaluate different methods for allocating scarce goods and services in the school and community.
- Theme 10: Civic Ideals and Practices
  - Evaluate positions about an issue based on the evidence and arguments provided, and describe the pros, cons, and consequences of holding a specific position.

C3 Framework

- Dimension 3: Developing Claims and Using Evidence
  - D3.3.3-5. Identify evidence that draws information from multiple sources in response to compelling questions.

UDL Toolkit

Multiple Means of Engagement	Multiple Means of Representation	Multiple Means of Action and Expression
Engaging students in activities that simulate real-world endeavors (digging for fossils, signing fossil permit, debating fossil ownership) increases motivation and sustains interest.	Using real fossils or fossil replicas, read-alouds, and interactive online simulations presents information in visual, auditory, and tactile ways and offers multiple entry points for students to process information.	Students express their learning via multiple means, including drawing, mapping, writing, speaking, completing interactive online simulations, and using an anticipation guide.
Using team roles, providing opportunities to work alone, in pairs, and in small groups, fosters collaboration and varies the social demands on students.	Using an anticipation guide to prime students for learning and tap prior knowledge and creating gridlines on the fossil dig to aid in mapping support students' comprehension and visualization.	Graphic organizers such as the four square template used in the dinosaur debate and the data recording chart for Is It a Fossil? support students' ability to organize and manage information.

Suggested Schedule and Sequence

- Day 1: **Engage** with Fossil Anticipation Guide ("Before" column) and *Fossils Tell of Long Ago* read-aloud

- Day 2: **Explore** and **Explain** with Is It a Fossil? activity and discussion

- Day 3: **Engage** with Fossil Collecting Permit, and **Explore** with Dig In! Fossil Bed Excavation

- Day 4: **Explain** with *Digging Up Dinosaurs* read-aloud, and **Elaborate** with Dino Dig simulation

- Day 5: **Elaborate** with Finders Keepers? article and Who Should Own Dinosaur Bones? four corners debate, and **Evaluate** with Fossil Anticipation Guide ("After" column)

Materials

For Is It a Fossil? Activity

(per class)

- Assortment of real fossils (a small fossil collection or sorting kit can be purchased inexpensively at *www.teachersource.com*, or fossils can be gathered from local schools, museums, and colleges)
- Rock
- Leaf or plant
- Stuffed animal or animal figurine
- Bone (such as a chicken bone)
- Shell

(per student)

- Clipboard and pencil

For Dig In! Fossil Bed Excavation

(per class)

- 2 Safari Ltd Ancient Fossils TOOBs and 1 Dinosaur Skulls TOOB (visit *www.safariltd.com*; for a class smaller than 20 students, order 1 of each)

- Small jar of petroleum jelly
- Mixing bowl
- Measuring cup
- Water
- Spoon
- String or yarn (optional)
- Scissors

(per group of 4–6)

- Plastic shoebox
- 2 cups plaster of paris
- 1 cup sand
- Plastic spoons
- Toothpicks
- Paper clips
- Toothbrushes or paintbrushes
- Pencils
- Dig In! Team Role Cards (p. 343), cut out

(per student)

- Indirectly vented chemical splash goggles, vinyl gloves, and a nonlatex apron

For Dino Dig Site Online Game

- Internet access

For Who Should Own Dinosaur Bones?

(per class)

- 4 pieces of paper on which are written 'Finders," "Landowners," "Government," and "Museums and Universities"

Lesson 10: Finders Keepers?

- Tape
- Internet access (optional)

Student Handouts

- Fossil Anticipation Guide
- Is It a Fossil?
- Fossil Collecting Permit
- Dig In!
- Finders Keepers?
- Who Should Own Dinosaur Bones?

Safety Notes

1. All students must wear safety goggles, vinyl gloves, and nonlatex aprons during the setup, hands-on, and takedown phases of the activity.
2. Use caution when working with plasticware, which can cut skin.
3. Immediately pick up any items dropped on the floor to avoid a slip-and-fall hazard.
4. Immediately wipe up any spilled water on the floor to avoid a slip-and-fall hazard.
5. Use caution when using hand tools or sharp materials to avoid cutting or puncturing skin.
6. Plaster of paris in powder form is a health hazard. Students must work with it only in liquid or semisolid form and should not sand or cut it.
7. Wash hands with soap and water after completing this activity.

Media

Books

- *Fossils Tell of Long Ago,* by Aliki
- *Digging Up Dinosaurs,* by Aliki

Websites and Apps

- TVO Kids Dino Dig (for PCs)
  https://tvokids.com/school-age/games/dino-dig

  or

- TVO Kids Dino Dan: Bone Caster (available for free at the App Store for iPads)
  https://tvokids.com/school-age/apps/bone-caster-dino-dan

- The Field Museum's website on Sue, the dinosaur (optional)
  http://archive.fieldmuseum.org/sue/index.html#behind-the-scenes

Background for Teachers

Fossils are fascinating diaries of Earth's history. Through careful and often painstaking study, fossil scientists, or paleontologists, have pieced together clues to what life looked like many thousands or millions of years ago and how Earth's environments have changed.

Fossils are evidence of organisms that lived at least 10,000 years ago. The evidence can be in the form of actual plant or animal parts that have been preserved, or "fossilized," such as bones, teeth, shells, or wood, or can be impressions that were left by the organisms, such as footprints or skin impressions. Fossils that were parts of the organisms are called body fossils, and fossils that are impressions left by the organisms are called trace fossils. Fossils may provide clues to how the organisms lived, including what they ate, if or how they raised young, and how they moved. Fossils can also provide clues to environmental and geological changes. For example, fossils of whales and other marine organisms that have been found high up in the Andes Mountains provide evidence that the Andes arose from the sea. Similarly, the Petrified Forest National Park in Arizona, now a desert environment, is known for its fossilized wood, ferns, and prehistoric amphibians, which together provide evidence that a lush prehistoric forest used to exist there. Fossils are truly storytellers, and there is no doubt that *A Framework for K–12 Science Education* anticipates that elementary teachers will help their students decipher and appreciate those stories. A key understanding that is expected for students at the end of fifth grade is that they recognize that some organisms that once lived are now extinct, with evidence of that coming from fossils. In addition, students need to understand the type of information fossils can illuminate about different organisms and their environments.

Lesson 10: Finders Keepers?

In this lesson, students examine fossil and nonfossil examples to solidify an understanding of what makes something a fossil. They then engage in a simulated fossil bed excavation to gain some understanding of the rigor of paleontologists' work and begin thinking about what fossil types and locations can tell scientists about Earth's past. They also engage in a virtual simulation game to reinforce nature of science understandings about the importance of recognizing patterns, being creative, and using different tools in science. Students then learn about the controversial case of Sue, the *Tyrannosaurus rex* whose fossilized remains were the subject of a five-year custody battle in the 1990s. This part of the lesson requires students to understand the perspectives of various parties in a fossil controversy, including the fossil finders, the landowners, the government, and museums and universities. Students will see that this issue is quite complex, but also quite fascinating, and explore the question of who should have access to important scientific discoveries.

Additional Resources

Articles on fossils

- Norell, M. A. 2003. Science 101: What is a fossil? *Science and Children* 40 (5): 20.

- Royce, C. A. 2004. Teaching through trade books: Fascinating fossil finds. *Science and Children* 42 (2): 22–24.

Articles on fossil hunting and ownership debates

- *www.theguardian.com/science/2016/jul/27/to-collect-or-not-to-collect-are-fossil-hunting-laws-hurting-science*

- *www.livescience.com/29331-dinosaur-smuggling-cases-t-rex.html*

- *www.smithsonianmag.com/smart-news/find-a-dinosaur-in-your-backyard-its-all-yours-19885792*

5E Lesson Plan

Engage: Fossil Anticipation Guide ("Before" Column) and *Fossils Tell of Long Ago* Read-Aloud

1. Begin by showing students the cover of *Fossils Tell of Long Ago,* by Aliki, and ask, "What do you think this book is about?" (students will likely infer that the book is about fossils, perhaps including dinosaurs). Tell students that you would like to know what they know about fossils. Distribute copies of the

Fossil Anticipation Guide (p. 339), and ask students to write a "T" for true or an "F" for false in the "Before" column for each sentence. Reassure students that it's OK to guess the answer now because they'll have a chance to answer in the "After" column after they have completed the lesson.

2. Read the book aloud. After reading, ask students:

    - "What is a fossil?" (the remains of a plant or animal that lived long ago)

    - "How do fossils form?" (when organisms die, parts of them can get pressed into the mud and over thousands of years become fossils; some fossils are the organisms' parts that become filled with minerals, and some fossils are just the imprints)

    - "Do all organisms become fossils?" (no, most don't because the soft parts can't make fossils and the hard parts are often destroyed)

    - Show the students pages 10–11. Ask, "What is the author trying to show in pictures 1 through 4?" (how a fish becomes a fossil) "Does this change happen over a short time or a long time?" (a long time; the author says "thousands of year went by," and the pictures show layers of land filling in the sea)

    - Show students page 22. Ask, "How do scientists know that the desert in the picture wasn't always a desert?" (fossils of a forest and forest organisms were found there) Ask, "Did this change happen over a short time or a long time?" (a long time; the picture says the forest was there 215 million years ago)

    - Show students page 23. Ask, "How did scientists know that there was once a sea where there are now mountains?" (because fossils of sea creatures were found there)

    - Finally, ask, "How do fossils tell us about Earth's past?" (they provide clues to what the environment and organisms were like a long time ago)

Explore and Explain: Is It a Fossil? Activity and Discussion

Advance Preparation: Set up eight stations around the room with one fossil or nonfossil (such as a rock, bone, or shell) at each, and number the stations from 1 to 8. Try to have at least three fossils represented.

1. Give each student a clipboard, a pencil, and a copy of the Is It a Fossil? handout (p. 340). Explain to students that their job is to examine each item and

decide whether they think it is or isn't a fossil, then explain why they think that. The handout includes a chart that asks students to describe the item at each station using words and/or pictures, record whether it is a fossil (yes or no), and explain their thinking. Students can visit the objects in any order; make sure they are distributing themselves well around the classroom. Allow students to confer if they wish about some of the trickier items (such as the shells and bones, which are remains of animals but not from long ago in terms of Earth's history).

2. When students have visited each of the items and recorded their findings, have a class discussion to review the results. Hold up each item and ask, "Is this a fossil?" Allow students to give a thumbs-up for yes or a thumbs-down for no. Scan the room to see whether there is consensus or disagreement. Ask students, "Who would like to explain their thinking to us?" Allow students to explain their ideas, and be sure to elicit opposing sides if students disagree. As students respond and discuss, try to focus on the definitions and criteria that they are using to make their decisions. If fossils are the remains of organisms (plants and animals) that lived a long time ago, then these two key criteria should be elicited:

    - Fossils must have once been alive or must be an imprint left by something once alive, which eliminates items such as a stuffed animal, figurine, or rock.

    - The organisms that formed fossils also must have died a very long time ago. This eliminates items such as a leaf, shell, and chicken bone.

3. Ask students to describe the items that were fossils. (responses will vary but will include descriptions of imprints or fragments of plants or animals in rocks; some may recognize certain fossils, like ammonites, that are no longer living)

4. Provide students with an opportunity to examine additional fossils in small groups so that they can appreciate the variety of shapes, textures, and appearances that fossils may exhibit. If your fossils came with a sorting guide or information, allow students to sort and identify the fossils. Note that the emphasis here is not on learning fossil names or types, but rather to provide students with experiences that reinforce their working definition of fossils as remains of organisms (living things) that lived long ago (many thousands or millions of years ago).

Engage: Fossil Collecting Permit

1. Explain to students that scientists who study fossils are called paleontologists. Ask, "How do paleontologists find fossils?" (they dig in places they think might have fossils) Share with students that they will now work in teams to simulate a fossil dig. However, before they get started, they need to complete a Fossil Collecting Permit. Permits are required when paleontologists go to dig sites because it is important to know who owns the land and whether the owner has granted permission for an excavation. Also, permits clarify how fossils can be removed and what the paleontologist can do with any fossils that are found (e.g., use them for research or education or sell them).

2. Divide the class into groups of four to six, and distribute a copy of the Fossil Collecting Permit (p. 341) to each student. Have students complete the top portion, which asks for their names and the names of their teammates. Then, read aloud the rules on the permit.

3. Have the paleontologists (students) sign and date the permits, and the fossil bed owner (you, the teacher) should sign the permits as well. This will set the tone for a realistic fossil dig and start students thinking about the complexities of fossil excavation!

Explore: Dig In! Fossil Bed Excavation

Advance Preparation: Rub a thin coating of petroleum jelly on the TOOB fossils. This will make the fossils easier to clean. For each shoebox fossil bed, mix 2 cups of plaster of paris with 1 cup of sand in a mixing bowl. (The sand makes the fossil bed a bit more like sandstone and easier to excavate.) Add water slowly while stirring until the mixture reaches a muddy consistency. Pour half of the mixture into the first shoebox. Place four to six TOOB fossils and dino skulls (Figure 5.36.) in various spots around the fossil bed. Cover with the remainder of the plaster mixture. Repeat for each of the shoeboxes. (*Note:* You can make all your plaster mixture at once if you prefer, but you will have to work quickly to pour the mixture into all the shoeboxes and place the fossils before

FIGURE 5.36.

An Assortment of TOOB Fossils

the bottom layer sets.) Assign each fossil bed a number (e.g., #1, #2, #3). Note that the fossils in the TOOB collections are as follows:

- Ancient Fossils TOOB: dinosaur footprint, giant crab, ammonite, raptor claw, fossilized frog, trilobite, *T. rex* tooth, fossilized fish, dino skin, and sea scorpion.

- Dinosaur Skulls TOOB: *Carnotaurus, Velociraptor, Brachiosaurus, Oviraptor, Nigersaurus, Diplodocus, Dilophosaurus, Dracorex, Triceratops, T. rex*, and *Parasaurolophus*.

1. Show a sample fossil bed. Explain that students will use various tools such as spoons, toothpicks, paper clips, and toothbrushes or paintbrushes to loosen and clean the fossils (Figure 5.37). They should be careful, as fossils may be fragile.

FIGURE 5.37.
Cleaning Fossils

2. Distribute a copy of the Dig In! handout (p. 342) to each student, and explain that paleontologists map where they find their fossils, so students need to be very careful about noting where they find each one. To help them map (and emulate some real paleontological digs), the handout asks students to lay down string or draw lines with a pencil to create a three-by-two grid on their fossil bed. Students should draw the fossils they find in the correct boxes on the handout and note whether they were deep or shallow (near the surface). They should also attempt to identify the fossils using books or websites.

3. You can assign roles as digger, cleaner, and inspector by distributing copies of the Dig In! Team Role Cards (p. 343), having students switch roles every five minutes or so. More than one student can perform the same role at one time.

Explain: *Digging Up Dinosaurs* Read-Aloud

1. When students have completed their excavations and fossil cleaning, have teams present their findings. Ask students:

    - "If you were told that your dig site was in a desert, would it surprise you? Why?" (answers will vary; teams with marine or freshwater aquatic animals such as fish, ammonites, giant crabs, sea scorpions, or frogs may be surprised, as this would indicate that the desert used to be underwater)

- "If I told you that your dig site was on top of a mountain, would it surprise you? Why?" (answers will vary; teams again may be surprised that fossils of marine or freshwater aquatic species could be found there)

- "What do the fossils tell us about the environment in which they're found?" (the fossils tell you what the environment used to be like, which may be very different from the present environment)

- "What does the depth of the fossil tell you?" (older fossils would typically be found deeper because the layers of sediment keep building up on top) Note to students that in your simulation, you weren't precise about putting the fossils from the earliest (oldest) species on the bottom.

- Ask, "How do you think our fossil dig might be similar to digs done by paleontologists?" (answers will vary; may include using tools, mapping where they are found, recording data, finding it hard to loosen and clean fossils, needing to be careful to avoid damaging fossils)

- "How do you think our fossil dig might be different from digs done by paleontologists?" (answers will vary; may include that it's indoors and on a smaller scale, the tools might differ, the fossils are not real and were put there by teacher)

2. Explain to students that now they are going to read a book by the same author as *Fossils Tell of Long Ago* to learn more about how paleontologists dig up fossils. Read aloud *Digging Up Dinosaurs*, by Aliki. Use some of these questions to guide reading:

    - After reading pages 10–11, ask, "What does *extinct* mean?" (there are no longer any of the organism alive on Earth)

    - After page 13, ask, "What does the author mean when she says, 'Fossils are a kind of diary of the past'?" (that fossils tell the story of what happened to the organism and what kind of environment it lived in)

    - After pages 16–17, ask, "Who are some of the experts who work together on an excavation, and what do they do?" (*paleontologists* study ancient plants and animals, *geologists* can determine the age of rocks

Lesson 10: Finders Keepers?

and fossils, *draftsmen* can draw fossils, *workers* carefully dig out the fossils, *photographers* take pictures of fossils, and *specialists* prepare the fossils for the museum)

- After pages 19–24, ask, "What are some of the challenges fossil hunters face, and how do they deal with them?" (fossils are fragile, so they need to dig carefully, use shellac to prevent crumbling, create casts, wrap and pack up carefully, keep big fossils in rock to be cleaned later at museum; they also need to keep fossils in the correct order to be reassembled, draw bones in their exact positions, take photos, and number fossils)

- After page 26, ask, "Why do paleontologists compare the fossilized bones of dinosaurs with the bones of other animals?" (to try to figure out a dinosaur's size and shape, how it stood, how it walked, what it ate, what it is related to)

- After pages 30–32, have students make an inference by asking, "Why do you think museums thought it was important to make copies of dinosaur skeletons?" (because seeing and studying dinosaurs is important work, and not everyone has access to real dinosaur fossils)

Tell students that they are going to have a chance to simulate some of the jobs they just learned about using an online game or an app!

Elaborate: Dino Dig Simulation

1. Two simulation options are included here, one focusing on dinosaur excavation and the other on dinosaur bone casting. Both connect students to nature of science understandings, including the use of various tools for inquiry, recognizing patterns, and the human and creative aspects of scientific investigation.

    #### Option A. Dino Dig, TVO Kids Online Simulation Game for PCs

    This simulation game (Figure 5.38, p. 334) allows students to use different tools in an excavation (a pick, a spade, and a brush), each with a specific job. Students select from five different dig sites on a world map. Once they choose a site, there is an introduction to a mystery dinosaur species that was found there (the species that are found at each site are historically accurate). As students dig, they collect fossils. Once they have found all the fossils at that site, they are taken to a lab where they reconstruct the dinosaur and learn its identity and some facts about its natural history. The purpose of this activity is to reinforce the challenges of fossil collection and reconstruction. Students can work in pairs at a computer station or laptop. The game has a voiceover that reads all text to aid students with reading and pronunciations. Students can choose to mute their devices if they find this distracting. Headphones can be worn by those who choose to listen to avoid distracting others.

FIGURE 5.38.

Screenshot from Dino Dig

Option B. Dino Dan: Bone Caster TVO Kids App for Apple iPads

This simulation (Figure 5.39) puts students in the role of a museum preparator, the specialist who prepares casts and molds of dinosaur bones for display and research. Students were introduced to this job in the book they just read, *Digging Up Dinosaurs*. Students choose a dinosaur and then follow the steps to frame the bones, fill the frame with liquid mold, and then fill the mold with plaster, racing against the clock to get the dinosaurs completed for the arriving visitors. This is a good app for introducing students to the steps taken to create the types of exhibits that they see in museums. They will also become familiar with various dinosaur species. While less directly related to

FIGURE 5.39.

Screenshot From Dino Dan: Bone Caster

the excavation itself than the Dino Dig game, it is nonetheless a clever and enjoyable simulation that introduces students to the final steps in dinosaur skeleton preparation.

2. When students are done, ask:

   - "What did you find interesting, enjoyable, or informative about this activity?" (answers will vary)

   - "What questions about dinosaur excavation or preparation do you still have?" (questions will vary)

   - "In what ways were the main characters creative?" (they had to use their imagination to envision the completed dinosaur)

   - "What were some of the tools the characters used for their work?" (e.g., chisels, shovels, resins, brushes)

   - "How does recognizing patterns help paleontologists or museum preparators with their jobs?" (they need to have seen different dinosaur skeletons and developed ideas about patterns of bones and how they might fit together)

   Explain to students that scientists try to recognize patterns in nature and use their creativity as well as different tools to do their job. That's part of the nature of science!

Elaborate: Finders Keepers? Article and Who Should Own Dinosaur Bones? Four Corners Debate

In this activity, students are introduced to the controversies surrounding fossil ownership through reading the Finders Keepers? handout (pp. 344–345). The reading discusses the legal battle that ensued over ownership of Sue, the famous *T. rex* that is now housed at the Field Museum in Chicago. The reading outlines the different arguments that fossil finders, landowners, governments, and universities and museums give for having the right to own fossils. Following the reading, students will complete the Who Should Own Dinosaur Bones? handout (p. 346) and then have a four corners debate.

1. Distribute copies of the Finders Keepers? handout. Also display it on a projector or document reader so that students can follow along in whichever manner is most comfortable for them. Tell students that they are going to be learning about an actual controversy that exists around fossil excavations, especially dinosaur fossil digs.

2. Read the Finders Keepers? article to students to model fluency, enlist student volunteers to read aloud, or begin with silent reading and have students pair up to discuss briefly afterward. The article ends with the questions "Who should own dinosaur bones? The finders? The landowners? The government? Universities and museums?" Invite students to share some of their initial thoughts on the question, but don't allow too long a discussion, as students will first be deciphering the arguments that each of these groups or parties gives for claiming ownership of fossils. Students will have a chance to share their own opinions later.

3. Distribute copies of the Who Should Own Dinosaur Bones? handout and also display it on a projector or document camera. The handout is like the four square template for evidence-based argumentation, with the question in the center, surrounded by four boxes.

4. Ask students to give one reason that people who find fossils might give for why they should own a dinosaur fossil, pointing to the first box on the chart. (responses might include because people who find things get to keep them, they did the work, they spent money to find it) Model how to complete the chart by writing one or more of these reasons in the first box.

5. Then, have students complete the chart, working individually or in pairs. As students are finishing, point out that they need to put a check mark on the group that they feel should own dinosaur fossils in a dispute and write their reasons at the bottom of the page. Reassure students that *there is no wrong answer to this question!* All the parties have valid arguments. You are asking for an opinion that is backed by information from the reading and any personal experiences that may inform their thinking.

6. As students are finishing, make four signs ("Finders," "Landowners," "Government," and "Museums and Universities") and post one in each corner of the room. When students are done, say, "Let's take a few minutes to discuss our ideas. I have posted signs for each of the four groups that might claim ownership of a dinosaur fossil in the four corners." (Show them the four signs.) "When I say, 'Go,' I'd like you to calmly go to the corner of the group that you checked on your sheet, the group that you feel should own the dinosaur bones. When you get to the corner, chat with the other people there about why you chose that group. You may take your sheet with you. Go!"

7. Allow students to go to their corners. If you notice a corner that has a single student, go to that corner and chat with the student about his or her answer.

Reassure the student that there is no wrong answer in this activity. It is about understanding different perspectives.

8. After students have discussed their ideas for a few minutes, inform them that in three to four minutes, you will ask for a spokesperson from each of the corners to share some of the reasons that were discussed. Allow students to work out who should be spokesperson, and encourage spokespeople to write ideas down if needed. Then, allow spokespeople to share their groups' ideas. After each spokesperson reports, ask the group if anyone has anything else to add, to ensure that students' ideas were represented.

9. Ask students:

    - "After hearing other groups' ideas, have any of your opinions changed? If so, how?" (responses will vary)

    - "What, if anything, did you find interesting or eye-opening about today's activities?" (responses will vary)

    - "Why do you think fossils, and dinosaur fossils in particular, cause people to argue (and even go to court) over fossil ownership?" (responses will vary but may include rarity of certain fossils, excitement about the age of fossils, importance to science and understanding Earth's history, money, dinosaurs in movies and media)

    - "Do you think that rules about whether people can keep fossils should differ depending on whether the fossil is very *rare* or whether it is *common*?" (accept all answers; it is worth noting that the Society of Vertebrate Paleontology has a policy statement that no private person should own *any* fossils of *any* kind!)

Evaluate: Fossil Anticipation Guide ("After" Column)

1. Have students revisit the Fossil Anticipation Guide, this time writing their answers in the "After" column. Briefly review answers and collect for assessment. The answers are as follows:

    1. Fossils are the remains of organisms that lived a long time ago. (T)

    2. Most prehistoric plants and animals became fossils. (F; most disintegrate or are washed away)

    3. It takes a long time for fossils to form. (T)

Lesson Plans

4. Fossils give us clues about where prehistoric plants and animals lived. (T)

5. Fossils of sea creatures can be found high up in mountains. (T)

6. Scientists who study fossils are called fossilologists. (F; they are paleontologists)

Going Deeper

- Students can create and curate a fossil museum from their excavated fossil finds.

- Classes can visit museums or local colleges and universities that house fossils.

- Students can create a timeline of the fossils that they excavated.

- Classes can learn about and contribute to citizen science projects about fossils at *www.myfossil.org/resources/citizen-science*.

Name: _____ Date: _____

Fossil Anticipation Guide

Before True or False		After True or False
	1. Fossils are the remains of organisms that lived a long time ago.	
	2. Most prehistoric plants and animals became fossils.	
	3. It takes a long time for fossils to form.	
	4. Fossils give us clues about where prehistoric plants and animals lived.	
	5. Fossils of sea creatures can be found high up in mountains.	
	6. Scientists who study fossils are called fossilologists.	

Name: _____   Date: _____

Is It a Fossil?

Complete the chart below as you visit the stations around the room.

Station #	Description (words and/or pictures)	Is It a Fossil? Yes or No	Explain Your Thinking
1			
2			
3			
4			
5			
6			
7			
8			

How can you tell if something is a fossil?

_____

_____

Fossil Collecting Permit

This permit grants permission for _____ to dig within
(paleontologist name)

fossil bed # _____. Other paleontologists in this group are _____

_____.
(other group members' names)

The following rules must be observed:

1. Digging must be done carefully, as fossils can be fragile.

2. Goggles must be worn by all paleontologists at all times.

3. Fossils may be excavated from the fossil bed *if and only if* the dig team has agreed on a fair and courteous division of labor for digging, cleaning, and inspecting.

4. Fossils that are extracted are to be used for educational purposes and public display only. No commercial use (selling) is allowed.

5. The fossil bed owner may stop the dig at any time if the above rules are not followed.

Signed:

_____   _____
          (paleontologist)   (fossil bed owner)

Date: _____

Paleontologist Name: _____ Date: _____

Team Members: _____

Dig In!

Create a three-by-two grid on your fossil bed that looks like the grid below by laying down string or drawing with a pencil. Then, carefully excavate your fossil bed using the tools provided. When you locate a fossil, record where you found it by drawing it in the corresponding box below. Also record the following: (1) What kind of fossil is it? (if known) (2) How deep was it? (deep or shallow)

Dig In!

Team Role Cards

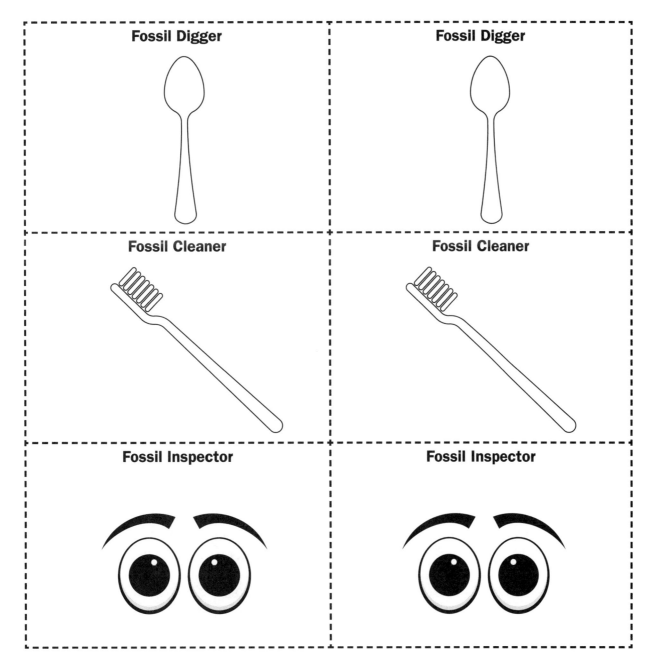

Finders Keepers?

Imagine this ...

Imagine that you are a paleontologist who has searched for dinosaur fossils for years. Then, one day, you happen upon an amazing find—the world's largest *Tyrannosaurus rex* fossil. It's around 67 million years old, and almost all the pieces are there! The company you work for is so excited it even names the dinosaur after you! But soon, things get a little messy. Your company thought that it had bought the rights to remove the dinosaur fossil from a rancher's land. But the rancher says that it's *his* fossil because he misunderstood the deal, and it's *his* land. Other groups, including a Native American tribe and the U.S. government, each think that *they* own the fossil because it was found on *their* lands! Scientists, museums, businesses, and the public begin to debate: Who owns the dinosaur bones? Does the old saying "Finders keepers, losers weepers" hold true?

And the owner is ...

Believe it or not, this was the true story of the famous *T. rex* named Sue, discovered in South Dakota in 1990 by Susan Hendrickson. Susan worked for a fossil-finding business. The debate about Sue went on for nearly five years. Ultimately, a judge decided that the rancher owned Sue because the fossil was part of his land, even though the fossil-finding business had invested so much time and money in the excavation. Two years later, the rancher sold Sue at an auction for $8.4 million! The buyer was the Field Museum in Chicago, which displays Sue to the public and allows scientists from around the world to study her.

This was a famous case, but it is just one of many that have arisen from dinosaur and other fossil finds. Often, the different groups (sometimes called parties in a dispute) have very strong arguments for why they should own the fossils. The people who find the fossils often say that they should own them because ... well, they found them! The concept of "finders keepers" dates back to very early times. This argument is especially convincing when the finder has spent a long time or a lot of money to find the fossil. But another concept that dates back a long time is the idea of *property*. Property is something you can own. It is also a word that is used synonymously with *land*. In the United States, if you find a fossil on land that you own (your "private property"), it is usually yours, and you have the right to keep it or sell it. But if you find a fossil on "public property," such as at a park or on public school grounds, then the government owns the fossil unless you have gotten permission to collect and keep it. In many other countries, the government owns *all* fossils that are found. They argue that when something is very important to the country's history or economy, or if it has high scientific value, like a fossil, then it should belong to the government.

In the United States, even universities and museums sometimes get into the debate and say that when something is *this* important to science, then it should be kept in publicly accessible collections such as university or public museums that are open for anyone with appropriate credentials to study. Scientists learn a great deal about Earth's history from fossils, and many worry that private fossil collectors may keep important fossils to themselves.

No Easy Answers

As you can see, the debate about who should own dinosaur bones and other fossils is very complex. All the different groups have persuasive arguments that support their rights to own fossils. What do you think? Who should own dinosaur bones? The finders? The landowners? The government? Universities and museums?

Who Should Own Dinosaur Bones?

Name: _____ Date: _____

In the boxes below, write a reason that each group might give for why *it* should own a dinosaur fossil in a dispute. Then put a check mark ✓ on the group that *you* think should own the fossil. Write your reasons at the bottom of the page.

(1) The people who find the fossils	(2) The owners of the land where the fossils are found
(3) The government of the country or state where the fossils are found	(4) Museums and universities that study and display fossils

Who Should Own Dinosaur Bones?

I think the (circle one) (1) finders, (2) landowners, (3) government, (4) museums and universities should own the dinosaur bones because _____.

Lesson 11: Blast From the Past

Lesson 11

Blast From the Past

Do We Still Need a Space Program?

Suggested Grade Levels

3–5

Driving Questions

- How do forces relate to rocket launches and landings?
- What are the arguments for and against maintaining a space program?

Lesson Overview

The flight of *Apollo 11* serves as a context for learning about forces and motion as students explore rocket launches and parachute landings. Students then examine the scientific, economic, political, and philosophical arguments for and against having a space program, culminating in a mock congressional hearing on whether to allocate funding to support one. Students also write opinion letters to members of Congress from their state in support of or against funding for a space program.

Connecting to the *NGSS*

(See full alignment in Table A.11 on pp. 513–514.)

- PS1.A: Structure and Properties of Matter
  - The amount (weight) of matter is conserved when it changes form, even in transitions in which it seems to vanish. (5-PS1-2)
- PS1.B: Chemical Reactions

- No matter what reaction or change in properties occurs, the total weight of the substances does not change. (Boundary: Mass and weight are not distinguished at this grade level.) (5-PS1-2)

- PS2.A: Forces and Motion
  - Each force acts on one particular object and has both strength and a direction. An object at rest typically has multiple forces acting on it, but they add to give zero net force on the object. Forces that do not sum to zero can cause changes in the object's speed or direction of motion. (3-PS2-1)

- PS2.B: Types of Interactions
  - Objects in contact exert forces on each other. (3-PS2-1)

- ETS1.B: Developing Possible Solutions
  - Tests are often designed to identify failure points or difficulties, which suggest the elements of the design that need to be improved. (3-5-ETS1-3)

- ETS1.C: Optimizing the Design Solution
  - Different solutions need to be tested in order to determine which of them best solves the problem, given the criteria and the constraints. (3-5-ETS1-3)

Societal Issues

Allocation of Resources, Planetary Protection, Pioneering/Manifest Destiny

Nature of Science

- Scientific Knowledge Is Based on Empirical Evidence
  - Science uses tools and technologies to make accurate measurements and observations.

- Science Is a Human Endeavor
  - Most scientists and engineers work in teams.
  - Science affects everyday life.

CCSS Connections

- English Language Arts
  - RI.3.1. Ask and answer questions to demonstrate understanding of a text, referring explicitly to the text as the basis for the answers.
  - W.3.7. Conduct short research projects that build knowledge about a topic.
  - W.5.7. Conduct short research projects that use several sources to build knowledge through investigation of different aspects of a topic.
- Mathematics
  - MP.2. Reason abstractly and quantitatively. (3-PS2-1) (5-PS1-2) (3-5-ETS1-3)
  - MP.4. Model with mathematics. (5-PS1-2)
  - MP.5. Use appropriate tools strategically. (3-PS2-1) (5-PS1-2) (3-5-ETS1-3)

NCSS Connections

- Theme 2: Time, Continuity, and Change
  - Use sources to learn about the past in order to inform decisions about actions on issues of importance today.
- Theme 7: Production, Distribution, and Consumption
  - Evaluate how the decisions that people make are influenced by the trade-offs of different options.
- Theme 8: Science, Technology, and Society
  - Ask and find answers to questions about the ways in which science and technology affect our lives.
- Theme 10: Civic Ideals and Practices
  - Evaluate positions about an issue based on the evidence and arguments provided, and describe the pros, cons, and consequences of holding a specific position.

C3 Framework

- Dimension 3: Developing Claims and Using Evidence
  - D3.4.3-5. Use evidence to develop claims in response to compelling questions.
- Dimension 4: Communicating Conclusions
  - D4.3.3-5. Present a summary of arguments and explanations to others outside the classroom using print and oral technologies.
- Dimension 4: Critiquing Conclusions
  - D4.4.3-5. Critique arguments.

UDL Toolkit

Multiple Means of Engagement	Multiple Means of Representation	Multiple Means of Action and Expression
Providing students with the opportunity to create and test rockets and parachutes gives them wide latitude in choices of materials and designs, thereby promoting autonomy.	Using a picture book, videos, and kinesthetic and hands-on activities to demonstrate laws of motion conveys information via multiple media.	Assessments (formative and summative) include a pictorial timeline, a diagram with a word bank, experimental data results and analyses, a T-chart, an oral discussion, and a letter, offering multiple opportunities for students to show what they've learned using varied means.
Scaffolded instruction on forces (through teacher demonstration followed by pair and small-group exploration) builds confidence and aids in motivation.	Explicit connections to social studies (through the study of a historical event, as well as civic action in letter writing), writing, and math aid in comprehension and generalization of material.	An *Apollo 11* sequence template, word bank–prompted diagram of forces, T-chart, and opinion letter template all support students' organizational skills.

Lesson 11: Blast From the Past

Suggested Schedule and Sequence

- Day 1: **Engage** with *Moonshot* read-aloud, *Apollo 11* video, and *Apollo 11* Timeline Foldable activity

- Day 2: **Explore** and **Explain** with Exploring Newton's Laws and **Evaluate** with Rocket Launches and Newton's Laws

- Day 3: **Explore** and **Explain** with Blast Off! rocket testing

- Day 4: **Explore** and **Explain** with Where Did the Bubbling Tablets Go?

- Day 5: **Explore** and **Explain** with Parachuting Pretzels and *How Do Parachutes Work?* read-aloud

- Day 6: **Evaluate** with Space CER Template and science conference

- Day 7: **Elaborate** with Do We Still Need a Space Program? Congressional Committee Meeting yes/no line debate, and **Evaluate** with Opinion Letter to a Member of Congress (this can be done as a take-home assignment)

Materials

For Blast Off! Rocket Testing

(per team of 2)

- Film canister with a snap-on lid (available at *www.amazon.com* or *www.walmart.com*)
- Tape
- Scissors

(per class)

- Glass
- Water
- Teaspoon
- Aluminum pan
- Effervescent tablets (such as Alka-Seltzer)
- Masking tape or chalk to mark height of launch (optional; alternatively, you can use existing landmarks)

Lesson Plans

(per student)

- Indirectly vented chemical splash goggles and a nonlatex apron

For Where Did the Bubbling Tablets Go?

(per team of 4)

- Small cup
- Quart- or gallon-size zipper-seal plastic bag

(per class)

- Water
- Effervescent tablets
- Triple beam balances or electronic scales

(per student)

- Indirectly vented chemical splash goggles and a nonlatex apron

For Parachuting Pretzels

(per team of 3–4)

- Plastic sandwich bag containing the following: paper towel, napkin, looseleaf paper, coffee filter, string (approx. 50 cm), large paper clip, tape or sticky dots, 2 pipe cleaners
- Scissors
- 1 pretzel
- Meter stick

(per class)

- 3–4 plastic garbage bags or tarps for landing sites

For Congressional Committee Meeting

(per class)

- 1 copy of "Yes" and "No" signs (pp. 152–153)

Lesson 11: Blast From the Past

Student Handouts

- *Apollo 11* Timeline Foldable
- Rocket Launches and Newton's Laws
- Blast Off! Rocket Testing Data Sheet
- Rocket Parts
- Where Did the Bubbling Tablets Go?
- Parachuting Pretzels
- Space CER Template
- Congressional Briefing
- Space Program Yes/No T-Chart
- My Opinion Letter Outline

Safety Notes

1. All students must wear safety goggles and nonlatex aprons during the setup, hands-on, and takedown phases of the activity.
2. Use caution when working with glass or plasticware, which can cut skin.
3. Immediately pick up any items dropped on the floor to avoid a slip-and-fall hazard.
4. Immediately wipe up any spilled water on the floor to avoid a slip-and-fall hazard.
5. Use caution when using hand tools or sharp materials to avoid cutting or puncturing skin.
6. This activity should be done outside or in a large indoor space like a gymnasium or cafeteria.
7. Be sure that students step away from the canister once it is capped.
8. Make sure the trajectory for projectiles is marked off, and do not allow anyone to stand in their path.
9. Never eat or taste any food, substance, or chemical used in the activity.
10. Use caution when using equipment such as ladders.
11. Wash hands with soap and water after completing this activity.

Media

Books

- *Moonshot*, by Brian Floca
- *How Do Parachutes Work?*, by Jennifer Boothroyd

Videos

- Short videos of *Apollo 11* takeoff, landing, and Moon walk
  www.nasa.gov/multimedia/hd/apollo11_hdpage.html

Background for Teachers

"We have liftoff!" These words have thrilled space enthusiasts and the general public for decades as humans have sought to explore beyond our planet Earth home. However, since the early days of space travel in the 1960s, debate has ensued about the importance of maintaining a space program, particularly in times of economic challenge, when some argue that the funds could be better spent on projects back home.

In this lesson, students read about the flight of *Apollo 11*, in which the United States became the first country to have astronauts walk on the Moon. On July 20, 1969, 600 million people (one out of five people on Earth) watched Neil Armstrong and Buzz Aldrin explore the lunar surface (with Michael Collins remaining in the *Columbia* command module) and heard Armstrong exclaim, "That's one small step for [a] man, one giant leap for mankind." The *Apollo 11* flight serves as an excellent example of putting forces to work for takeoffs and landings. Newton's first law (and *NGSS* performance expectation 3-PS2-1) essentially states that an object at rest will remain at rest (and an object in motion will remain in motion at a certain velocity) as long as the forces acting on it are *balanced*; when *unbalanced* forces are introduced, the object's motion will change. When spacecraft are on the launch pad, they are not moving because the force that they are exerting down, due to gravity, is balanced with the force that is pushing up from the ground (called the normal force). To get them moving, spacecraft use rockets to provide the unbalanced force needed to propel them into space.

During a rocket launch, the fuel that is burned by the rockets releases gas that creates a pushing force down toward the ground. So then, why do rockets go *up*? That's an example of Newton's third law of motion: For every action, there is an equal and opposite reaction. Forces always occur in pairs. When the gas creates a pushing force toward the ground (the action), the equal and opposite force from the ground (the reaction) pushes the rocket in the other direction upward, allowing the rocket to take off. In the case of a rocket launch, the upward pushing force from this reaction is called *thrust* (Figure 5.40). Of course, since *gravity* is a force that pulls objects, including the rocket, toward Earth, the thrust of the rocket must be strong enough to overcome the gravitational force. The heavier the spacecraft (the load), the stronger the force would have to be to lift it. To send astronauts to the Moon, the National Aeronautics and Space Administration (NASA) had to develop a rocket that was powerful enough to lift the spacecraft that would carry the astronauts into space but could shed parts so as not to be too heavy once the launch had taken place. This is because heavy objects require more force to lift up (and overcome the force of gravity) than lighter ones (Newton's second law). You can observe this yourself by simply trying to lift a heavy object versus a light object with one hand; it takes more force to lift the heavy one.

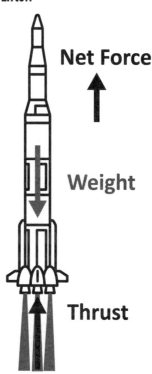

FIGURE 5.40.

The Forces Involved in Rocket Liftoff

Spaceships also have to deal with *friction*, the force that slows objects down as they rub against each other. Specifically, spaceships need to deal with *air resistance* or *drag*, friction with the air, as they climb through Earth's atmosphere. To minimize air resistance, spaceships are built to be streamlined, without any unnecessary objects protruding from the exterior. The force of air resistance is also at work when returning astronauts to Earth. In the case of *Apollo 11*, parachutes were used to increase air resistance in order to slow the descent of the command module *Columbia* and allow it to gently splash down in the Pacific Ocean.

Note: To summarize, a rocket takes off because the upward pushing force (thrust) is greater than the downward pulling force (weight due to gravity). The result of these unbalanced forces is liftoff!

In this lesson, students experiment with rocket launches by using effervescent (bubbling) tablets mixed with water as the "fuel." Film canisters decorated with fins and nose cones serve as the rockets. When the tablets are mixed with the water in the capped film canister, bubbles (gas) build up until the force from the gas pops the cap off. The canister can be placed with the cap side down so that the canister (decorated like a rocket) is lifted into the air. Students then test the effects of adding weight to the rocket on the launch height. By seeing that lighter, more streamlined rockets fly higher, they will understand why the *Apollo 11* engineers needed to keep the spacecraft as light as possible. Students also dig deeper into the reaction of the bubbling tablets in water to gain an understanding of *conservation of mass*; specifically, they design an experiment that uses a scale or balance to determine that the total weight of substances that are combined is the same as the total weight after the substances are combined, even if one of the substances seems to disappear! This is important for students to understand because it clarifies that even when substances like rocket fuels are mixed or burned, they don't simply vanish; they are still somewhere, possibly contributing to pollution. Students are also challenged to create a parachute that can allow a pretzel to be dropped from a high height without breaking. This helps them see that lightweight materials that form a large canopy to catch the air (create resistance) work best, similar to the parachute used by *Apollo 11* astronauts during their splashdown. Finally, students engage in a mock congressional debate about the need for continued space exploration. Congress is responsible for determining whether projects such as the space program are funded, so this activity provides students with a glimpse of what it is like to make decisions with competing interests. They then put this knowledge into action by writing opinion letters to their state legislators.

Additional Resources

NASA's resources for teachers

- *www.nasa.gov/audience/foreducators/index.html*

Smithsonian National Air and Space Museum information on *Apollo 11*

- *https://airandspace.si.edu/explore-and-learn/topics/apollo/apollo-program/*

Article on air pollution produced by rockets

- *www.scientificamerican.com/article/how-much-air-pollution-is-produced-by-rockets*

Note: The Blast Off! Rocket Testing activity is based on the activity at *http://spaceplace.nasa.gov/pop-rocket*.

5E Lesson Plan

Engage: *Moonshot* Read-Aloud, *Apollo 11* Video, and *Apollo 11* Timeline Foldable

1. While showing students the cover of the book, tell them that you will be reading a book called *Moonshot*, by Brian Floca, which tells the true story of the

Apollo 11 space mission. Draw a KWL chart (see pp. 15–16) on the board and ask students, "What do you Know about the *Apollo 11* space mission?" Enter what students Know under the "K" column. Then ask, "What do you Wonder about the *Apollo 11* space mission?" Enter students' Wonderings under the "W" column. After the read-aloud, ask students, "What did you Learn about the *Apollo 11* space mission?" and fill in the "L" side. Figure 5.41 shows an example of what the completed chart might look like.

FIGURE 5.41.
Sample KWL Chart for *Moonshot*

K	W	L
It happened a long time ago.	When was the *Apollo 11* mission?	The mission was in July 1969.
It landed on the Moon.	Was it the first time people landed on the Moon?	*Apollo 11* was the first manned spaceship to land on the Moon.
	Who was on the spaceship?	There were three astronauts. Two of them walked on the Moon!
	What did the astronauts do?	The astronauts collected rock and soil samples and took pictures and videos.

2. After the read-aloud, discuss the following Wonderings with students if they haven't already been discussed:

   - "What was the purpose of the *Saturn* rocket?" (to launch the *Columbia* and *Eagle* spaceships into space)

   - "Why does the rocket shed parts?" (to make it lighter so it can fly more quickly)

   - "Why did they use Velcro onboard the spaceships?" (because there was no gravity, so objects needed to be held down)

   - "Did everything on this flight go smoothly?" (no, the spaceship overshot its original landing site)

   - "What were some of the questions that the astronauts hoped to answer by collecting rocks and taking pictures and videos?" (Where did the Moon come from? What is it made of? How old is it?)

- "What was the purpose of the parachutes?" (to slow down the landing in order to make it more gentle)

3. Show students the "Apollo 11 Introduction" video montage of the *Apollo 11* launch, Moon walk, and landing found at *www.nasa.gov/multimedia/hd/apollo11_hdpage.html* (bottom of the page). Invite students to share their thoughts on this remarkable video, which includes an iconic newscaster, Walter Cronkite, becoming visibly emotional when Neil Armstrong sets foot on the Moon.

4. Ask students to think about the order or *sequence* of steps that took place during the *Apollo 11* flight story and share their ideas. Write student responses on the board or chart paper. The steps should be written in this order:

Launch → *Saturn V* Rocket Separates → Journey to Moon → Lunar Landing → Moon Walk → Lunar Liftoff → Journey Home → Splashdown

5. Open the *Moonshot* book to the inside front cover and review the steps with students. Note to students that this timeline provides much more technical information than is covered in the book, such as the times the different modules needed to turn around, dock, eject, and so on. The story in the book is a simplified version of the actual sequence.

6. Give each student a copy of the *Apollo 11* Timeline Foldable handout (p. 377). Explain that they are going to create a timeline foldable booklet that they can keep to help them remember the *Apollo 11* journey. Students should cut out the foldable booklet and the titles, then paste the titles in the correct order on the booklet. Then, they should illustrate each step with a picture and a short sentence caption. For example, the "Launch" section could have a picture of the launch with the caption "3-2-1 … Blast Off!" The front cover should be titled and have an illustration of the students' choosing. The correct order for the foldable, which has five pages, is as follows:

Launch → *Saturn V* Separates → Lunar Landing and Moon Walk → Lunar Liftoff → Entry and Splashdown

Explore and Explain: Exploring Newton's Laws

1. Begin by reviewing some of the key points from the *Moonshot* read-aloud. Ask the following think-pair-share questions (elicit student responses but accept all answers at this point):

Lesson 11: Blast From the Past

- "Why did *Apollo 11* need engines to launch it?' (the engines push it into space)

- "How did the engines make the rocket go up?" (burning fuel produced gas that pushed down against the launch pad)

- "Why did the *Saturn V* rocket have to shed parts?" (to make it lighter so it could go faster and farther)

- "Why do you think the spaceship needed a huge rocket engine to leave Earth but didn't need as big an engine to return?" (the spaceship was going against gravity heading away from Earth, but it was being pulled toward Earth on the way back)

2. Tell students, "To better understand how rockets work, we need to understand forces. We can do this with our chairs and our bodies!" Place a chair in front of the room. Ask students:

    - "What is a force?" (a push or a pull)

    - "Are there any forces acting on this chair?" (students will probably say no)

    - "Is the chair moving or standing still?" (standing still)

3. Ask a volunteer to come up and gently push the chair a short distance. Then ask:

    - "Why did the chair move?" (the child exerted a pushing force on the chair)

4. Invite a second child to come up to the chair, and have both children gently push the seat of the chair from opposite sides so that it doesn't move. Ask:

    - "Why isn't the chair moving?" (because the two children are exerting the same forces from the opposite directions)

    - Introduce the term *balanced*. Explain to students, "We can say that the forces that are being exerted are balanced. When something doesn't move, it is because the forces acting on it are balanced."

    - Give the chair a gentle push so that it moves. Ask, "Why does the chair move now?" (because you are exerting a force, but there's no one pushing back)

- Introduce the term *unbalanced*. Explain to students, "When something moves, it is because there is a force acting on it that isn't balanced by another force. It is unbalanced."

5. Point to the chair and ask again, "Are there forces acting on this chair? How do we know?" (there must be forces because it's staying still) Explain to students that gravity is pulling the chair down toward the center of Earth, but the floor is pushing back to keep the chair in place! The force that pushes up is called the *normal force*. Ask:

    - "What would happen if the floor weren't strong enough to exert a force equal to gravity?" (the chair would fall through the floor because the normal force wouldn't be enough to push back against gravity!)

6. Allow students to repeat this test themselves with a partner. Explain to students that they must use gentle forces during the test.

7. Then, explain that they have just demonstrated Newton's first law, which basically says that objects at rest, like the chair, have balanced forces acting on them (gravity and the normal force from the floor or two children pushing equally from opposite sides). The objects tend to stay at rest unless an unbalanced force (such as a push from one child) is introduced, causing them to move.

8. Have students stand up in place. Ask:

    - "Why are we able to stand still in place?" (because there are balanced forces acting on us: gravity pulling down and normal force from the ground pushing up)

    - "How can we create a force that is stronger than the force from gravity pulling us down?" (by jumping!) Have students stand up and jump several times, including small jumps and big jumps.

    - "When you want to jump, which way do your legs push?" (down against the ground)

    - "How does pushing down make you go up?" (when a force pushes in one direction, there is a force pushing against it in the opposite direction; when our legs push down on the floor, the floor pushes back up)

    - "How do you make small versus big jumps?" (a small force down makes a small jump; a big force down makes a big jump)

- Explain that this is called action-reaction. When one forces pushes against an object (the action), the object pushes back with an equal force (the reaction). When we jump, we push against the ground (action), and the ground pushes back (reaction). This is Newton's third law. The reaction force that pushes you up is called *thrust*.

9. Ask students to think back to the *Moonshot* book. (You can show page 8, which depicts *Apollo 11* on the launchpad, as a reminder. The second question below is covered on pages 12–15.) Ask:

    - "Why did *Apollo 11* stand still on the launchpad before the engines were started?" (the rocket had balanced forces acting on it: gravity pulling it down toward the center of Earth and a normal force pushing up from the ground)

    - "How did the engines make the rocket launch?" (The gas released from the burning fuel builds up in the engine and eventually creates a strong downward pushing force [the action]. This force is met by an equal upward pushing force [the reaction], called thrust. This thrust is stronger than the downward pull of gravity, so the rocket takes off!)

Evaluate: Rocket Launches and Newton's Laws

1. Distribute copies of the Rocket Launches and Newton's Laws handout (p. 378). Allow students to work in pairs to discuss and complete the handout. They can fill in details on the pictures (e.g., the launchpad, engine gases) as they finish. The answers are as follows:

    - In the two pictures, A = Normal Force, B = Gravity (used twice), C = Thrust

    - In the first picture, the forces are **balanced**; the normal force (upward push from the ground) is equal to gravity (downward pulling force). That is why the spacecraft is at rest.

    - In the second picture, the forces are **unbalanced**; the thrust (the upward pushing force, which is a reaction to the engine gases pushing down) is greater than gravity (downward pulling force). That is why the spacecraft is in motion.

    - When an object is at rest, the forces acting on it are balanced.

    - When an object is in motion, the forces acting on it are unbalanced.

Explain to students that next they are going to get to build and test rockets that have their action force powered by bubbles!

Explore and Explain: Blast Off! Rocket Testing

Safety Alert: This activity should be done outside or in a large indoor space such as a gymnasium or cafeteria. Safety goggles should be worn by students and teachers. Be sure that students step away from the canister once it is capped. Also, remind students that they are not to eat or taste the effervescent tablets. To ensure safety, hold tablet pieces until teams are ready to test so that you can keep track of all tablets.

1. Demonstrate bubbling for students: Fill a glass halfway with water and drop an effervescent tablet in it. Have students observe. Ask:

    - "What is in the bubbles?" (a gas called carbon dioxide, or carbonation)
    - "Where does the gas go when the bubbles pop?" (into the air)
    - Show students a film canister. Ask, "What do you think would happen if I put the tablet and water in this canister and put the cap on?" (the gas will build up until it pops)

2. Put ¼ of a tablet in the canister and add 1 teaspoon of water. Quickly cap the canister, put it in an aluminum pan on the floor *with the cap side up,* and step aside. After a few seconds, the cap will fly up into the air with a loud noise. Ask:

    - "Why did the cap fly up?" (the gas in the canister pushed it up when the force inside became stronger than the force holding the cap down)
    - "Why did it make noise?" (the quick pushing of air by the gas makes noise)
    - "How can I make the canister fly rather than the cap?" (turn the rocket over)

3. Do a second demonstration with the same materials, but this time, place the canister in the aluminum pan *with the cap side down* and step away. The canister will fly up and make a noise. Ask:

    - "Why did the canister fly up?" (the gas in the canister built up, and when it was strong enough to push the cap off, it pushed against the ground)

Lesson 11: Blast From the Past

4. Distribute copies of the Blast Off! Rocket Testing Data Sheet and Rocket Parts handouts (pp. 379–380) to students. Explain that they are going to work in pairs to build and test a rocket. The design variables include the number of fins, presence or absence of a nose cone, whether the nose cone is pointy or flat, and whether the rocket is tall or short. Some possible designs would be a tall rocket with two fins and a pointy nose cone, a tall rocket with no fins and a pointy nose cone, a short rocket with two fins and a pointy nose cone, a short rocket with two fins and a flat nose cone, and so forth. To make rockets, students wrap the paper around the canister and tape it down. The lip of the canister should extend slightly past the paper so that they can easily cap the canister. *The open end of the canister must be at the bottom of the rocket.* They can then add fins and a nose cone.

5. Show students the launchpad, which is an aluminum pan. As a class, establish markers for high, medium, and low launches. You can mark an interior or exterior wall with tape or chalk (low is below 3 feet, medium is 3–6 feet, and high is over 6 feet) or use existing landmarks such as playground equipment (e.g., low is below the swing, medium is between the swing and the top bar, and high is above the top bar) or trees (e.g., low is below a low branch, medium is between a set of branches, and high is above the tree).

6. Have pairs take turns launching their rockets from the launching pad using ¼ of a tablet for each launch. Give each pair three chances to test the rocket with different designs. Rocket bodies and fins should be reused if they haven't gotten too wet.

7. If time allows, students can test additional designs by recording data on the back of their data sheets. You can also have students test other variables including the amount of effervescent tablet (¼, ½, or 1 tablet), the amount of water, and even the temperature of the water—the sky's the limit!

8. Draw a whole-class data table on the board, and have teams report their highest and lowest launches. Then, ask:

    - "Which combination of fins, rocket heights, and nose cones gave you the highest launch? Which combination gave you the lowest?" (While the findings may vary, student will likely find that short, lighter rockets with fewer added parts fly the highest. Too many added pieces weigh the rocket down and create more air resistance, or drag. However, having a pointy nose cone helps the rockets fly higher than a flat nose cone or no nose cone at all, as it reduces drag on the rocket.)

- "What made your rockets fly up?" (the gas building up from the tablets and water eventually creates a downward force; there is an equal and opposite force pushing the rocket up—Newton's third law)

- "What made your rockets come back down?" (the force of gravity pulls objects toward Earth)

- "Why did *Apollo 11* shed parts of its engines after liftoff?" (to make the rocket lighter so that it could go farther)

- "Why didn't *Apollo 11* need the same engines to come back home?" (gravity can be used during reentry to bring the spaceship back home)

Explore and Explain: Where Did the Bubbling Tablets Go?

1. This activity illustrates conservation of mass. Ask students to remind you how they fueled their rockets. (using gas bubbles from the effervescent tablets) Repeat the demonstration of ¼ of an effervescent tablet being dropped in half a glass of water, then ask students, "Where does the tablet go?" (accept all answers; students may suggest that the tablet dissolves in the water or that it disappears)

    - Ask students, "If the tablet has dissolved in the water, is it still there?" (accept all answers; students may suggest that dissolving makes the tablet too small to see, or they may say that the tablet has become a gas and gone into the air)

    - "If we can't see the tablet anymore, can you think of a way to tell if it's still there?" (accept all answers)

2. Show students the electronic scale or triple beam balance and ask, "How can we use the scale (or balance) to determine whether the effervescent tablet is still in the water or not?" (accept all answers; students may suggest weighing the cup, water, and tablet before mixing and then after the tablet dissolves to see if they weigh the same) *Note:* At this grade level, no distinction is made between weight and mass. Ask, "If the materials weigh the same before and after the tablet dissolves, what does that tell us?" (that the tablet materials are still in the water or the air)

Lesson 11: Blast From the Past

> **Misconception Alert**
>
> This activity addresses a commonly held misconception that when materials dissolve, they simply disappear. One way to counter this misconception is to weigh substances before and after they are mixed to demonstrate that the weight is the same. For example, you can weigh sugar cubes and water (and the container the water is in) before and after mixing to show that the total weight is the same, even when the sugar is completely dissolved and "invisible." However, things get a bit more complicated when gases are produced from a reaction, such as when effervescent tablets are mixed with water, because the weight will *not* be the same before and after the reaction if the gas is released into the air before it is weighed. Gas is *matter*—it takes up space and has weight. So in this experiment, students need to figure out that they must capture the gas and weigh it to make sure that they are considering *all* the substances that go into and come out of the reaction. To do this, they can enclose all the materials in a sealed plastic sandwich bag that captures the gas that is produced. But don't tell them! Allow your students to design and test the reaction and compare results with other teams. You can then facilitate discussion of why it's important to trap the gas to see that the weight before and after the reaction is the same, or conserved. By doing this, you can address yet another common misconception—that gas isn't matter; gas is indeed matter (it takes up space and has weight) even if we can't see it!

3. Divide students into groups of four. Distribute copies of the Where Did the Bubbling Tablets Go? handout (pp. 381–382) Show students the available materials, which include cups, water, effervescent tablets, and quart- or gallon-size zipper-seal plastic bags. Explain to teams that they are challenged to design an experiment that uses weight to test whether effervescent tablets simply disappear when they are mixed with water. Teams must develop their own stepwise procedures and determine what data to collect.

4. Allow student teams to discuss their experiments and write out their procedures. Remind teams that their procedures should be detailed enough that any team could follow them!

5. Circulate around the room to check on team procedures and answer questions. Once teams have completed their procedures, provide each team with one or two effervescent tablets.

6. After teams have completed their data sheets, discuss them by asking the following questions:

    - "How did you determine where the bubbling tablet went after it was mixed with water?" (Have students describe their procedures.)

    - "What do you think happens to the bubbling tablets after they are mixed with water? How do you know?" (Have teams report their findings. Elicit the idea that the tablet materials don't just disappear—they are still in the water or released into the air as gas. We know this because the total weight of the materials before mixing was about the same as the weight of the materials after they are mixed. If the tablets just disappeared, the weight after mixing would be much lighter, by the amount of the tablets' weight.)

    - "Does gas have weight? How do you know?" (yes; the teams that captured the gas given off by conducting the experiment in a sealed plastic bag likely found that their products of the reaction weighed the same or nearly the same as the starting materials, and these results were closer than those of teams that didn't capture the gas)

    - "If rockets use fuels that release gas when mixed or burned, does the gas just disappear?" (no, the gas still exists in the atmosphere; in fact, you may wish to note that scientists are currently studying the pollution that is released when rockets are launched to assess the impact of the emissions and debris on climate and the ozone layer)

    - Review the final question on the data sheet by asking the students to complete this sentence:

        The total weight of the substances before they are combined is about equal to / less than / greater than (circle one) the total weight of the substances after they are combined.

Most teams will find that the before and after weights are about equal, particularly if they captured and weighed the gas in the bag. Discussion of why different teams got different results may include allowing the gas to be released and not including the weight of the cup or bag before and after.

Lesson 11: Blast From the Past

Explore and Explain: Parachuting Pretzels and *How Do Parachutes Work?* Read-Aloud

1. Show students the splashdown page from *Moonshot*. (*Note:* The book does not have page numbers, but it is the next-to-last page.) Ask:

   - "What force is pulling the *Columbia* module down toward Earth?" (gravity)

   - "Why did the module use parachutes for landing?" (to slow the descent of the module to make a gentler landing)

   - "How do you think the parachutes work?" (accept all answers; students will have a general idea that the parachute "catches the air" but may not recognize this as air resistance, or drag, which is friction with the air that slows objects down)

2. Introduce the following vocabulary to students:

   - *Canopy:* the main part of the parachute that catches the air. Canopies come in many shapes, sizes, and configurations.

   - *Suspension lines:* the cords that attach the canopy to the harness and hold the load (a person or anything else that is parachuting).

3. Explain to students that they are going to have the opportunity to build and test different parachutes to see what attributes create a gentle landing. To do this, they will parachute pretzels, which are very fragile!

4. Distribute premade parachute materials packets to teams of three or four students. Distribute a copy of the Parachuting Pretzels handout (pp. 383–384) to each student. Teams should also have access to a meter stick. Teams should also have access to a meter stick and scissors.

5. Explain to students that their team challenge is to develop a parachute that can safely allow the pretzel to drop from a high point without breaking.

6. Review meter stick measurements with students. Each meter stick is 100 cm. Students will be measuring drops from 50 cm, 100 cm, and 150 cm (approximating 50 cm above 100 cm).

7. Show students some of the materials that are in the packets, including paper towel, napkin, looseleaf paper, coffee filter, string, large paper clip, sticky dots or tape, and pipe cleaners. Write a list of the materials on the board so that students have a visual reference for recording data.

8. Explain the rules:

    - The pretzel must be hooked onto the paper clip. (To do this, simply pull open the paper clip slightly and hook the pretzel on.)
    - The pretzel can't be covered with anything.
    - Attempt to land your pretzel on a plastic garbage bag (or tarp) for easy cleanup.
    - Don't eat the pretzel!

9. Encourage teams to examine and test different materials before drawing their plans. They can simply try dropping items such as the napkin, paper towel, or plastic bag to see how these different materials catch the air.

10. Remind teams that they must draw their plans before testing. The drawings should include labels for the various items. Circulate the room to check students' drawings. Distribute one pretzel per team once drawings are complete.

11. Students should record their drops on the handout. They will start with the pretzel at 50 cm. If that drop is successful, they can try 100 cm (1 meter). If they are unsuccessful, they can revise their parachute and retest. Teams should continue this mode of testing until they have their best design.

12. If time allows, have a 3-meter drop with the teacher dropping the pretzel with parachute. *Safety Alert:* This may require a small stepladder. Use caution.

13. Have students share their answers to the parachute discussion questions on the handout. Answers are as follows:

    1. Which materials worked best for making a parachute? Why do you think these materials worked well? (answers will vary; thin, light materials work best, as less weight means less downward force)

    2. What are the characteristics of a successful parachute? (a big canopy creates more drag, or resistance, slowing the descent; longer strings allow more air to be caught under the canopy, creating more drag; a small hole in the top of the canopy may help the parachute come down straighter, as the air is released evenly through the hole, releasing some pressure)

    3. What improvements would you make if you had more time and more materials? (answers will vary)

    4. Using words and pictures, explain how a parachute works to slow a falling object down. Include the words *gravity* and *drag* in your answer!

Lesson 11: Blast From the Past

(gravity pulls the object down toward Earth, but the parachute creates drag, a force that pushes up on the parachute, thereby slowing the fall)

14. In the remaining time, read aloud *How Do Parachutes Work?*, by Jennifer Boothroyd, which reinforces many of these ideas in a simple narrative that takes students from parachute jump preparation right through the jump and navigation back to Earth. The book reiterates the way that the parachute creates drag, which slows the descent of the pilot.

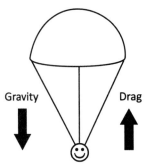

Evaluate: Space CER Template and Science Conference

1. Begin this activity with a think-pair-share. Ask students to take a few minutes to discuss some of the experiments that they've conducted on rockets and parachutes over the last two days. What were some of the things that they learned? After giving students several minutes to discuss, ask for volunteers to share.

2. Explain to students that when scientists have results from their experiments, they make claims, or statements, that answer the question that they were researching. Ask:

    - "What question were we researching when we tested rockets?" (What rocket design would give the highest launch?)

    - "What is a claim that you could make to answer that question?" (students will suggest different answers based on their findings)

    - "How are you making those claims? In other words, how do you know that those claims are true?" (our experiment results) Explain to students that those results are considered *evidence* to support their claim. Evidence is data that support a scientific claim. Data can be results from experiments, information from texts, observations, and so on.

    - "What is an example of evidence that can support a claim about the best rocket design?" (accept answers that give data from experiments)

3. Distribute copies of the Space CER Template handout (p. 385), and display one in the room. Have students read the question: "What is the best design for a (circle one) rocket/parachute?" Model for students that you are going to start with the rocket (circle the word *rocket*).

IT'S STILL DEBATABLE! USING SOCIOSCIENTIFIC ISSUES TO DEVELOP SCIENTIFIC LITERACY, K–5

4. Explain to students, "You are going to have a chance to complete these handouts with a partner or partners from your rocket or parachute experiments. Then, just as scientists do, we're going to have a conference where we share and discuss our findings. But first, I'm going to do an example for you."

5. Say, "Let's pretend that I found that a short rocket with a pointy nose flew the highest. What is my claim?" (that short rockets with pointy noses are the best design) Ask, "What's my evidence?" (students will say that you had data to support that in your results) Model writing as evidence: "I measured the heights of launches with different noses (flat or pointy) and different heights. The flat-nosed rocket flew medium." (*Note:* This is intentionally not evidence of that claim. It is to get students thinking and talking.) Students will likely point out that your evidence about the flat-nosed rocket doesn't support the claim about the pointy-nosed rocket! Thank your students for correcting you, and let them know that the evidence needs to be *appropriate* for the claim; in other words, it needs to match the claim that you're making. Erase your first evidence statement and write, "The pointy-nosed rocket went high." Ask students, "What do you think?" Allow students to discuss; they will likely point out that without knowing how it compared with other rockets, we still don't know if the evidence supports your claim. Perhaps the other rockets also went high. Thank students again and share that the evidence needs to be *sufficient*, or enough, to support the claim. Now write, "We ran three trials of rocket tests, each with different nose types (pointy, flat, and no nose) and the same height rocket (short). The pointy-nosed rocket went high. The flat-nosed rocket went medium. And the no-nose rocket went low. Is that appropriate and sufficient?" (probably yes)

6. Tell students, "Now, we need to tie it all together with reasoning. The reasoning explains how the evidence connects to the claim. It usually includes some sort of science principle, like the different things we learned about forces such as thrust, drag, and gravity, to help explain. What might my reasoning be in this example?" (allow students to make different attempts; answers will likely include connecting the idea that flying high is the way we measure a good rocket launch; for a rocket to go high, it needs to have enough thrust to overcome gravity and not get slowed down by drag, and the pointy nose may reduce drag, so, pointy noses may be the best design)

7. Ask students, "Is it possible for us to make different claims from the same experiment?" (yes!) "Why might that happen?" (different tests, different people testing, only one trial per test, interpreting the evidence differently, error in recording) Explain to students that this is why scientists share their

results—so that they can discuss them, challenge them, and see if the claims make sense based on the evidence and reasoning.

8. Have teams of students (based on their experiment groups) choose whether they'd like to work on rocket or parachute claims. See how many students want to work on each; try to divide students fairly evenly. Explain that they are going to simulate a science conference where scientists present their research to other scientists who have an opportunity to ask questions. Have the groups work on their Space CER Templates and then have other groups working on the same topic (rockets or parachutes) visit so that the team can present to them. The visiting groups should ask questions about the evidence to make sure it's appropriate and sufficient and that the reasoning makes sense. Then, the visiting groups will present.

9. Allow students to collaborate and discuss. While they're working, circulate around to the various groups to monitor and support their discussions.

10. With 10 to 15 minutes left in class, have the parachute groups visit the rocket groups, who will do brief presentations of their CER templates. Then, switch and allow the rocket groups to visit and present to the parachute groups.

11. Assess CER templates using the rubric on page 390. Figure 5.42 shows an example of a model (full credit) CER.

FIGURE 5.42.

Sample Completed Space CER Template for a Rocket

Question: What is the best design for a (circle one) (rocket) / parachute?

Claim: The best design for a rocket is to have a long pointy nose and small fins.

Evidence:

In our experiment, we tested flat noses, pointy noses, and no noses, all with small fins. We found that the rocket with the long pointy nose flew "high," the flat nose flew "medium," and the no nose flew "low."

Reasoning:

The rocket with the pointy nose and small fins flew highest. High flying means a good rocket design. The pointy nose probably had less drag. Drag is the force that slows things down in the air. That's probably why it was best.

Elaborate: Do We Still Need a Space Program? Congressional Committee Meeting

1. Show students the inside back cover of *Moonshot,* which gives a great deal of background information on the *Apollo 11* mission. Allow students to respond to the following questions using think-pair-share:

    - "Astronaut Neil Armstrong famously said as he stepped onto the Moon, 'That's one small step for [a] man, one giant leap for mankind.' What did he mean by that? Do you agree?" (answers will vary)

    - "Why do you think so many people on Earth watched the Moon landing?" (answers will vary)

2. Inform students that although space travel captured, and still captures, the attention and imagination of people, there is some debate as to whether a space program is still necessary. Ask students if they can imagine any reasons why someone might oppose a space program. (answers will vary)

3. Explain to students that they are going to have the opportunity to learn about and discuss this issue—as members of Congress! Inform students that one of the responsibilities of Congress is to determine whether various government-run programs should be funded. One such program is the space program, which is overseen by NASA. Explain that the question of whether or not to fund NASA has arisen, and it is their job as members of Congress to discuss and debate this question. To do so, they will need to learn a bit more about the issues.

4. Distribute copies of the Congressional Briefing and the Space Program Yes/No T-Chart handouts (pp. 386–388). Have students read the Congressional Briefing, either aloud as a class or with partners. Instruct students to complete the T-chart with a partner by writing the arguments for and against a space program ("yes" and "no") on the appropriate sides of the chart. Then, they should explain their own decision at the bottom of the page. Circulate around the room as students are working to ensure that they understand the task. Completed T-charts should include many of the answers shown in Figure 5.43.

Lesson 11: Blast From the Past

FIGURE 5.43.

Sample Completed Space Program Yes/No T-Chart

Yes	No
• It helped us to understand Earth's history as well as flight, weather, satellite communications, climate change, medical breakthroughs, and so on. • Research has also advanced inventions like cell phones, freeze-dried food, prosthetic limbs, and so on. • Natural resources might be found in space. • As Earth's population keeps growing, we need more places to live. • Humans should be pioneers and explorers. • Space exploration can help us find other life forms. • Exploring space can protect us from asteroids and hostile nations.	• Space programs are too expensive. • Money could be better spent on solving problems on Earth such as ridding the world of hunger or solving environmental issues. • Scientists who work on space programs could be working on big Earth-bound problems instead. • We haven't even explored most of the oceans on Earth; why are we exploring space? • Space exploration creates pollution and waste in space, including space junk. • Humans can bring germs to other planets and harm other life there. • It can be dangerous—the Challenger and Columbia disasters killed all of the crew.

5. When students are done, let them know that they are going to use a yes/no debate line to discuss this issue. Post the "Yes" sign (with the thumbs-up picture) on one side of the room and the "No" sign (with the thumbs-down picture) on the other. Put a line of masking tape approximately 3 feet long on the floor in the center of the room, so that it divides the room into two sides. (For young students, you can put the posters on the floor on either side of the line to provide a closer visual cue.) The room should look similar to Figure 5.14 (p. 140).

6. Write the following sentence frames on the board:

   • "I think that _____ because _____."

   • "I disagree with you because _____."

   • "According to the article/book/my personal experience, _____."

Lesson Plans

- "I think you should also consider_____."

7. Tell the members of Congress that you suggest organizing their discussion or debate in an efficient and enjoyable way. Remind them that there is *no wrong answer*. There are good reasons on both sides of the issue; the important thing is that they share their ideas in a thoughtful and respectful manner.

8. Explain that the question you'd like them to discuss is whether we still need a space program: either yes or no. If they think the answer is yes, they should go to the "Yes" side of the line. If they think the answer is no, they should go to the "No" side of the line. They may not straddle the line by putting one leg on either side. For now, you are asking them to choose a side.

9. When students are on their chosen sides, show them the sentence frames and ask them to share their opinions in that format. Remind them that they need to take turns and listen hard to other people's opinions. *Note:* If all the students go to one side of the room, you can take the other side and debate with them! See if you can persuade students to switch sides. You can ask students to do this as well. Let them know that they'll have the opportunity to write about their real opinions later, but for now, you'd appreciate them arguing for the other side—it will help them improve their debating skills!

10. Have students share their opinions, making sure that students justify their opinions (claims) using evidence from either the article, the *Moonshot* book, or their personal experiences. Also pay careful attention to whether you see anyone switching sides or scooting toward or away from the line. If you notice that, you can say, "I see that some members of Congress may be rethinking their positions on the issue, and that's fine. Would anyone like to share their thoughts on that?" Ask, "What arguments persuaded you to change your mind?"

11. When all students who wish to share their opinions have shared (and all students will have the opportunity to share their opinions in a letter as well), ask students to move to their final spots so that they can see where their classmates stand on the issue. Ask:

    - "What other information might be helpful for you to know about the NASA space program to help you make your decisions?" (students may be interested in learning about specific projects, budgets, or what other countries' space programs are like; encourage additional research!)

Lesson 11: Blast From the Past

- "Would it matter to you if the question were 'Do we still need a *crewed* space program?' In other words, do we still need to send people up in space or can we let robots do the flying and exploring?" Allow students to discuss this option.

Tell your Congress members that now they are going to have the opportunity to share their opinions with real members of Congress!

Evaluate: Opinion Letter to a Member of Congress

1. Distribute copies of the My Opinion Letter Outline handout (p. 389), and tell students that they are going to be using this form to organize their letters, which they are actually going to send to members of Congress. They can use their T-charts to assist them.

2. After their outlines are completed, have each student write a formal letter, either handwritten or using a computer. You and your students can identify your senators and representatives on the following websites: *www.senate.gov* and *www.house.gov*. Once you locate the senators (there are two per state) or the representatives you would like to contact, visit their websites to find their addresses. You may wish to write to their local addresses rather than their addresses in Washington, DC. The proper saluations and addresses are as follows:

To a Senator
The Honorable (full name)
(room #) (name) Senate Office Building
United States Senate
Washington, DC 20510
or local address

Dear Senator (last name):

To a Representative
The Honorable (full name)
(room #) (name) House Office Building
United States House of Representatives
Washington, DC 20515
or local address

Dear Representative (last name):

Note: Because of increased security around letters to members of Congress, I suggest putting all the letters together in one envelope (or divided among a few members of Congress), with a cover letter from you that explains your class project. Include a clear return address with your name (or Mr./Ms. _____'s third-grade class) and school name on the envelope. Maybe your class will even hear back from one or more members of Congress!

3. Before mailing, evaluate student letters by making sure all have the following components:

- A proper salutation (Dear Senator/Representative _____)
- An introductory statement expressing the student's opinion
- Three reasons and the source for each of those reasons (e.g., the article, the book, discussions)
- A closing statement reiterating the student's opinion
- A signature line (e.g., Sincerely, Respectfully)

Going Deeper

- Students can interview family and school community members about their memories of space missions or events. The class can create a documentary of the interviews!

- Students can investigate NASA's student website (*www.nasa.gov/audience/forstudents/index.html*) and identify topics of particular interest. They can then create displays for a gallery walk to share what they've learned.

- Another issue regarding space exploration is the idea of private citizens or companies traveling to space. Your class can research this issue and discuss the advantages and disadvantages of public versus private space travel.

- Students can create a full timeline of space exploration in your school hallway to educate the school community about space exploration history.

Apollo 11 Timeline Foldable

July 16, 1969 9:32 AM EDT	July 16, 1969 9:34–9:41 AM EDT	July 20, 1969 4:18/10:56 PM EDT	July 21, 1969 1:54 PM EDT	July 24, 1969 12:50 PM EDT
Lunar Liftoff	Lunar Landing and Moon Walk	Entry and Splashdown	Saturn V Separates	Launch

IT'S STILL DEBATABLE! USING SOCIOSCIENTIFIC ISSUES TO DEVELOP SCIENTIFIC LITERACY, K–5

Rocket Launches and Newton's Laws

Name: _____ Date: _____

Write the names of the forces acting on the spacecraft next to the letters.
(Hint: One of the forces is used twice!)

Word Bank
Gravity
Normal Force
Thrust

1. *Apollo 11 on the launchpad before liftoff.*

A
B

The forces are (circle one) balanced unbalanced

2. *Apollo 11 lifting off the launchpad.*

B
C

The forces are (circle one) balanced unbalanced

When an object is at rest, the forces acting on it are _____.

When an object is in motion, the forces acting on it are _____.

Name: _____ Date: _____

Blast Off!

Rocket Testing Data Sheet

Today we are going to test the effects of different rocket designs on launch height. You and your partner can build and test three different rocket designs. Enter your data in the chart below. Use H for high, M for medium, and L for low flights.

Safety First:

- Keep your goggles on.
- Step away from the launch pad once your rocket is set.
- Do not eat or taste the effervescent tablets.

Rocket Drawing	Rocket Description (fins, nose cone, rocket height)	Launch Height (H, M, or L)	Observations/ Notes
Design #1			
Design #2			
Design #3			

1. Which design flew the highest? Why?

   _____

   _____.

2. Which design flew the straightest? Why?

   _____

   _____.

Rocket Parts

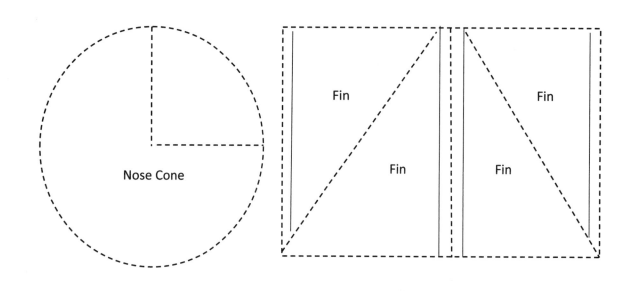

Use this space for the body of the rocket. Try rockets of different heights.

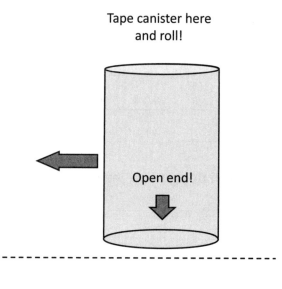

Scientist Name: _____   Date: _____

Where Did the Bubbling Tablets Go?

When an effervescent (bubbling) tablet is dropped in water, it seems to disappear ... but does it? Work with your teammates to design an experiment that uses weight to determine what happens to a bubbling tablet when it reacts with water. You can use any or all of the materials below:

Water
Small Cup
1 or 2 Bubbling Tablets
Zipper-Seal Plastic Bag
Balance or Electronic Scale

Write a stepwise procedure below. Your procedure should be detailed enough that any other group can follow your directions!

Our Procedure:

I think that _____ because
(what you think you will find out from your experiment)

_____.
(why you think so)

IT'S STILL DEBATABLE! USING SOCIOSCIENTIFIC ISSUES TO DEVELOP SCIENTIFIC LITERACY, K–5

Use this box to record your data. Remember, you might need to record data **before** and **after** you add the tablets to the water!

Our Data:

Questions:

1. Where do the bubbling tablets go when they are added to water? How do you know?

   _____

   _____

2. Did your group or another group use the plastic bag in the experiment? Was it helpful? Explain.

   _____

   _____

3. Circle the answer:

   The total weight of the substances before they are combined is

   about equal to / less than / greater than (circle one)

   the total weight of the substances after they are combined.

Name: _____ Date: _____

Parachuting Pretzels

Challenge

Your challenge is to build a parachute that allows you to drop a pretzel from a high height without breaking upon landing.

The pretzel should not be covered with any materials. It must hook onto a paper clip and swing freely.

Procedure

1. Design and build your first parachute. Test it from 50 cm high (with the bottom of the pretzel at 50 cm). Enter your result in the table below.

2. If your test is successful (the pretzel lands safely), try dropping it from 100 cm high. Continue this procedure until you've reached your maximum height for a successful drop.

3. Revise your parachute to see if you can go even higher!

Record whether your pretzel landed safely (☺) or not (☹).

Diagrams of Our Parachutes	50 cm	100 cm	150 cm
1.			
2.			

IT'S STILL DEBATABLE! USING SOCIOSCIENTIFIC ISSUES TO DEVELOP SCIENTIFIC LITERACY, K–5

Questions about Your Parachutes

1. Which materials worked best for making a parachute? Why do you think these materials worked well?

   _____
   _____
   _____

2. What are the characteristics of a successful parachute? (Hints: big or small canopy, thin or thick material, long or short strings, holes or no holes in parachute?)

   _____
   _____
   _____

3. What improvements would you make if you had more time and more materials?

   _____
   _____
   _____

4. Using words and pictures, explain how a parachute works to slow a falling object down. Include the words *gravity* and *drag* in your answer!

   _____
   _____
   _____

Name: _____

Date: _____

Space CER Template

Question: What is the best design for a (circle one) rocket/parachute?

Claim:

Evidence:

Reasoning:

IT'S STILL DEBATABLE! USING SOCIOSCIENTIFIC ISSUES TO DEVELOP SCIENTIFIC LITERACY, K–5

Congressional Briefing

Dear Member of Congress,

You will be discussing the following question at our next meeting:

Do We Still Need a Space Program?

Please use the article below, and any other resources that are permitted by the committee chairperson (your teacher), to help you prepare.

Thank you,

Congressional Secretary

In 1958, Congress established the National Aeronautics and Space Administration (NASA). Its mission was to explore peaceful (nonmilitary) applications of space science and technology. When President John F. Kennedy challenged the United States to "choose to go to the Moon," NASA led the way with Project Apollo, achieving its goal of reaching the Moon in 1969, with Apollo missions continuing until 1972.

A few of NASA's many other notable projects have included the Space Shuttle program, which began in 1981 and continued for nearly 30 years. The Space Shuttle helped build the International Space Station, which continues to conduct experiments in space today. NASA's Hubble Space Telescope provided data that helped us understand how our universe formed, while several of NASA's discoveries have helped us with many things on Earth. For example, NASA has contributed to our understanding of flight, weather, communication satellites, electronics, global climate change, and life sciences. NASA's research has also helped create new inventions such as cell phones, freeze-dried foods, and prosthetic limbs.

Another reason for supporting the space program is to find natural resources that might run out on Earth. As the human population on Earth keeps increasing, looking to space for materials and possible places to colonize will become increasingly important. The space program also can protect us from asteroids by warning us and redirecting them away from Earth. And some people feel that the space program protects us by watching out for nations that might be hostile toward the United States. Finally, many people feel that NASA should be supported because they believe that humans are driven to understand their world and be pioneers. They say that space inspires us.

However, a space program comes with great cost. The annual budget in 2018 was over $20 billion. While this is actually a very small percentage of the government's budget, some people feel that it could be better spent on challenges here on Earth, such as medical research, solving environmental problems, and ridding the world of hunger. Some also argue that the scientists who work on the space program could be better used to solve these Earth-bound problems. Critics also point out that there are places on Earth, such as oceans, that have barely been explored. They wonder, "Why are we going into space when we've got plenty to explore on Earth?"

Space exploration also creates waste. The rocket fuels that are used to send rockets into space contribute to pollution, and the industries that create the rocket materials themselves create pollution on Earth. There are over 500,000 pieces of "space junk," including small pieces of spacecraft, discarded launch vehicles, and other human-made items, orbiting Earth as a result of space travel. It is also possible that when humans enter the environments of other planets, they will bring germs that could harm any life that might be there. While NASA has an Office of Planetary Protection, which works on issues like avoiding contamination (spreading germs), there is always a possibility of causing harm to other environments. Some people think that humans would simply create many of the same environmental challenges on other planets that they have caused on Earth.

And finally, another argument against space travel is that it can be dangerous; the explosions of the space shuttles *Challenger* in 1986 and *Columbia* in 2003, both of which killed all seven crew members onboard, are examples of this. One compromise solution is to focus on robotic rather than human space travel, which lowers costs (since the robots don't need to return to Earth, the flight is only one-way), eliminates the risks to astronauts, and minimizes the possibility of causing harm to other planets.

Congress Member's Name: _____   Date: _____

Space Program Yes/No T-Chart

Question: Do we still need a space program?

Yes	No

My conclusion is that we do / don't (circle one) need a space program because

_____

_____

_____

_____

_____

Name: _____   Date: _____

My Opinion Letter Outline

I think that we do / don't (circle one) need the space program. Here are my reasons.

Reason #1	Reason #2	Reason #3

That is why I think that we _____

_____

_____

_____

_____

_____

_____

Rubric for Space CER Template

	0 pts.	1 pt.	2 pts.
Claim	Student does not make a claim.	Student makes an inaccurate or incomplete claim.	Student makes an accurate and complete claim.
Evidence	Student does not provide evidence.	Student provides inappropriate or inadequate evidence.	Student provides appropriate and adequate evidence.
Reasoning	Student does not explain reasoning.	Student provides inappropriate, illogical, or inadquate reasoning.	Student provides appropriate, logical, adequate reasoning.

Total: ___ / 6 pts.

Lesson 12

"Mined" Your Own Business

Was the California Gold Rush Good for the United States?

Suggested Grade Levels

3–5

Driving Questions

- What are properties of matter?
- What are the impacts of human development on natural resources and people?

Lesson Overview

Students simulate panning for gold within the context of the California gold rush to examine the properties of pyrite, or fool's gold, and compare them with the properties of real gold. Additionally, they analyze the impacts of the gold rush on the people and environment of California and decide whether the overall impact of the event was good or bad (or both) for the United States. In a culminating activity, students create a journal written from the perspective of a gold prospector to review and assess what they learned.

Connecting to the *NGSS*

(See full alignment in Table A.12 on p. 515.)

- ESS3.C: Human Impacts on Earth Systems
  - Human activities in agriculture, industry, and everyday life have had major effects on the land, vegetation, streams, ocean, air, and even

outer space. But individuals and communities are doing things to help protect Earth's resources and environments. (5-ESS3-1)

- PS1.A: Structure and Properties of Matter
    - Measurements of a variety of properties can be used to identify materials. (5-PS1-3)

Societal Issues

Environmental Concerns, Discrimination, Property Rights

Nature of Science

- Scientific Knowledge Is Based on Empirical Evidence
    - Science uses tools and technologies to make accurate measurements and observations.

CCSS Connections

- English Language Arts
    - W.5.7. Conduct short research projects that use several sources to build knowledge through investigation of different aspects of a topic.
    - W.5.9. Draw evidence from literary or informational texts to support analysis, reflection, and research.
- Mathematics
    - MP.2. Reason abstractly and quantitatively. (5-ESS3-1)
    - MP.4. Model with mathematics. (5-ESS3-1)
    - MP.2. Reason abstractly and quantitatively. (5-PS1-3)

NCSS Connections

- Theme 2: Time, Continuity, and Change
    - Describe how people in the past lived, and research their values and beliefs.
- Theme 7: Production, Distribution, and Consumption

- Examine and evaluate different methods for allocating scarce goods and services in the school and community.
- Theme 8: Science, Technology, and Society
    - Research a scientific topic or type of technology developed in a particular time and place, and determine its impact on people's lives.

C3 Framework

- Dimension 3: Developing Claims and Using Evidence
    - D3.4.3-5. Use evidence to develop claims in response to compelling questions.
- Dimension 4: Communicating Conclusions
    - D4.2.3-5. Construct explanations using reasoning, correct sequence, examples, and details with relevant information and data.

UDL Toolkit

Multiple Means of Engagement	Multiple Means of Representation	Multiple Means of Action and Expression
The mining activity and pyrite observations allow for varied demands of complexity, letting students participate while learning from peers.	Using a concept map, Venn diagram, and realistic budgeting activity guides information processing and visualization.	Data recording, math, and writing in a journal give students multiple opportunities to express what they have learned.

Suggested Schedule and Sequence

- Day 1: **Engage** with Gold Concept Map and *The California Gold Rush* read-aloud (part 1), and **Explore** with Stake Your Claim!
- Day 2: **Explain** with Stake Your Claim! data, Venn diagram to compare gold and pyrite, *The California Gold Rush* read-aloud (part 2), and What Could You Buy During the Gold Rush?
- Day 3: **Elaborate** with *The California Gold Rush* read-aloud (part 3) and Weighing the Evidence graphic organizer

- Day 4: **Evaluate** with California Gold Rush Journal

Materials

(per class)

- Pasta (6–8 pounds of different sizes)
- Bedsheet or 2 large plastic garbage bags (to place under the mine)
- Pyrite (approx. 1 lb) (available at *www.nature-watch.com*)
- Small vial of gold flakes (available at *www.walmart.com* or *www.nature-watch.com*)
- Sticky notes
- Balances with gram weights
- Steel nail

(per team of 3–4)

- 3 feet of yarn, tied in a loop (use a different color for each team)
- Pie pan
- Penny
- Pencils
- Rulers
- Magnifiers

(per student)

- Indirectly vented chemical splash goggles and a nonlatex apron

Student Handouts

- Stake Your Claim!
- What Could You Buy During the Gold Rush?
- Weighing the Evidence
- California Gold Rush Journal

Lesson 12: "Mined" Your Own Business

Safety Notes

1. All students must wear safety goggles and nonlatex aprons during the setup, hands-on, and takedown phases of the activity.

2. Use caution when working with glass or plasticware, which can cut skin.

3. Immediately pick up any items dropped on the floor to avoid a slip-and-fall hazard.

4. Immediately wipe up any spilled water on the floor to avoid a slip-and-fall hazard.

5. Never taste any food, substance, or chemical used in the activity.

6. Use caution when working with sharp tools or materials to avoid cutting or puncturing skin.

7. Wash hands with soap and water after completing this activity.

Media

Book

- *The California Gold Rush,* by Mel Friedman

Background for Teachers

The California gold rush was triggered on January 24, 1848, by the discovery of gold in a river in Northern California by a carpenter named James Marshall. He and his sawmill business partner, John Sutter, tried to keep their discovery a secret, but that didn't last long! Soon, people from all around the United States and the world came to try to find gold and strike it rich. Many became known as "forty-niners" because of the swell of people who arrived in 1849. The gold rush helped bring new cities and industrialization to the western United States. In addition, many innovations, from the Wells Fargo wagons that carried mail and gold to Levi's blue jeans, were developed. Yet, while stories about the gold rush are sometimes romanticized as being a time of adventure and fortune, most prospectors who went to California didn't become rich, and the gold rush was a difficult time for many different groups of people. For example, many Native Americans became prospectors but were often driven off the land. Many Spanish-speaking settlers from Mexico, called Californios, were driven out as well. Chinese prospectors often needed to pay special mining taxes, and anti-Chinese laws arose. And although California did not allow slavery, many African American slaves were brought from states that did still allow slavery and were forced to mine.

There were many methods for mining gold, but panning was one of the simplest. Miners would put dirt that they found along rivers in a shallow pan and gently rinse it under the water. While lighter sediment like sand and gravel would wash away, any gold pieces that were present would remain behind, as gold is very dense and heavier than many other materials for its size. Miners who found gold by panning were said to have "hit pay dirt." The weight of gold was an important property to miners, as it helped them identify it. Scientists use the term *specific gravity* to describe heaviness in this way. The specific gravity of gold is 19.3. This means that it is 19.3 times as heavy as an equivalent volume of water, and gold is twice as heavy as pure lead! This "heavy" property of gold enabled panning to work.

Once miners thought they had found gold, they would sometimes use another property of gold to make sure it was real: They would bite it! Gold is a soft, malleable mineral, which makes it ideal for shaping into jewelry, coins, and other items. Most other rocks and minerals that the miners would find, including pyrite, or fool's gold, were much harder. Teeth are harder than pure gold, so miners could tell if the piece they found was real gold if they could bend or mark it by biting. Today, gold is typically mixed with other metals, such as copper, nickel, or zinc, to make it harder so that it can withstand usage. When you see a designation like "14 karat gold," this tells you how pure the gold is. A karat is a unit equal to 1/24 part of pure gold by weight, so 24 karat gold is 100% pure gold, and 12 karat gold is 50% pure gold. Fortunately for miners, pure gold is very soft (2.5–3.0 on the Mohs Hardness Scale), so they could do a quick and easy test for authenticity!

For a few years, panning and digging for gold sufficed for the miners, but in 1853, a new technology called hydraulic mining was introduced. In this type of mining, hoses were used to shoot strong streams of water at canyon walls and hillsides. The strong force from the water separated the various materials so that miners could extract the gold. But this type of mining did tremendous damage to the environment by breaking down hillsides and sending sediments into rivers. Clearly, the California gold rush was important in bringing people, industry, and attention to California, but it also had some negative impacts on the environment and on groups of people who were subject to discrimination.

During this activity, students compare the properties they observe in pyrite with the ones they observe and read about gold in order to distinguish between the two (in alignment with *NGSS* 5-PS1-3) in a manner similar to the miners. They also consider the environmental and human impacts of the gold rush and how those impacts may help us with decision making today.

Lesson 12: "Mined" Your Own Business

Additional Resources

Information on the gold rush

- *www.pbs.org/wgbh/americanexperience/features/goldrush-california*
- *www.history.com/news/8-things-you-may-not-know-about-the-california-gold-rush*

5E Lesson Plan

Advance Preparation: Pour the pasta on top of the bedsheet or garbage bags. Carefully hide the pyrite under and inside the pasta so that it isn't visible. Cover the mine so that students can't see it when they come into the room.

Engage: Gold Concept Map and *The California Gold Rush* Read-Aloud (Part 1)

1. Pass a small vial of gold flakes around the class and write the word *gold* on the whiteboard. Ask students to brainstorm words that come to mind when they think about gold. Write each word students suggest on a sticky note, and stick these on the board. Some common terms that students may suggest include *valuable, jewelry, coins, rare, glittery, mines, nugget, mineral, yellow, shiny, gold rush*, and *24 karat*.

2. Create a concept map about gold on the whiteboard. Ask students, "Are there ways that we could group these words into categories?" Create categories such as "Appearance," "Uses," and "History," and group the sticky notes underneath the appropriate headings (see Figure 5.44 on p. 398). Let students know that as they learn more about gold, they can add more notes. Ask, "Which of our words relate to gold's *properties,* the ways in which different materials appear or behave?" (students will likely suggest that the yellowish color and the shiny luster are its main properties) Read aloud pages 1–17 and 30–31 of *The California Gold Rush*, by Mel Friedman. Allow students to add new words to the concept map.

Lesson Plans

FIGURE 5.44.

Sample Gold Concept Map

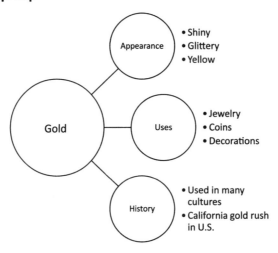

Explore: Stake Your Claim!

1. Explain to students that, working in teams, they are going to become prospectors and pan for gold in the class mine. Show students the pasta mine and explain that gold is hidden in the mine. Distribute a copy of the Stake Your Claim! handout (p. 404) to each student.

2. Divide students into teams of three or four. Ask one prospector from each team to come up to the mine with the team's yarn loop and place it where the team would like to mine. Tell students that this is called staking a claim (Figure 5.45).

FIGURE 5.45.

Staking a Claim

3. When all the groups have staked their claims, have a second miner from each team come up and dip the pie pan into the field within the team's claim (yarn loop) to find gold. This is called prospecting. Give each prospector a maximum of three chances (dips) to get a pan full of the pasta within his or her team's claim and look through it to find gold. *Tip:* This step works best if you have one student at a time come up to the mine so that everyone can watch to see if he or she

Lesson 12: "Mined" Your Own Business

finds any gold. Students can return the pasta to the mine and carry any gold they have found back to their team tables.

4. Rotate jobs and have teams repeat by staking claims and prospecting until all students have had a chance to be prospectors and claim stakers.

5. Once all the mining is completed, have teams observe their gold nuggets. Ask students to draw detailed pictures of their finds, using magnifying glasses. Then have them complete their data sheets, which include spaces in which to describe the listed properties. Teams should use a balance to determine the weight of each of their nuggets in grams.

6. Then, have teams use the Mohs Hardness Scale (Figure 5.46) to perform a simplified test. Students should test whether a nugget of their gold can be scratched by a fingernail. (It can't.) Next, have them check to see if it can be scratched by a penny, which is harder than a fingernail. (It can't.) You can then demonstrate that it takes something as hard as a steel nail to scratch pyrite (6.5 on the Mohs Hardness Scale). Scratch some pyrite with a steel nail to show that it does scratch. Real gold, because it is so soft (2.5–3.0 on the Mohs Hardness Scale), can be scratched by all three, even the soft fingernail. Therefore, we know that our gold isn't real gold—it's pyrite!

FIGURE 5.46.
Mohs Hardness Scale

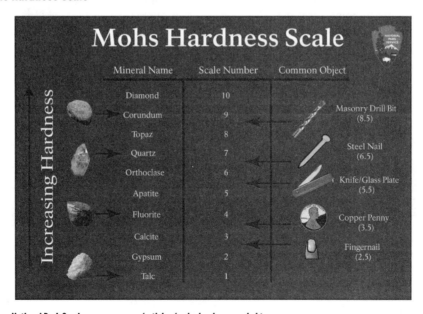

Source: National Park Service, www.nps.gov/articles/mohs-hardness-scale.htm.

Explain: Stake Your Claim! Data, Venn Diagram to Compare Gold and Pyrite, *The California Gold Rush* Read-Aloud (Part 2), and What Could You Buy During the Gold Rush?

1. Have teams report their results by sharing the data from their Stake Your Claim! handouts.

2. Then, draw a Venn diagram on the board. Write "Gold" in one circle and "Pyrite" in the other (Figure 5.47). Ask:

    - "How are pyrite and gold similar? How are they different?" Add to the Venn diagram as students respond.

FIGURE 5.47.

Sample Venn Diagram to Compare Gold and Pyrite

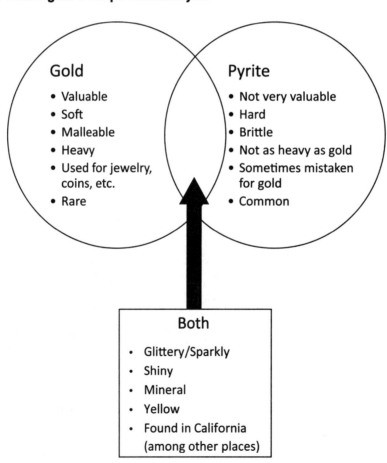

Lesson 12: "Mined" Your Own Business

- "How did the prospectors use their knowledge about the properties of gold to identify it?" (they knew how much it should weigh for a given nugget size, and they could check to see how soft it was)

3. Have each team report the weight of its total find. Write the weights on the board. Ask students, "If you had found that much weight in real gold, do you think it would be valuable? Let's find out *how* valuable!"

4. Read aloud pages 23–29 of *The California Gold Rush*, in which students learn about the cost of living during the gold rush, as well as the way miners lived.

5. Tell students that to understand the value of their find, they are going to play a game called What Could You Buy? Distribute copies of the What Could You Buy During the Gold Rush? handout (p. 405). Explain to students, "For our game, we're going to assume that 1 gram of pyrite equals $1. Your handout shows some of the prices during the gold rush, according to our book and some other sources."

6. Instruct students to work with their teammates to determine what they'd like to buy as a group with their pyrite. They should mark how many of each item they want in the third column and the total cost of each in the fourth column. The idea is to collaboratively make decisions without going over the team's budget! When teams are done, remind them that most of the forty-niners made between $20 and $30 per day during the high point of the gold rush, but much less in later years. The goods were very expensive to buy, and for the most part, only the people who sold goods to the miners got rich!

Elaborate: *The California Gold Rush* Read-Aloud (Part 3) and Weighing the Evidence Graphic Organizer

1. Read aloud pages 33–end of *The California Gold Rush*, in which students learn about the positive and negative impacts the gold rush had on the development of California, the environment, and various ethnic groups. The book describes hydraulic mining and notes that it has continuing effects on the environment even today. Ask:

   - "What is hydraulic mining? How does it work?" (high-pressure jets of water are used to break rocks and dislodge minerals; it was very good at getting gold, especially when much of the placer gold in the rivers was gone)

   - "What are some of the environmental impacts of hydraulic mining?" (it causes increased flooding, erosion, and sediment in the water)

- "Why are technologies like hydraulic mining used if they harm the environment?" (they also have good impacts; in the case of hydraulic mining, the new technology helped people earn money and stay in California) Introduce the term *trade-offs*. Explain that every new technology has trade-offs: There are both positive aspects and negative aspects.

- "What were some of the *positive* effects of the gold rush?" (quick development of California; many people, especially business people, became rich; new groups of immigrants came to the United States, adding to diversity)

- "What were some of the *negative* effects of the gold rush?" (harm to the environment from mining, especially hydraulic mining; African American slaves were brought in from states where slavery was legal and forced to mine; there was discrimination against Native Americans, Californios, and Chinese immigrants)

2. Distribute copies of the Weighing the Evidence graphic organizer handout (p. 406). Ask students to fill in the boxes on the balance to indicate the reasons they think that the gold rush was good or bad for the United States. Then, they should write and explain their decision whether it was good or bad at the bottom of the page.

Evaluate: California Gold Rush Journal

1. Distribute copies of the California Gold Rush Journal handout (pp. 407–411). Ask students to complete the journal from the perspective of a gold rush miner.

2. Evaluate students' journals using the rubric on page 412.

Going Deeper

- Students can compare and contrast the California gold rush with other resource booms, such as the Pennsylvania oil rush (1859), the Texas oil boom (1901–1940s), or the Klondike gold rush (1896–1899).

- Students can develop and perform a play about the California gold rush for other classrooms to reinforce the key concepts and various perspectives illuminated in this study.

Lesson 12: "Mined" Your Own Business

- Students can work in groups to investigate the impacts of the gold rush on Native Americans and Chinese immigrants by reading the PBS article at *www.pbs.org/wgbh/americanexperience/features/goldrush-value-land* and *www.pbs.org/wgbh/americanexperience/features/goldrush-chinese-immigrants*.

Prospector Name: _____   Date: _____

Team Name: _____

Stake Your Claim!

How many nuggets did your team find? _____

Choose one nugget and complete this data table, observing your nugget with a magnifier, weighing it with a balance, and performing a hardness test.

Property	Description
Color	
Texture	
Luster	
Shape	
Size (length and width in cm)	
Weight (in grams)	
Hardness (Does your fingernail or a penny scratch it?)	
Draw your nugget here:	

Prospector Name: _____   Date: _____

Team Name: _____

What Could You Buy During the Gold Rush?

Use this sheet to determine what your team could buy during the gold rush! We are going to assume that

1 gram of pyrite = 1 dollar ($1)

How many dollars did your team earn? _____

Enter the number of each item you would like to purchase in the "How Many?" column, then calculate the cost of that purchase and write it in the last column.

(Cost of Item × How Many = Cost of Purchase)

Item	Cost During the Gold Rush	How Many?	Cost of Our Team's Purchase
Meal for one person	$20		
Rice	$8/pound		
Coffee	$4/pound		
Eggs	50 cents each		
Butter	$20/pound		
Cheese	$25/pound		
Flour	$13/bag		
Beef	$10/pound		
Shovel	$36		
Blanket	$5		
Boots	$20/pair		

1. **List what your team was able to purchase with your find. Be sure to include how many of each you purchased.**

2. **Were you able to buy more or less than you thought? Explain.**

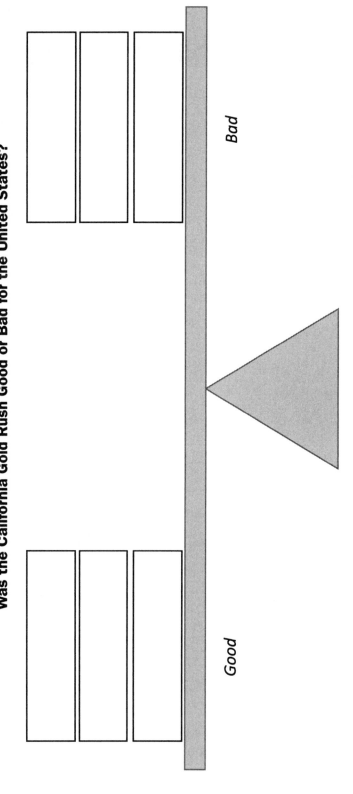

California Gold Rush
Journal

Miner: _____

Date: _____, 1849

A bit about me:

Some things I know about gold:

I'm glad/not glad I came to California because …

Here's how I pan for gold:

3

A typical day goes like this …

6

I can tell the difference between real gold and fool's gold because ….

4

People here represent many cultures, including ….

5

Rubric for California Gold Rush Journal

Criteria	Not Yet (1 pt.)	Emerging (2 pts.)	Secure (3 pts.)
Describes properties of gold	Student unable to record or communicate about the properties of gold.	Student records or communicates one property of gold.	Student records and communicates multiple properties of gold.
Describes how to distinguish gold from pyrite	Student cannot describe how to distinguish the two minerals.	Student makes inaccurate claims about how to distinguish them.	Student makes accurate claims about how to distinguish them backed by evidence.
Uses perspective writing to incorporate learnings from the lesson about the lives of the miners	Student is unable to write from the perspective of the miner.	Student is able to write from the miner's perspective, but inconsistently.	Student demonstrates consistent ability to write from the miner's perspective.

Total: ___ / 9 pts.

Lesson 13
Fueling Around

Which Energy Sources Are Best?

Suggested Grade Levels

3–5

Driving Questions

- What is energy, and how is it related to forces and motion?
- What are the pros and cons of different renewable and nonrenewable energy sources?

Lesson Overview

Students begin an exploration of energy by reading the true story of a Danish island that became completely energy independent, relying only on renewable energy sources including wind, solar, and biomass. They then make connections between energy, forces, and motion by building and testing pinwheels and Puff Mobiles, which use air puffs to simulate wind. This engineering challenge leads students into a study of renewable and nonrenewable energy sources, in which they learn about the pros and cons of each source. Finally, students become experts in and advocates for a particular energy source as they conduct a congressional debate to determine national energy recommendations. Issues of technological trade-offs, economics, and environmental justice are all examined.

Connecting to the *NGSS*

(See full alignment in Table A.13 on pp. 516–517.)

- PS3.A: Definitions of Energy

- The faster a given object is moving, the more energy it possesses. (4-PS3-1)
- ESS3.A: Natural Resources
  - Energy and fuels that humans use are derived from natural sources, and their use affects the environment in multiple ways. Some resources are renewable over time, and others are not. (4-ESS3-1)
- ESS3.C: Human Impacts on Earth Systems
  - Human activities in agriculture, industry, and everyday life have had major effects on the land, vegetation, streams, ocean, air, and even outer space. But individuals and communities are doing things to help protect Earth's resources and environments. (5-ESS3-1)
- ETS1.A: Defining and Delimiting Engineering Problems
  - Possible solutions to a problem are limited by available materials and resources (constraints). The success of a designed solution is determined by considering the desired features of a solution (criteria). Different proposals for solutions can be compared on the basis of how well each one meets the specified criteria for success or how well each takes the constraints into account. (3-5-ETS1-1), (secondary to 4-PS3-4)

Societal Issues

Environmental Costs/Benefits, Economics of Energy Production and Consumption, Environmental Justice

Nature of Science

- Science Is a Human Endeavor
  - Science affects everyday life.
  - Creativity and imagination are important to science.
- Influence of Science, Engineering, and Technology on Society and the Natural World
  - People's needs and wants change over time, as do their demands for new and improved technologies.

CCSS Connections

- English Lanugage Arts
    - RI.4.1. Refer to details and examples in a text when explaining what the text says explicitly and when drawing inferences from the text.
    - RI.4.3. Explain events, procedures, ideas, or concepts in a historical, scientific, or technical text, including what happened and why, based on specific information in the text.
    - RI.4.9. Integrate information from two texts on the same topic in order to write or speak about the subject knowledgeably.
    - W.4.7. Conduct short research projects that build knowledge through investigation of different aspects of a topic.
- Mathematics
    - MP.2. Reason abstractly and quantitatively. (4-ESS3-1)
    - MP.4. Model with mathematics. (4-ESS3-1)

NCSS Connections

- Theme 7: Production, Distribution, and Consumption
    - Evaluate how the decisions that people make are influenced by the trade-offs of different options.
- Theme 8: Science, Technology, and Society
    - Identify examples of the use of science and technology in society as well as consequences of their use.
- Theme 10: Civic Ideals and Practices
    - Evaluate positions about an issue based on the evidence and arguments provided, and describe the pros, cons, and consequences of holding specific positions.

C3 Framework

- Dimension 3: Developing Claims and Using Evidence

- D3.4.3-5. Use evidence to develop claims in response to compelling questions.
- Dimension 4: Communicating Conclusions
    - D4.1.3-5. Construct arguments using claims and evidence from multiple sources.
- Dimension 4: Taking Informed Action
    - D4.6.3-5. Draw on disciplinary concepts to explain the challenges people have faced and opportunities they have created, in addressing local, regional, and global problems at various times and places.

UDL Toolkit

Multiple Means of Engagement	Multiple Means of Representation	Multiple Means of Action and Expression
Having clear goals in the Puff Mobile Challenge, RAN chart research, and research project promotes sustained effort.	A picture book read-aloud, hands-on investigations, jigsaw readings, icons/symbols, and debate serve as multiple media for communicating information.	Use of team roles, word banks, RAN chart, and the jigsaw strategy facilitates managing information and resources.
Use of the jigsaw strategy allows students to become "experts" in a topic, increases motivation, and promotes collaboration and a sense of community.	The use of a congressional hearing makes explicit connections to social studies, thereby making transfer and generalization of information easier.	Drawings of Puff Mobiles, research on energy, and a debate provide different opportunities for students to express what they have learned.

Suggested Schedule and Sequence

- Day 1: **Engage** with *Energy Island* read-aloud, and **Explore** and **Explain** with Pinwheel Power
- Day 2: **Explore** and **Explain** with Puff Mobile Engineering Challenge
- Day 3: **Elaborate** and **Evaluate** with Get Energized! article and energy jigsaw activity with RAN chart
- Day 4: **Elaborate** with Energy Consumption Pie Chart think-pair-share and Congressional Committee Debate introduction and research

- Day 5: **Evaluate** with Congressional Debate rubric and My Energy Choice Statement

Materials

For Pinwheel Power

(per student)

- Pencil with eraser (in addition to the one used to write with)
- Straight pin
- Scissors
- Bead (optional)
- Safety glasses or goggles

For Puff Mobiles

(per class)

- Meter sticks
- Masking tape
- Marker

(per team)

- 4 Life Savers candies
- 3 straws
- 2 paper clips
- 1 piece of paper
- Tape
- Scissors
- Set of Puff Mobile Team Role Cards (p. 440) cut apart

(per student)

- Safety glasses or goggles

Lesson Plans

For Congressional Committee Debate

(per class)

- Set of Alternative Energy Expert Group Cards (p. 452) cut apart
- Plastic sandwich bag

Student Handouts

- Pinwheel Power
- Pinwheel Template
- Calling All Engineers! Puff Mobile Challenge
- Puff Mobile Challenge Data Sheet
- Puff Mobile Challenge Questions
- Get Energized!
- Get Energized! Questions
- Energy RAN Chart
- Expert Group Readings: Solar Energy, Nuclear Energy, Wind Energy, Geothermal Energy, Biomass Energy, Hydropower Energy
- Congressional Committee Debate on Alternative Energy
- My Energy Choice Statement

Safety Notes

1. All students must wear safety glasses or goggles during the setup, hands-on, and takedown phases of the activity.
2. Immediately pick up any items dropped on the floor to avoid a slip-and-fall hazard.
3. Use caution when working with sharp tools or materials to avoid cutting or puncturing skin.
4. Do not eat any food items used in the lab activity.
5. Wash hands with soap and water after completing this activity.

Lesson 13: Fueling Around

Media

Book

- *Energy Island,* by Allan Drummond

Background for Teachers

Energy makes things happen. It makes forms of transportation move, light bulbs (and the Sun) emit light, and fuels provide heat, and it even keeps our bodies alive. While scientists typically define energy as the ability to do *work,* this definition is quite confusing for students, as the type of work that scientists are talking about is different from what children think of as work (e.g., homework, schoolwork, parents' jobs). Work, in the scientific sense, is the ability to move something over a distance. So, it makes sense that when we sweep dirt from the floor, drive a nail into wood with a hammer, or carry a heavy bag of groceries into the house, we are doing work. But what is less obvious is that when wind blows a sailboat along the water, when a rocket blasts off into space, or even when your fingers press down on a keyboard, work is also happening and therefore energy is needed. It is also apparent that forces are closely related to energy, because when two objects interact (such as the wind and the sailboat), each one exerts a force on the other. These forces can transfer energy between the objects and lead to motion.

While energy has always been necessary for human existence, the last century has seen exponential growth in the energy needs of countries, towns, and even individuals. Industrialization, globalization, transportation, and communication have all created tremendous demand for energy. Most of these energy needs have been met by burning *fossil fuels.* Fossil fuels are formed from the decomposition of organic matter when plants and animals die and get trapped under Earth's surface. Over millions of years, under the pressure of the surface along with heat from Earth's core, the matter is compressed into coal, oil, or natural gas. These fossil fuels contain tremendous stored energy within their chemical bonds; these bonds are broken and release energy when the fuels are burned. Unfortunately, pollutants such as carbon dioxide, methane, and sulfur dioxide are also released, contributing to greenhouse gases, climate change, and acid rain. These fuels are becoming much less available because, although fossil fuels are constantly being formed, the time it takes for them to be produced is so much longer than the time it takes to extract and use them; therefore, they are considered *nonrenewable* energy sources.

Renewable energy sources, such as the Sun, wind, water, and biofuels, are able to replace themselves and can't be depleted. These renewable sources are referred to as alternative energies, since they are alternatives to fossil fuels. Solar, wind, biomass, geothermal, and hydropower energy are all considered alternative energies. Interestingly, there is some debate within the scientific community as to whether nuclear energy (which comes from uranium) should be considered renewable or nonrenewable. Uranium deposits on Earth are finite, which is an argument for nuclear energy being considered nonrenewable. Nuclear energy also has tremendous environmental impacts due to harmful nuclear waste. However, proponents of nuclear energy suggest that breeder reactors, which are nuclear reactors that are able to create more source material than they consume, are essentially renewable. Proponents also point to the low carbon emissions of nuclear power plants to argue that they are a form of alternative energy.

In this lesson, students learn through a read-aloud about an island in Denmark that became completely *energy independent,* meaning that the island uses only renewable energy. By relying on wind, solar, and biomass energy, the island no longer needed to have oil shipped to it or electricity carried to it from the mainland. The book also shows the way that individuals and communities can work together to solve energy and resource issues, a disciplinary core idea in the *NGSS*. Students then have the opportunity to build and test pinwheels and to design and test their own wind-powered cars, working with material constraints. The latter activity not only reinforces the concept of wind as a source of energy but also provides an engineering challenge and demonstrates key concepts in forces and motion. Students will see that by applying more wind to their cars by blowing on them harder, they are able to change the speed of the cars and the distance traveled. This is because the energy from the wind exerts a pushing force on the car due to the collision between the air and the cars. Students will also see that changing the direction of the wind changes the direction of the cars. While this activity highlights only one form of alternative energy, it sets the stage for students' research as they prepare for a congressional debate on alternative energies. The jigsaw strategy, which divides students into "expert groups" who read together and then teach their "home groups," is used for this research. Students are challenged to develop and present recommendations for alternative energy investment by the United States. By analyzing the pros and cons of different sources of energy, students will see that no energy source is without challenges and that decision making on this topic requires deep knowledge of the issue.

Additional Resources

National Energy Education Development Project

- *www.need.org/educators*

Lesson 13: Fueling Around

U.S. Energy Information Administration's Energy Kids page (students use this resource for their research)

- *www.eia.gov/kids*

National Renewable Energy Laboratory's educational resources

- *www.nrel.gov/about/energy-education.html*

Article on pinwheel science

- *www.scientificamerican.com/article/strong-wind-science-the-power-of-a-pinwheel*

Note: This lesson is based in part on the following:

> AIMS Education Foundation. 2004. Puff mobiles. In *Popping with power,* 114–119. Fresno, CA: AIMS Education Foundation.

5E Lesson Plan

Engage: *Energy Island* Read-Aloud

Show students the cover of the book *Energy Island,* by Allan Drummond, which depicts children holding pinwheels while large wind turbines stand in the background. Ask students, "What do you think this book might be about?" (responses will vary) Tell students that the book tells the true story of an island in Denmark. (Note that there is a map showing the location of Denmark in the book.) Read the book aloud, then ask:

- "At the beginning of the story, how did the island of Samsø get its energy?" (it got electricity from the mainland by using a cable, and oil was delivered by a tanker)

- "What kinds of things did people on the island use energy for?" (lights, heaters, hot water, cars)

- "What was Søren Hermansen selected to do?" (lead the project to make the island of Samsø energy independent)

- "What does energy independence mean?" (making their own energy on the island and using renewable energy)

- "What does renewable energy mean?" (energy from resources that will never run out, such as the Sun, wind, or water)

- "What was the problem with nonrenewable energy?" (coal, oil, and natural gas produce gases that trap heat in the atmosphere, which contributes to climate change; nonrenewable energy can run out and also was expensive for the island to buy)

- "How did the islanders on Samsø respond to the idea of becoming energy independent?" (the children were excited, but the adults were skeptical)

- "How did Søren Hermansen persuade people to make changes in their energy sources?" (he spoke to everyone on the island and met with different groups, until finally someone decided to build a wind turbine)

- "What happened on the night of the storm?" (all the electricity went out, except at the house with the wind turbine, so the islanders became interested in becoming energy independent, too!)

- "At the end of the book, how did the island of Samsø get its energy?" (it got electricity from wind turbines and solar panels to run cars, bicycles, and lights; burned biomass such as straw for heat; used canola oil for tractor fuel oil; and could even send electricity back to the mainland of Denmark!)

Explore and Explain: Pinwheel Power

1. Explain to students that they are going to explore how wind-powered devices, such as windmills, wind turbines, and pinwheels, use the wind as a source of energy. To do this, they are going to make and test pinwheels like the ones they saw in *Energy Island*. Students can work in pairs to assist each other, but they should each be provided their own set of materials to make their own pinwheels.

2. Distribute one copy of the Pinwheel Power handout (p. 435) and the Pinwheel Template (p. 436) to each student.

3. Students should follow the directions provided at the bottom of the template sheet. They will be cutting and folding the blades, poking holes through the dots, and using a pin to attach the folded blades to the pencil eraser. A bead (optional) may be used as a spacer to keep the blades from rubbing against the eraser.

4. Once students have all assembled their pinwheels, have them test the effects of wind direction and wind speed on the pinwheel using the Pinwheel Power data sheet.

5. After students test their pinwheels, discuss their findings as a group. Ask:

- "What effect does wind strength have on the speed of the pinwheel?" (stronger wind makes the pinwheel spin faster)

- "Why do you think this is so? Try to use the words *energy* and *force* in your answer." (the stronger wind has more energy than the gentle wind; the energy gets transferred to the pinwheel when the force of the wind touches and pushes the pinwheel)

- "What effect does wind direction have on the speed of the pinwheel?" (blowing into the blades of the pinwheel from the front makes the pinwheel turn more quickly than blowing from the sides or the back)

- "What combination of wind strength and direction gave the fastest spin?" (strong wind blown into the blades of the pinwheel from the front)

- "If you were setting up a wind farm, what should you think about in terms of *where* and *how* you set up the windmills or turbines?" (set up the wind farm in a place where there are a lot of strong winds; set up the windmills or turbines so that they are facing the prevailing wind or can turn into the wind)

Explore and Explain: Puff Mobile Engineering Challenge

1. Explain to students that today they are going to apply their knowledge of energy, forces, and motion to design, build, and test cars that run only on wind energy: Puff Mobiles!

2. Divide students into groups of three or four. Distribute copies of all three handouts (Calling All Engineers! Puff Mobile Challenge, Puff Mobile Challenge Data Sheet, and Puff Mobile Challenge Questions, pp. 437–439) to each student.

3. Read the Calling All Engineers! Puff Mobile Challenge sheet aloud or have student volunteers read. If possible, display the sheet using a document reader to provide students with an additional visual cue. You can also point out the various parts of the handout as the class is reading along.

4. After reading the handout, reinforce that the challenge is to make a car that can travel 3 meters with the fewest puffs to test the *efficiency* of the car. Explain to students that they will draw their designs and record their results on the Puff Mobile Challenge Data Sheet. Lay out three meter sticks on the floor or tabletop to show students the official start and finish lines for the testing. (You can also use one meter stick and show students how to measure 3 meters by marking the end of each meter and flipping the meter stick to the next meter.) *Tip:* You may wish to set up three or four sets of start and finish lines around the room as practice areas. You can simply place strips of masking tape 3 meters apart on the floor, and write "Start" or "Finish" on each piece. Older students can do this task themselves. Give teams the choice of their test tracks being on the floor or a tabletop. This allows for working at different levels, which can be helpful for students who may have difficulty sitting or kneeling on the floor.

5. Inform teams how much time is allotted for this activity (approximately 30 minutes), and show them the materials that they may use, although teams *do not* have to use all the materials. You can either prepare the materials for each group in advance or allow a Materials Manager from each group to use the list on the handout to go "shopping" for his or her group's materials.

6. Ask students:

    - "What is the *criterion,* or goal, for this challenge?" (to build a car that can travel 3 meters with the fewest puffs)

    - "What are the *constraints,* or limitations, for this challenge?" (teams can only use the materials given and need to finish in 30 minutes)

7. Give each student a Team Role Card (p. 440, cut apart in advance). The team roles for this challenge are Materials Manager, Test Manager, Puffer, and Spokesperson. Go over the responsibilities of each role with the class. The roles are described in detail on the cards.

8. Have students come up with team names and work collaboratively to design their Puff Mobiles and complete the Puff Mobile Challenge Data Sheet and Puff Mobile Challenge Questions handouts. As students work, remind them that they will likely need to revise their plans and retest their Puff Mobiles several times. A few minutes before the time is up, tell students how much time is remaining and remind them to answer all the questions on the handouts.

Lesson 13: Fueling Around

9. When you are ready for the Puff Mobile Challenge Derby, inform students that you will be calling up the Spokesperson and Puffer from each team, one team at a time, to describe their Puff Mobiles and test their best designs on the track.

10. Draw a chart like the one in Figure 5.48 on the board.

FIGURE 5.48.

Sample Chart for Puff Mobile Challenge Derby

Team Name	Number of Puffs to Reach 3 Meters

As the Spokesperson and Puffer come up, write their team name in the chart. Then ask the Spokesperson to do the following:

- Show and describe the team's final design
- Explain any changes the team made to the car to create their best design
- Describe any challenges that the team experienced while building and testing its car

11. Then, have the Puffer demonstrate the team's Puff Mobile (he or she may ask other teammates to assist or substitute). See Figure 5.49 on page 426. Enter the number of puffs it takes for the car to reach 3 meters. Repeat until all the teams have tested their Puff Mobiles and results have been recorded.

12. Congratulate all engineers, and ask students, "Did you notice any commonalities among the teams' best designs and challenges?" (answers will vary, but students will likely notice that lighter materials and a large surface to catch the wind were very helpful; getting the wheels to turn and having the car go straight are both common challenges)

13. Review the following answers to the questions on the Puff Mobile Challenge Questions handout:

- What makes your car move? Try to use the word *energy* in your answer. (our puffs provided the energy that was transferred to the car when the wind hit it, and the energy made the car move)

- What changes did you make to your car to make it travel farther on fewer puffs? (review the answers above; changes may include using lighter materials, using a surface to catch the wind, adjusting wheels so that they turn well)

- How did you keep your car traveling straight or straighten it out if it went sideways? Why do you think this worked? (puffing straight onto the car helps make it go straight; puffing from the sides helps move it back to the center if needed because changing the direction of the force changes the motion of the car)

- What other materials might have been helpful in building a Puff Mobile? (answers will vary)

- Complete the following paragraph using each term in the word bank only once:

   Puff Mobiles use <u>wind</u> energy to make them move. When we blow on Puff Mobiles, we are providing a pushing <u>force</u> that hits the car and transfers energy to it. When a car moves fast or far, it has more <u>energy</u> than a car that goes slowly or for a short distance. We can make our Puff Mobiles go straight or sideways by changing the <u>direction</u> of the puffs.

FIGURE 5.49.
Testing Puff Mobiles

Elaborate and Evaluate: Get Energized! Article and Energy Jigsaw Activity With RAN Chart

1. This activity consists of a whole-class read-aloud with questions, followed by a jigsaw activity (p. 36) in which students collaborate in "expert groups" to learn about alternative energy sources and then teach their "home groups" about their topic. Students complete RAN charts (p. 18) as they learn.

2. Inform students that they are going to work at becoming experts on energy today. Distribute copies of the Get Energized! article (pp. 441–442) and the accompanying questions handout (p. 443) to each student. If possible, display the article using a projector or document camera so that you can point to key phrases, model fluency as students follow along, and highlight key passages.

3. Read the article aloud to students, or call on student volunteers to read aloud. This article introduces the concept of energy and the pros and cons of fossil fuels. After reading the article, give students the option of working on their own or with a partner to complete the questions. Then, review the answers:

    - How does energy make things move? Include an example from the article in your answer. (energy is transferred from one object to another when they collide, and the energy creates a force that makes things move; examples include wind or moving air hitting sailboats, water, pinwheels, or Puff Mobiles)

    - How does the amount of energy an object has relate to the speed of the object? (the more energy an object has, the faster it moves)

    - How do fossil fuels provide energy? (fossil fuels contain stored energy from the Sun; when they are burned, the energy is released in the form of heat, light, and fast-moving air)

    - What are the advantages of fossil fuels? (fossil fuels are plentiful, fairly easy to get, and efficient at providing energy; also, the fossil fuel industry provides jobs for people)

    - What are the disadvantages of fossil fuels? (when burned, they release pollutants that contribute to climate change and acid rain; they also pollute through oil spills, pipe leaks, and oil rig accidents; newer ways of getting fossil fuels, such as fracking and mountaintop removal mining, can be very harmful to the environment; they're nonrenewable, so they will eventually run out)

    - What are alternative energy sources? List them in your answer. (alternative energy sources are sources other than fossil fuels; they include solar, wind, hydropower, nuclear, geothermal, and biomass)

4. Inform students that they are going to be working together to become experts in a certain alternative energy source and then teach their teammates using a strategy called jigsaw (because they are like the pieces of a jigsaw puzzle that will come together to form a whole picture).

Lesson Plans

5. Divide the class into "home groups" of six students. Don't worry if your class can't be divided evenly by six—just divide the class as evenly as you can with groups as close to six as possible. For example, if you have 17 students, make two groups of 6 and one group of 5.

6. Have students in each group number off from 1 to 6. If there are more than six in a group, have students double up (e.g., have two students designated as 6). If there are fewer than six in a group, give students an extra number as necessary (e.g., designate one student as both 5 and 6).

7. Explain to students that they are going to be assigned a particular alternative energy to research. To learn about that energy, they are going to join with classmates from other home groups to form new "expert groups." After reading together and becoming knowledgeable about the alternative energy, they will return to their home tables to teach their home groups about it. By doing this jigsaw, they will be learning about six different alternative energies in a fun and efficient way!

8. Assign expert group topics based on students' number designations as follows:

    - 1: Wind Energy
    - 2: Solar Energy
    - 3: Nuclear Energy
    - 4: Biomass Energy
    - 5: Hydropower Energy
    - 6: Geothermal Energy

9. Distribute a copy of the Energy RAN Chart (pp. 444–445) to each student. Tell students that this is a graphic organizer that will assist them in gathering and sharing their research.

10. Project the Energy RAN Chart so that the enlarged version is visible to them. Point out to students that the rows in the chart correspond to the different alternative energy sources (the same as their expert groups), while the columns of the chart relate to what students know, learn, and wonder about each source.

11. Ask students to take a moment and try to fill in anything they think they know about each of the alternative energy sources. You can model the first

one by saying, "In this first box, I need to write what I think I know about wind energy. Hmm … Do I know anything about wind energy?" Then invite students to suggest some answers. For example, students might say, "Wind energy is what moves sailboats." You could then add "Moves sailboats" in the first box. If students offer that "wind energy is a renewable resource," you could add "Renewable" in the box. Give students a few minutes to write what they think they know about each of the energy sources, reassuring them that it's fine if they don't know anything about it because they are going to learn today! You can also allow students to turn and talk to a partner and discuss their ideas with each other.

12. Once students have completed the first column, explain that they will be completing the other columns when they do some reading research at their expert groups and when their classmates teach them afterward at their home groups. Take a moment to explain the columns as described in Table 5.3.

TABLE 5.3.

Explanations of the Columns in Energy RAN Chart

Confirmed	This means that what you thought you knew was correct. Write the information that was confirmed through your research or from the experts' lessons in this column.
Misconceptions	This means that something you thought you knew wasn't correct. Being aware of our misconceptions is very important because this helps us learn better. Learners often find out that things they thought were true aren't. Write any misconceptions you had in this column. For example, if you thought wind energy was nonrenewable, but you learned it was renewable, you would write, "Wind is renewable, not nonrenewable."
New Information	Write new information that you learned from your research or from the experts' lessons in this column.
Wonderings	Write things that you still want to know after you're done with the jigsaw activity in this column. We will share these at the end of the class so that we can follow up with further research to find out the answers.

13. Let students know that they will have 20 minutes to work with their expert groups. When they arrive at their expert group, they should read the passage silently to themselves first. They can ask an expert group buddy to assist

with any questions. When everyone is done reading, they should discuss the key ideas in the expert group reading, complete the row for their alternative energy source on their RAN charts, and practice what they would like to say when they return to their home groups to teach their groupmates about the resource. Stress that teaching does not mean simply telling students what to write on their charts; they should think about sharing the information in an interesting and informative way. Some "Big Ideas" to guide them in thinking about what they have learned about their energy sources are listed at the end of each expert group handout.

14. Send students to their expert groups by indicating which table will represent which group. For example, say, "This table is for all students who were assigned as number 1. You will be researching wind energy." Place copies of the Wind Energy Expert Group Reading handout on the table, enough for each student in the group. (For all Expert Group Reading handouts, see pp. 446–451.) Do the same for the remaining expert groups. Have students get started by reading silently. Once discussions begin, circulate around the room and check that students are sharing ideas, entering information on their RAN charts, and practicing the lessons they will teach their home groups.

15. At the end of the 20 minutes, have students return to their home groups. Students will then take turns teaching their peers about their alternative energy sources. Reinforce the "Big Ideas," and remind students that they do not have to write everything they hear; just listen carefully, ask questions if they wish, and complete their RAN charts. As students are sharing, circulate around groups to listen to presentations, clarify any misunderstandings, and informally assess the RAN charts.

16. If time allows, ask students to share one item from their "New Information" column with the class. Encourage students to listen carefully so that items aren't repeated. Items from the "Wonderings" column can also be shared in this manner. Reassure students that they will have more time to research alternative energy sources and answer their questions. Have students keep their RAN charts for reference during the next activity.

Elaborate: Energy Consumption Pie Chart Think-Pair-Share and Congressional Committee Debate Introduction and Research

1. Begin by showing students the pie chart in Figure 5.50, either by drawing it or using a document camera. *(Note:* The word *petroleum* refers to crude oil that is extracted from the ground, as well as the products that are made from

Lesson 13: Fueling Around

refining it, which include diesel fuel, gasoline, heating oil, and jet fuels. This does *not* include vegetable or cooking oils. Vegetable oils *can* be used for fuel, as students learned in the book *Energy Island*. When you discuss fossil fuels and use the word *oil*, be sure to specify that you mean petroleum (oil from the ground), as students may be more familiar with cooking oils than with crude oil or its products.)

FIGURE 5.50.

U.S. Energy Consumption by Source in 2017

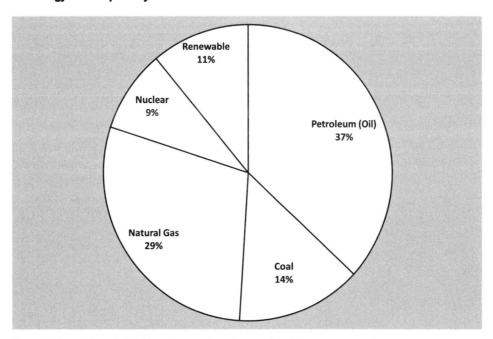

Source: U.S. Energy Information Administration, www.eia.gov/energyexplained/?page=us_energy_home.

2. Point out each of the "slices" of the pie and read the percentages for each of the energy types. Then say, "I'd like you to take a few moments to think about this pie chart. Then I'd like you to pair up with a partner and discuss, 'What information does it convey? Does anything catch your attention?'" Give students a few minutes to share with a partner. Afterward, ask students to volunteer what they understand or find interesting about the pie chart. Responses will differ, but here are some the follow-up questions you might want to include (if they aren't addressed by students):

- "What is the subject of this pie chart?" (energy consumption in the United States in 2017) "What does *consumption* mean?" (use)

- "What was the source of most energy used in the United States in 2017? What percentage was it?" (petroleum [oil], 37%)

- "What was the next biggest source, and what percentage was it?" (natural gas, 29%)

- "Which energy source had the smallest percentage?" (nuclear, 9%)

- "What does *renewable* energy mean?" (energy sources that can replace themselves)

- "How does energy use in the United States compare with the island of Samsø that we read about in *Energy Island*?" (Samsø became energy independent, using 100% renewable energy; the United States only gets 11% of its energy from renewable sources)

3. Next, introduce the Congressional Committee Debate by informing students that they are now members of Congress. One of the jobs of Congress is to advise the president on issues of national importance. Tell students that the president has expressed concern about the United States' dependence on fossil fuels (petroleum, coal, and natural gas) and has asked for their advice on which alternative energy might be best for the country to invest in. Since they are now experts in this area, thanks to the jigsaw activity, they are to debate the issue among themselves and then prepare a report for the president.

4. Distribute copies of the Congressional Committee Debate on Alternative Energy handout (p. 453). The handout describes the purpose, procedure, and time frame, and it provides additional suggested websites that students might wish to use to prepare.

5. Divide students by their expert groups from the RAN chart research. However, for this debate, each group will be representing a *different* source of energy. You can use the Alternative Energy Expert Group Cards (p. 452, cut apart in advance) to randomly assign each group a new energy source (perhaps by putting the cards in a plastic sandwich bag and having a representative from each of the groups pick a card). *Note:* While students might want to stick with the sources they already became experts about, it is helpful to push students a bit to learn new information and perhaps use their other knowledge from the RAN chart activity to develop counterarguments for other teams.

Lesson 13: Fueling Around

6. Give students time to prepare for their debate. When you are ready to begin the debate, follow the sequence described on the handout. After all teams have presented their opening statements, allow 5 to 10 minutes for students to ask questions and politely challenge each other's energy sources. Provide students with questioning frames such as "Can you please explain why you think _____?" "Did you consider _____?" or "Can you please clarify _____?" (You can also provide each team with a copy of Sentence Frames for Arguments on p. 26.)

7. To promote students' use of evidence in their arguments challenging other team's plans, you can write the following frame and example on the board, or you might wish to use the CES template in Figure 3.9 (p. 22):

    Claim + Evidence + Source

    Example: We think that wind power is the best alternative energy for the country (claim). Wind energy is renewable, is very clean, and is efficient at creating electricity (evidence), according to our article (source).

    (*Note:* You can remind students of the many possible sources from which they can glean information by referring to the Sources of Evidence template in Figure 3.14 on p. 27.)

8. At the conclusion of the debate, have Congress (all students) vote on the energy that they now feel most strongly about as being a good choice, except that they can't vote for the energy source they just represented. This prompts students to be a bit more thoughtful about their choices.

9. Afterward, congratulate your members of Congress on their fine efforts. Ask students:

    - "What did you like about this activity?" (accept all answers)
    - "What didn't you like about the activity?" (accept all answers)
    - "Were there any arguments that were particularly persuasive for you? Why?" (accept all answers, but focus on the use of evidence as being persuasive)
    - "Did any of you change your mind about any energy sources through this study?" (accept all answers)

Evaluate: Congressional Debate Rubric and My Energy Choice Statement

1. Assess each team using the rubric on page 455.

2. Distribute copies of the My Energy Choice Statement handout (p. 454). Have students complete the handout by choosing the alternative energy they feel is best for the future, writing a statement that includes at least one evidence-based argument to support their opinion, and drawing their energy choice at work. Post these around the school so that other school community members can learn about energy!

Going Deeper

- Students can interview the school maintenance staff about energy usage and efforts at sustainability in the school.

- Students can create maps depicting where in the country and the world different energy sources are prevalent.

- Students can do additional research on *Energy Island* to see what has been happening on Samsø since the book was written.

Name: _____   Date: _____

Pinwheel Power

1. What effect does *wind strength* have on the speed of your pinwheel? Blow directly onto the front of the pinwheel to find out, and fill in the chart below.

Wind Strength	Pinwheel Spin Speed (Fast, Medium, or Slow)
Strong wind	
Gentle wind	

2. What effect does the *wind direction* have on your pinwheel's speed and the direction of its spin? Try to keep the wind strength the same for each of the tests. Fill in the chart below with your findings.

Wind Direction	Pinwheel Speed (Fast, Medium, or Slow)	Pinwheel Direction (Clockwise C or Counterclockwise ↻)
Blowing onto the **front** of pinwheel		
Blowing from the **left side** of the pinwheel		
Blowing from the **right side** of the pinwheel		
Blowing from the **back** of the pinwheel		

What combination of wind strength and wind direction made your pinwheel spin the fastest?

_____

Why do you think this is so?

_____

_____

Name: _____ Date: _____

Pinwheel Template

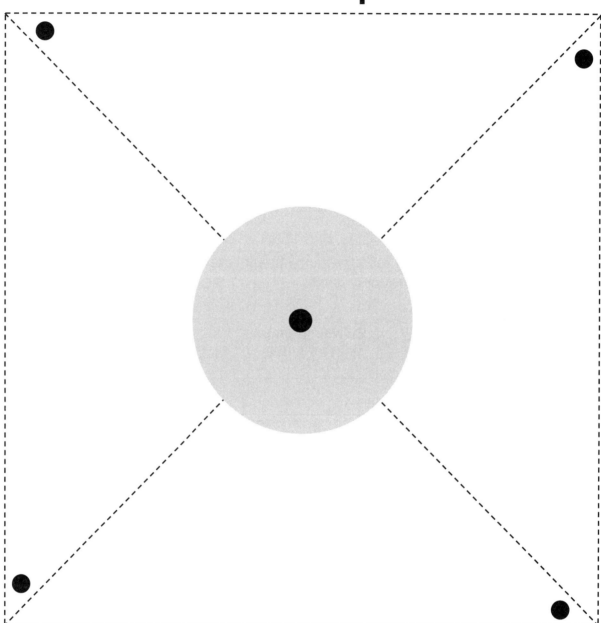

1. Cut along the dotted lines.
2. Poke holes through the black dots with a pin or pencil.
3. Bring the corners of the triangles with the dots to the center. Try to line up the holes.
4. Push a pin through the holes and then into a pencil's eraser.
5. Try blowing on the pinwheel. If it doesn't spin easily, try putting a bead on the pin between the pinwheel and the eraser.

Calling All Engineers!

Puff Mobile Challenge

Dear Engineers:

Your challenge is to make a car using only the materials listed below. Your car needs to be wind-powered so that it will be environmentally friendly. Therefore, you can only make your car move by blowing on it!

Materials

4 Life Savers candies

3 straws

2 paper clips

1 piece of paper

Tape

Scissors

We will test your car by seeing how many puffs it takes for your car to travel 3 meters. The most *efficient* cars will be able to travel 3 meters with the *fewest* puffs!

Use your Puff Mobile Challenge Data Sheet to plan your car. Then test your Puff Mobile at one of the designated testing areas. Keep revising and testing your design until you get your final, best design. Your best design will compete in the Puff Mobile Derby!

Complete the Puff Mobile Challenge Data Sheet and Questions when you are done testing. Remember these rules:

1. You must use only the materials on the list to make your Puff Mobile. (You do not have to use all them.)

2. Your Puff Mobile must travel using only wind (puffs).

Good luck!

Engineer Name: _____   Date: _____

Team Name: _____

Puff Mobile Challenge Data Sheet

1. Plan your Puff Mobile in the box below. Be sure to draw and label all the parts.

Our First Puff Mobile Design

---

2. Build and test your Puff Mobile! How many puffs does it take to make your Puff Mobile travel 3 meters? _____

3. Can you make your Puff Mobile more *efficient?* In other words, can you make changes so that it travels farther on fewer puffs? Draw and label your final, best design here:

Our Best Design Traveled 3 Meters in _____ Puffs

Engineer Name: _____   Date: _____

Team Name: _____

Puff Mobile Challenge Questions

1. What makes your car move? Try to use the word *energy* in your answer.

   _____

   _____

2. What changes did you make to your car to make it travel farther on fewer puffs?

   _____

   _____

3. How did you keep your car traveling straight or straighten it out if it went sideways? Why do you think this worked?

   _____

   _____

4. What other materials might have been helpful in building a Puff Mobile?

   _____

   _____

5. Complete the following paragraph using each term in the word bank only once:

Puff Mobiles use _____ energy to make them move. When we blow on Puff Mobiles, we are providing a pushing _____ that hits the car and transfers energy to it. When a car moves fast or far, it has more _____ than a car that goes slowly or for a short distance. We can make our Puff Mobiles go straight or sideways by changing the _____ of the puffs.

Word Bank

direction	force
energy	wind

Puff Mobile Team Role Cards

Materials Manager

Make sure your team has all the materials on the list. Also, monitor your teammates' recording sheets to make sure everyone is recording his or her plans and test results. Collect all materials from your team at the end of the activity and help clean up your work area.

Test Manager

Set up the Puff Mobile at the starting line for every test. Count the number of puffs it takes for the car to reach 3 meters for each test. Help your teammates stay on task and help resolve any disputes.

Puffer

Use your puffing power to move the car from the start to the finish line. Try to keep your car as straight as possible using only puffs. Assign an additional puffer or puffers if you need assistance.

Spokesperson

Show and describe your team's final design at the Puff Mobile Derby. Explain any changes you made to the car to create your best design. Describe any challenges that your team experienced during Puff Mobile design or testing.

Get Energized!

What Is Energy?

Energy makes things happen. It makes cars and trains move, turns on light bulbs, and even keeps our bodies alive! Energy makes a baseball take flight from a batter's swing, a sailboat move in the wind, and music travel to our ears. Energy is everywhere something is moving or changing. So whether you are kicking a ball, warming up soup, banging on a drum, flying in a plane, or turning on a light, energy is making it happen.

Energy Moves ... and Can Make Things Move!

Energy is able to move from object to object. When the wind blows, the moving air has energy. When the wind hits a sailboat's sails, the wind's energy is transferred to the boat, making the boat move. Waves on the water are also caused by the wind's energy. When the wind blows and comes in contact with the water, it transfers its energy to the water. The waves now have energy to move and crash up on the shore. When you blow on a pinwheel, your breath has energy that is transferred to the pinwheel, making it move. The same thing even happens when you blow on your Puff Mobile! In all these examples, the energy provided by the moving air (either the wind or your breath) creates a force that pushes on the object (sailboat, water, pinwheel, or Puff Mobile) and makes it move. The more energy that is transferred, the faster the object moves!

Energy Changes ... and Can Change Things!

Energy can also change form. For example, energy that was originally in the form of sunlight can be changed into stored energy in fossil fuels, which include coal, petroleum (oil), and natural gas. The stored energy doesn't make anything move or change until the fuels are burned. When that happens, energy from the fuels is released in the form of heat, light, or fast-moving air, which allows us to heat our homes, light our grills, power the engines of our cars, or convert water into steam to create electricity in power plants.

Energy Ups and Downs: Fossil Fuels

Humans have always needed energy to survive, but industrialization in the 18th and 19th centuries created much greater energy demands. The invention of engines that used coal, oil, or natural gas to provide power for transportation, factories, and power plants resulted in many important amenities of modern life, including cars, trains, heat and hot water, and electricity. Fossil fuels were critical to meeting increased energy demands, and their use made sense because they were plentiful, fairly easy to find, and very efficient at providing energy.

Today, fossil fuels still provide most of the world's energy, and the fossil fuel industry provides jobs for many people. These are important advantages of fossil fuels. But fossil fuels also add to pollution, as they give off pollutants when they are burned. Carbon dioxide, methane, and sulfur and nitrate gases, among others, contribute to global environmental challenges such as climate change and acid rain. Fossil fuels can also cause pollution from oil spills when tankers carrying oil crash into rocks or reefs, offshore oil rigs have accidents, or fuel pipelines leak. In addition, some of the newer ways that fossil fuels are collected, such as fracking (injecting high-pressure water, sand, and chemicals into the ground to force oil or natural gas out) or mountaintop removal mining (removal of the summits or high ridges of mountains to access coal), have very serious environmental impacts. Fossil fuels are also nonrenewable, which means they can't replace themselves … or at least, they can't replace themselves nearly as quickly as they are used by humans. So, although fossil fuels have been a critical part of history and everyday life, it is important to consider some alternative sources of energy to try to minimize harm to the environment and ensure that we have energy for the future.

Alternative Energy

Alternative energy sources, which are energy sources other than fossil fuels, include solar, wind, hydropower, biomass, geothermal, and nuclear energy. Like fossil fuels, each of these energy sources has advantages and disadvantages. This makes decision making about energy very challenging. In the next activity, you will learn more about each of these alternative energies so that you can decide for yourself which energies you think are best for the future.

Name: _____ Date: _____

Get Energized! Questions

Answer these questions after you have read the Get Energized! article.

1. How does energy make things move? Include an example from the article in your answer.

   _____

   _____

2. How does the amount of energy an object has relate to the speed of the object?

   _____

   _____

3. How do fossil fuels provide energy?

   _____

   _____

4. What are the advantages of fossil fuels?

   _____

   _____

5. What are the disadvantages of fossil fuels?

   _____

   _____

6. What are alternative energy sources? List them in your answer.

   _____

   _____

Name: _____ Date: _____

Energy RAN Chart

Source of Energy	What I Think I Know	Confirmed (Yes, I was right)	Misconceptions (Oops!)	New Information	Wonderings
Wind					
Solar					
Nuclear					

Energy RAN Chart (*continued*)

Source of Energy	What I Think I Know	Confirmed (Yes, I was right)	Misconceptions (Oops!)	New Information	Wonderings
Biomass					
Hydropower					
Geothermal					

Solar Energy

Expert Group Reading

Solar energy is produced from the energy that comes from the Sun. The Sun's energy is considered renewable because the Sun is able to create its energy by continuously forming and breaking bonds between different gases. Solar energy can be used directly to provide heat or light for our homes. It can also be used to heat or cool homes and water through the use of solar thermal technology. In addition, by using special solar cells that can convert solar energy to electrical energy, solar energy can be used to meet all of a home's energy needs. Solar energy can also power cars, and solar panels can be placed on top of large buildings or out on solar farms to collect sunlight, convert it to electricity, and power communities. Solar energy is a very clean, renewable fuel, and there are many promising new technologies that are making it more efficient. But solar energy can be very expensive to install, and of course, it's not always sunny outside, so reliable systems must be available to store the energy for cloudy days.

Big Ideas to think about:

- How does this energy source work?
- What are the advantages of using it?
- What are the disadvantages of using it?
- Any other interesting facts?

Nuclear Energy

Expert Group Reading

Nuclear energy comes from the splitting of atoms, which are the building blocks of all matter, into smaller atoms. When the atoms are split apart, they release heat and radiation energy, which can be captured and used to change water into steam. The steam is then used to create electricity. The process of splitting the atoms is called fission. Atoms of the element uranium are used for the fission reaction, which takes place in nuclear reactors in nuclear power plants. Nuclear energy is very efficient, as one small fingertip-sized pellet of uranium can produce as much energy as 150 gallons of oil can produce. But nuclear energy also has disadvantages. Because radioactive wastes are produced by nuclear power plants, the safe disposal of those wastes is difficult. Also, accidents at nuclear power plants can happen as a result of mechanical malfunctions or natural disasters such as earthquakes. The release of radioactive materials into the air or water can be very dangerous or even deadly. Although uranium is a common element in Earth and nuclear power plants are very efficient, nuclear power is considered nonrenewable because it relies on the mining of uranium, which can eventually run out.

Big Ideas to think about:

- How does this energy source work?
- What are the advantages of using it?
- What are the disadvantages of using it?
- Any other interesting facts?

Wind Energy

Expert Group Reading

Wind is formed by the movement of air. Air moves when the Sun heats up the air and causes it to rise. Cooler air comes in to take the place of the warmer air. All this movement of air creates wind. Wind has been used for centuries for sailboats, windmills, and to dry clothes on clotheslines. Modern wind power relies on large wind turbines with propellerlike blades that turn when the wind hits them. The turning of the blades causes the main shaft to spin a generator that creates electricity. Wind power is renewable because the Sun keeps making the air move around Earth. Wind power is also very clean, as no pollution is created. Wind turbines can be placed on land or in water. And in places where there is a lot of wind, wind energy is very efficient, as the savings from cheap electricity generation make up for the initial costs of the turbines. But wind energy also has disadvantages. Wind turbines are noisy, and some people find the large turbines to be unattractive. Wind turbines can kill birds and bats that accidentally crash into the blades. Wind farms also create a "shadow flicker" effect, which is similar to a light flashing on and off, when the turning blades create alternating shadows and then light. Some people find this effect to be annoying. Aside from wind, air can also be used as a source of energy when it is squeezed, or compressed, into a small container under pressure. This compressed air can be used to power cars and hovercrafts. Compressed air is a clean energy, and it is cheaper than gasoline. But there are not many compressed air filling stations, so until this technology is expanded, it may not be practical.

Big Ideas to think about:

- How does this energy source work?
- What are the advantages of using it?
- What are the disadvantages of using it?
- Any other interesting facts?

Geothermal Energy

Expert Group Reading

Geothermal energy comes from heat below Earth's surface. Volcanoes, which can release hot magma (melted rock), and geysers, which release steam, are both evidence of Earth's heat below the surface. Hot springs, which are formed when heat from inside Earth reaches surface water, are enjoyed by many people for recreation and therapeutic purposes (to return to or maintain good health). But Earth's heat can also be used to heat homes and businesses. By placing water-filled pipes underground, Earth's heat can be used to heat the water. Similarly, pipes can be used to capture naturally occurring hot water or steam. The hot water or steam can then be used to create electricity by turning turbines in generators. Electricity generation from geothermal energy is very clean, as little pollution is created. Geothermal energy is renewable because Earth is constantly generating heat. Geothermal power plants use less land to make the same amount of energy as power plants using other sources of energy. But geothermal energy also has disadvantages. Geothermal energy requires a large investment of money to get set up, as drilling and piping are costly. The location of geothermal power plants is usually limited to places where heat reaches close to Earth's surface. In the United States, for example, geothermal power plants are primarily found in the western part of the country. Drilling for geothermal energy can trigger earthquakes. Also, the liquids that are emitted from geothermal systems can contain very dangerous chemicals that can pollute the air and water and can release smelly gases.

Big Ideas to think about:

- How does this energy source work?
- What are the advantages of using it?
- What are the disadvantages of using it?
- Any other interesting facts?

Biomass Energy

Expert Group Reading

Biomass is material that comes from plants or animals. It contains stored energy from the Sun. Examples of biomass include crops such as soybeans, corn, and sugarcane; wood from trees; food and yard wastes; and even animal manure and human sewage. Some forms of biomass, such as wood and garbage, can be burned directly for heat. Other forms of biomass, such as food, yard, and paper wastes, produce biogas as they break down in landfills. The biogas can be captured and burned for fuel. Animal and human sewage can create biogas when it is processed in special containers called digesters. Biodiesel fuel is made from vegetable oils and can be used to power cars and heat homes. Ethanol is a biofuel that comes from fermenting (breaking down) crops like corn or sugarcane into a liquid that can be burned for energy. It is often added to gasoline to reduce the pollution that is given off by cars. Some cars, called flex-fuel vehicles, are able to run on fuel that is either mostly gasoline or mostly ethanol. Biomass fuels are renewable, since plants and animals are constantly replacing themselves. Biomass also releases less carbon dioxide than fossil fuels when it is burned, so it contributes less to climate change and could possibly be used to replace fossil fuels. However, using biomass on a large scale may lead to loss of forests or farms. If many farms begin growing biofuels instead of other crops, foods may become very expensive. Biofuel plants are expensive to build and maintain, and transporting biomass to the biofuel plants often requires trucks that burn fossil fuels.

Big Ideas to think about:

- How does this energy source work?
- What are the advantages of using it?
- What are the disadvantages of using it?
- Any other interesting facts?

Hydropower

Expert Group Reading

Hydropower, or water power, is energy that comes from moving water. Hydropower is one of the oldest sources of energy used by people, as humans have been using the force of water flowing in rivers and streams to turn paddle wheels to grind grain for thousands of years. Hydropower is also used to make electricity through the use of hydroelectric dams. These dams create electricity by using swiftly flowing water to turn turbines that are connected to generators. Hydropower is the largest renewable source for electricity in the United States. It is renewable because water is continuously moved through the water cycle. The water on Earth's surface evaporates up into the sky, where it condenses to form clouds, then returns to Earth as precipitation (rain, snow, sleet, or hail). Hydropower is a very clean and plentiful source of energy. Dams are generally reliable, and they create reservoirs for boating and swimming. But dams can also change the environment for wildlife, especially for fish that migrate along rivers. Dams can cause flooding if they fail to hold water back. They can lose the ability to produce much energy in years of drought, when less water is available. Another form of hydropower is called tidal power. Tides, which are regular changes in water levels caused by the Moon's gravitational pull, can provide energy. By using damlike structures called barrages, people harness the energy of tides as they move in and out to create electricity. Ocean waves can also be used to provide energy. Yet tidal and wave power are both relatively new energy sources compared with hydroelectric energy, so more research has to take place before they can be used more widely.

Big Ideas to think about:

- How does this energy source work?
- What are the advantages of using it?
- What are the disadvantages of using it?
- Any other interesting facts?

Alternative Energy Expert Group Cards

Group 1 Wind	**Group 2** Solar
Group 3 Nuclear	**Group 4** Biomass
Group 5 Hydropower	**Group 6** Geothermal

Congressional Committee Debate on Alternative Energy

The Scenario

When Congress makes laws or decides whether to spend money on different projects, it first has smaller groups of members, called congressional committees, research and debate the issues. Members of the committees are experts on these issues. You are a member of the Energy Committee. Your committee has been asked to advise the full Congress on which alternative energy is best for America's future.

The Task

You will collaborate with other committee members to research a particular alternative energy source. You will then participate in a debate to try to convince other committee members that your alternative energy source is best. The committee will then vote on the best energy source and develop a report to share with the full Congress.

The Format

- Opening statements: Present arguments for why your energy is best (1 min.)

- One round of questions and discussion: Committee members can respectfully ask questions, clarify statements, or raise discussion points with other groups (time to be determined by the committee chair, your teacher)

- Closing argument: Present your final arguments for your energy (1 min.)

Research Materials

Prepare for the debate using your RAN Energy Chart and these websites:

- *www.eia.gov/kids* (click on "Energy Sources")

- *www.need.org/Energyinfobooks* (open an "Elementary" book about your energy)

Name: _____ Date: _____

My Energy Choice Statement

I think that _____ is the best energy source for the future.

I think this because _____

_____

_____

_____

_____ .

Here is a picture of my energy source at work!

Rubric for Congressional Debate

Each team should make a two-minute clear, organized, and evidence-based opening statement that describes the following: (1) why they think their energy source is best, (2) the advantages of the energy source, (3) the disadvantages of the energy source, and (4) the resources they used to do research.

	Early (1 pt.)	Emerging (2 pts.)	Sophisticated (3 pts.)	Points and Comments
Use of Evidence	Students use opinion without evidence to back their claims.	Students use tenuous or incomplete evidence to back claims.	Students demonstrate complete and accurate use of evidence to back claims.	
Source and Quality of Evidence	Students are unable to identify sources of evidence.	Students demonstrate some effort in identifying and evaluating the sources of evidence.	Students thoughtfully identify and evaluate the sources of evidence.	
Science Content Understanding	Students demonstrate minimal understanding of science content.	Students demonstrate a moderate degree of understanding of science content.	Students demonstrate strong understanding of science content and consistently apply it to their arguments.	
Clarity and Organization of Presentation	Presentation is unclear and disorganized.	Presentation is somewhat clear and organized.	Presentation is clear, organized, and compelling.	
Response to Questions	Students are unable to respond to questions.	Students respond in inappropriate manner or with inaccurate information.	Students respond appropriately, thoughtfully, and accurately.	

Total: __ / 15 pts.

Lesson Plans

Lesson 14
Watch Your Step

Should Distracted Walking Be Illegal?

Suggested Grade Levels

3–5

Driving Questions

- How do our senses work with our brain to process and respond to information?
- How do these systems and structures aid in survival?

Lesson Overview

Students are introduced to the problem of "distracted walking," which occurs when people use their phones for calling, texting, or taking photos while walking. After being introduced to the problem through videos of news reports, students explore reaction time and distraction through a ruler drop investigation. Students then learn about the role of the brain and nervous system in processing and responding to information, with particular attention paid to how distraction affects response time. Then, students develop final projects of their choice, such as public service announcements or position papers to send to local government officials expressing their views about whether distracted walking should be illegal.

Connecting to the *NGSS*

(See full alignment in Table A.14 on p. 518.)

- LS1.A: Structure and Function

- Plants and animals have both internal and external structures that serve various functions in growth, survival, behavior, and reproduction. (4-LS1-1)
- LS1.D: Information Processing
    - Different sense receptors are specialized for particular kinds of information, which may be then processed by the animal's brain. Animals are able to use their perceptions and memories to guide their actions. (4-LS1-2)

Societal Issues

Government Control, Individual Freedom Versus Public Good

Nature of Science

- Scientific Knowledge Is Based on Empirical Evidence
    - Science findings are based on recognizing patterns.
    - Science uses tools and technologies to make accurate measurements and observations.
- Science Is a Human Endeavor
    - Science affects everyday life.
    - Creativity and imagination are important to science.

CCSS Connections

- English Language Arts
    - W.4.1. Write opinion pieces on topics or texts, supporting a point of view with reasons and information.
    - SL.4.5. Add audio recordings and visual displays to presentations when appropriate to enhance the development of main ideas or themes.
- Mathematics
    - 4.G.A.3. Recognize a line of symmetry for a two-dimensional figure as a line across the figure such that the figure can be folded across the line

into matching parts. Identify line-symmetric figures and draw lines of symmetry. (4-LS1-1)

NCSS Connections

- Theme 6: Power, Authority, and Governance
  - Examine issues involving the rights and responsibilities of individuals and groups in relation to the broader society.
- Theme 8: Science, Technology, and Society
  - Identify examples of the use of science and technology in society as well as the consequences of their use.
- Theme 10: Civic Ideals and Practices
  - Develop a position on a school or local issue and defend it with evidence.

C3 Framework

- Dimension 3: Gathering and Evaluating Sources
  - D3.1.3-5. Gather relevant information from multiple sources while using the origin, structure, and context to guide the selection.
- Dimension 4: Communicating Conclusions
  - D4.3.3-5. Present a summary of arguments and explanations to others outside the classroom using print and oral technologies (e.g., posters, essays, letters, debates, speeches, and reports) and digital technologies (e.g., internet, social media, and digital documentary).

UDL Toolkit

Multiple Means of Engagement	Multiple Means of Representation	Multiple Means of Action and Expression
Allowing students to develop their own tests for distraction and providing choices for final projects sustain interest by optimizing personal choice and autonomy.	Illustrating the problem of distracted walking through various media, such as video, text, graphics, and a physical demonstration, aids in comprehension of language and symbols.	Providing a choice of tools for the ruler drop, developing two- and three-dimensional models of the nervous system, giving options of having cues or no cues on the model, and providing choices for final projects allow for options in physical action, expression, and communication.
Focusing on a topic that is connected to students' everyday lives stimulates and sustains interest by providing an authentic, relevant experience.	Connecting to prior knowledge through visual imagery, discussion, and explicit connections to social studies aids in comprehension.	Providing data sheets and graphic organizers to maintain information and a checklist for final project tasks supports students' executive functions and enhances self-monitoring.

Suggested Schedule and Sequence

- Day 1: **Engage** with Distracted Walking News Report 1, and **Explore** with Ready, Set, React! Checkpoint Lab

- Day 2: **Explain** with Distracted Walking News Report 2, Reaction/Distraction reading and questions, and Wikki Stix Nervous System Modeling

- Day 3: **Elaborate** with Distracted Walking News Report 3 and Should Distracted Walking Be Against the Law? article and Agree/Disagree T-Chart

- Day 4: **Elaborate** with Distracted Walking Choice Project

- Days 5+: **Evaluate** with Presentation of Distracted Walking Choice Project

Materials

For Ready, Set, React! Checkpoint Lab

(per team of 2)

- Ruler and meter stick (teams have their choice of measurement instrument)

(per student)

- Safety glasses or goggles

For Wikki Stix Nervous System Modeling

(per class)

- Assortment of Wikki Stix (wax-covered yarn that is moldable and sticks to paper; available at *www.amazon.com, www.walmart.com,* and most craft stores)

(per student)

- Safety glasses or goggles

For Distracted Walking Choice Project

(per class)

- Internet access for research
- Additional materials depending on student project choices, such as paper (for pamphlets, letters, or comic strips), trifold boards (for poster presentations), video devices such as iPads or smartphones (for recording public service announcements), and craft supplies (for sculptures, dioramas, or mobiles)

Student Handouts

- Ready, Set, React! Checkpoint Lab
- Evaluating Media Sources Template (p. 28)
- Reaction/Distraction
- Wikki Stix Nervous System Modeling (2 versions available, one with picture cues and one without)
- Should Distracted Walking Be Against the Law?

Lesson 14: Watch Your Step

- Should Distracted Walking Be Against the Law? Agree/Disagree T-Chart
- My Distracted Walking Project Checklist

Safety Notes

1. All students must wear safety glasses or goggles during the setup, hands-on, and takedown phases of the activity.

2. Use caution when working with meter stick or other sharp materials to avoid cutting or puncturing skin or eyes.

3. Wash hands with soap and water after completing this activity.

Media

Videos

News Report 1: "New Research Shows Danger of Distracted Walking," from *CBS This Morning*

- *www.youtube.com/watch?v=wdW3lC67llQ*

News Report 2: "Texting and Walking? Jeff Rossen Explains How It Could Get You Killed," an experiment on distracted walking and brain imagery from the *Today Show*

- *www.youtube.com/watch?v=G4WAhH_buFQ*

News Report 3: "California City Bans Texting and Walking," from *Good Morning America*

- *www.youtube.com/watch?v=5cODio26wLM*

Background for Teachers

Studies say that most of us use our cell phones while walking. Over the last decade, injuries from distracted walking, which includes texting, talking, video calling, and taking photos while walking, has increased steadily. Most alarming is that a 2013 study by Safe Kids Worldwide found that one in five high school students and one in eight middle school students cross the street while distracted (*www.safekids.org/research-report/research-report-teens-and-distraction-august-2013*). Moreover, crossing the street isn't the only time this practice is dangerous; a quick perusal of videos about distracted walking on the internet shows people walking into walls, falling into fountains, or even more deadly, falling onto railroad tracks or over cliffs.

The issue of distracted walking relates to the way the human brain and nervous system work. Humans, like other animals, use their senses to detect information from the environment. Sights, sounds, smells, and tactile cues all help us sense danger, which is critical for survival. Our sense organs, such as our eyes and ears, perceive information through *sensory receptors*. Once information is perceived, it is transmitted via sensory neurons (nerve cells) to the brain. The brain and the spinal cord process the information and send out signals about how to respond to the muscles in our arms, legs, and other parts of our bodies. The brain and spinal cord together make up the *central nervous system*, essentially the central processing area for information. The sensory organs, such as eyes, ears, nose, tongue, and skin, along with the nerves that carry information to and from the central nervous system, form the *peripheral nervous system*. Together, these systems help us perceive stimuli, determine whether to respond to them, and respond if needed. This amazing coordination of systems and information helps us survive!

Unfortunately, the time that it takes for this reaction to happen, or *reaction time*, increases when we are distracted. Distraction occurs when the brain receives too much information at once and is forced to divide attention between different stimuli. When, for example, we are texting while walking, the brain is forced to dedicate attention to texting, which reduces the attention to walking. This reallocation of brainwave activity has been documented using brain scans and is demonstrated in a video included in this lesson.

Because of the increase in distracted walking injuries, some towns and cities have passed laws against using a cell phone while walking or crossing the street. Proponents of such legislation cite the importance of government regulation when it comes to the safety of its citizens. Proponents also cite the cost to society when emergency rooms, doctors, and insurance companies must bear the weight of increased accidents. Opponents of such legislation cite personal rights, arguing that people should have the right to do what they want with their cell phones, even if it means they're putting themselves at greater risk. Opponents also feel that having police enforce these laws is a poor use of their time and will take them away from more pressing issues. They point to the fact that many of the injuries attributed to cell phones happen at home, where enforcement would be impossible.

In this lesson, students are introduced to the issue of distracted walking through a news video. They then explore their own reaction time by trying to catch a ruler or meter stick with their fingers while doing different tasks that distract them. Next, they read about the human nervous system and create a tactile model using Wikki Stix (wax-covered string), to align with the *NGSS* standards that anticipate students recognizing the organs and systems that aid in survival and the manner in which information from various stimuli is transmitted between various organs and systems of the body. The controversial question of whether distracted walking should be illegal is then introduced to students, who answer the question by creating a project of their choice that meets the requirements set out in a rubric.

Additional Resources

Information on distracted walking by teens

- *www.nsc.org/home-safety/safety-topics/distracted-walking/teens*
- *www.safekids.org/press-release/1-5-high-school-students-crosses-street-while-distracted-technology*

Information on kids' health for educators, parents, and students

- *https://classroom.kidshealth.org*

5E Lesson Plan

Advance Preparation: Create a KLEW chart like the one in Figure 3.2 (p. 17).

Engage: Distracted Walking News Report 1

1. Show students the first 15 seconds of the *CBS This Morning* news report titled "New Research Shows Danger of Distracted Walking" at *www.youtube.com/watch?v=wdW3lC67llQ*. Have students share what they think they know about distracted walking. Enter this information in the "Know" column of the KLEW chart.

2. Continue showing the video to the end. After the video, have students think-pair-share to discuss the question "What have you learned about distracted walking from the video?" Record student ideas in the "Learned" column of the KLEW chart. (These may include that injuries from distracted walking are on the rise, people fall or get hit by cars because of distracted walking, research suggests people veer off course when distracted, and millennials ages 21 through 25 are most likely to be injured by distracted walking.) Be sure to ask students what the evidence of each claim is, such as "a study in the *CBS This Morning* video" or "the orthopedic surgeon in the video," and include it under "Evidence." Invite students to share their questions about the topic, and add them to the chart under "Wonderings."

3. Ask students, "What does *distracted* mean?" (not focusing on what you are supposed to; trying to do too many things at once) "What does distraction have to do with accidents?" (accept all answers)

4. Explain to students that they are going to do an experiment to learn more about distraction.

> **Misconception Alert**
>
> Students often think that their brains are only working when they are thinking about something. This misconception is understandable because we often talk about the brain in terms of intelligence, thought, and learning. But the brain is always working, even when we are completely unaware! The brain controls our voluntary nervous system, which controls things like standing up or singing a song, and our involuntary nervous system, which controls things like our heartbeat, breathing rate, and even our stomach growling! In the next activity, students are going to test their reaction times as they try to catch a ruler that is dropped. It is worth noting that scientists use the term *reaction* for voluntary activities (like trying to catch a ruler or a ball) and *reflex* for involuntary activities (like blinking when something comes near your eyes).

Explore: Ready, Set, React! Checkpoint Lab

1. Divide the class into pairs, distribute copies of the Ready, Set, React! Checkpoint Lab handout (pp. 471–473) to each student, and read through the lab with students. Demonstrate a simple ruler drop test with one student to show the class how it is done. Ask students, "Why are we measuring how far the ruler travels before it is caught?" (because this gives us an idea of how quickly the person reacted; the shorter the distance, the quicker the reaction) Be sure that students are clear that a smaller measurement means a quicker reaction time.

2. Next, review why and how the Average Catch Distance will be calculated. Ask, "Why are we looking at averages when comparing the numbers from the different tests?" (because they allow us to compare groups of several numbers using only one number; they help us to get a rough idea of what a typical test result will be) "How do we find the Average Catch Distance?" (add the three trial distances and divide by three)

Lesson 14: Watch Your Step

3. Explain to students that because this is a checkpoint lab, they need to have you check their work at each of the lab's three checkpoints before they move on. Then, tell students they may begin the lab, following the instructions on the handout.

4. After the lab, review the answers to the questions:

    - Does a larger catch distance mean that reaction time was faster or slower? Explain. (larger catch distance means slower reaction time because the ruler was able to travel farther before getting caught)

    - Which of your three tests had the slowest reaction time? Why do you think this was so? (results will vary, but typically, one of the distraction tests will have the slowest reaction time; this is because when people are distracted, the brain is not able to attend to the task as quickly, so reaction is slower)

    - How does distraction affect reaction time? (in general, greater distraction leads to slower reaction time)

    - What are some examples of when a quick reaction time is important? (playing sports, walking, running, riding a bike, driving, gaming, protecting yourself from something coming toward you)

5. Allow students to add their Learnings, Evidence, and Wonderings from the lab to the KLEW chart.

Explain: Distracted Walking News Report 2, Reaction/Distraction Reading and Questions, and Wikki Stix Nervous System Modeling

1. Distribute copies of the Evaluating Media Sources template (Figure 3.15 on p. 28) to students. Inform them that they are going to be watching another video about distracted walking, but this time, you'd like them to think about some of the questions on the template in addition to the information that the video communicates on distracted walking. Review the template with them before showing the video, and tell students to jot down answers during the video.

2. Watch the *Today Show* news report titled "Texting and Walking? Jeff Rossen Explains How It Could Get You Killed" at *www.youtube.com/watch?v=G4WAhH_buFQ*. The video shows an experiment in which the news reporter has his brain scanned while just walking and again while walking and texting.

3. After the video, ask students:

- "What new information have you learned from this video?" (Add responses to the KLEW chart in the "Learnings" and "Evidence" columns.) (answers will vary; may include that people walk into walls or fall into fountains while walking and texting; the brain changes when we're focused on more than one thing at a time; the brain can't really multitask as well as we sometimes think it can; experiments show that people overshoot their goal and walk sideways while texting; we can't see far beyond our phones when we're looking at them)

- "What new questions do you have about distracted walking?" (Add student questions to the "Wonderings" column.)

- Think-pair-share: "The video showed two experiments: one to see how well people walk while texting, and another showing the reporter's brain scan first while just walking and again while walking and texting. Do you think these were good experiments? Talk to a partner to discuss the strengths and weaknesses of these experiments." (strengths: both demonstrate the challenges of walking and texting very clearly, the first walking test was repeated with several people and got the same results, and the brain test used advanced technology; weaknesses: the first walking test used mostly older people, but younger people may be better at walking and texting, and the second test used a treadmill for walking, which may not give realistic results)

- "Using the Evaluating Media Sources template, what do you think about this video as a media source?" (answers will vary but may include that the *Today Show* is very well known, the experiments seemed generally trustworthy, the information mostly agreed with the first video, the doctor had good credentials, it is a fairly recent video, perhaps there is some bias because the reporter seems eager to prove that distracted walking is bad) *Note:* There is no right or wrong answer in this analysis. The goal is to help students begin to question and compare media sources and understand why this is important.

- "How did this video compare to the first one we watched?" (answers will vary but may include that both are from reputable news sources, both interviewed doctors, both gave similar statistics, the first one was older but more or less agreed with the second)

- "Why is it important to think about our sources of information?" (some are more trustworthy than others; we don't want to rely on information that isn't accurate). *Note:* Have students keep the template at school so that they can refer to it when watching another video during the next class.

4. Distribute copies of the Reaction/Distraction handout (pp. 474–475). Have students read in pairs and answer the questions. Review the answers as follows:

    1. Why is the brain sometimes called the command center of the body? (it receives all the information coming into the body and then sends out instructions to different parts of the body; it controls all major functions including breathing, moving, thinking, and responding to stimuli)

    2. How does your nervous system keep you safe? (the nervous system detects stimuli that might cause you harm and makes your body respond)

    3. Complete the following paragraph using each term in the word bank only once:

    The brain, spinal cord, and nerves are all part of the nervous system. When information is picked up by sensory receptors in our sense organs, the information travels along nerves to the brain. The brain processes the information and sends instructions back out through the spinal cord and nerves to our body so that we can respond and survive.

5. Distribute copies of the Wikki Stix Nervous System Modeling handout (pp. 476–477). (*Note:* Two versions of the handout are included: one with picture cues and the other with just a list of the required parts. You can allow students to choose whether they would like cues or would rather design the model themselves.) Explain to students that they are going to be making a tactile (touchable) model of the nervous system that shows how the body responds to stimuli. Show them how Wikki Stix (wax-covered pieces of yarn) work by pressing one onto the paper. Wikki Stix can also be braided, wrapped, or woven to create different patterns and shapes. Explain that students can use any design scheme they wish that helps them remember the parts of the nervous system. Their model should include a sensory organ (e.g., eyes, nose, tongue, skin, ears), nerves, the brain, the spinal cord, and a part of the body that responds to the stimulus (legs, arms, hands, or feet). Students who choose not to use the picture cues can draw on the paper before they use the Wikki Stix if they'd like.

6. As students work, circulate around the room to check for student understanding of the way that the different parts of the nervous system work together to control responses.

7. When students are done, allow them to circulate around the room to see their classmates' work. Models can be displayed on tables or, if pressed sufficiently onto paper, hung on the walls.

8. Assess student work by checking written answers to questions on the reading sheet and making sure that students have developed tactile representations of all required features of the model.

Elaborate: Distracted Walking News Report 3 and Should Distracted Walking Be Against the Law? Article and Agree/Disagree T-Chart

1. Have students take out their Evaluating Media Sources templates. Briefly review the purpose of the template and their evaluations of the two videos that they have viewed.

2. Show the *Good Morning America* news report titled "California City Bans Texting and Walking" at *www.youtube.com/watch?v=5cODio26wLM*. This video introduces the idea of distracted walking laws.

3. Ask students:

    - "Did you hear any new information in this video that we could add to our KLEW chart?" (some cities are passing laws against distracted walking; this isn't just happening in the United States; in London, they've wrapped padding around lampposts to protect people if they walk into them; the term *cell phone zombies* is used in Europe)

    - "Using your Evaluating Media Sources template, what do you think about this video as a media source?" (video is recent, data are similar to those in the other videos; may be biased because no opposition to the laws is mentioned)

4. Explain to students that the question of whether to make distracted walking illegal is controversial; some people support it and others oppose it. Distribute copies of the Should Distracted Walking Be Against the Law? article and T-chart handouts (pp. 478–479).

5. Read the article as a class, encouraging students to underline or highlight the arguments for and against the laws.

Lesson 14: Watch Your Step

6. Introduce the T-chart to students. The claim is that distracted walking should be against the law. Ask students to think about the different videos as well as this reading, and list arguments supporting laws against distracted walking in the "Agree" column and arguments opposing laws under "Disagree." If you have time, allow students to research additional newspaper articles and videos on the topic. As students work, circulate around the room to make sure they understand the T-chart. Some sample arguments appear in Figure 5.51.

FIGURE 5.51.

Sample Should Distracted Walking Be Against the Law? Agree/Disagree T-Chart

Claim: <u>Distracted walking should be against the law.</u>

Agree	Disagree
• There are more than 6,000 pedestrian deaths per year.	• Not all of the pedestrian accidents are due to distracted walking.
• These have increased along with cell phone use.	• The government doesn't have a right to control what I do on my phone.
• People who get injured fill up emergency rooms and cost society money.	• A fine won't change people's behavior.
• There's no reason to be on your phone when you are walking.	• Some of the studies (like the walking test in the video) were done on older people; younger people can walk and text.
• Millennials are most at risk of injury.	• Police have better things to do than enforce this.
• People can't walk straight when they are looking at their phones.	• A lot of cell phone injuries happen at home so this law won't help.
• The brain changes when you try to multitask.	• My cell phone is my business.
• The government makes a lot of laws to protect us.	

7. Explain to students that they will be using these T-charts for a project on distracted walking.

Elaborate: Distracted Walking Choice Project

1. Distribute copies of the My Distracted Walking Project Checklist handout (p. 480). Inform students that it's time to put their expertise on the topic of distracted walking to use! Explain that their challenge is to develop a project that can both educate people about the issue of distracted walking and

communicate their opinion on whether distracted walking should be against the law.

2. Explain to students that their project can take many forms, including a pamphlet, a letter to a local government official, a comic strip, a poster presentation, a video, a sculpture, a diorama, a mobile, or any other approved medium. However, all projects must meet the requirements in the checklist and will be shared with the school community.

3. Review the checklist for the project with students, and give them a few minutes to think about what they'd like to develop. Have students sketch out their proposals on the back of the checklist sheet to ensure that you (and they) are clear about what they'd like to do.

4. Have students create their projects. As students work, circulate around the room to answer questions and ensure that students are adhering to the requirements on the checklist.

Evaluate: Presentation of Distracted Walking Choice Project

1. Have students present their projects to the school community. This can be done by visiting other classes or having a small assembly, a Distracted Walking Convention, or a class open house.

2. Use the checklist as a rubric to evaluate student work. You can determine whether students have indeed completed each of the tasks and can assign a CAN (creativity, accuracy, and neatness) score as well.

Going Deeper

- Students can research local distracted walking laws and accident statistics.

- Students can research international laws on distracted walking; many countries are dealing with the same issue.

- Students can develop STEM solutions for distracted walking. Ideas that have already been put into place include electric lights in the sidewalk at crosswalks, clear cell phones, and padded signposts.

Name: _____   Date: _____

Ready, Set, React! Checkpoint Lab

In this lab, you and a partner will take turns testing each other's reaction times by catching a ruler as it is dropped. The partner who drops the ruler is called the dropper and the partner who catches the ruler is called the catcher. You will have an opportunity to be a dropper and a catcher.

Part A: React!

1. Decide who is going to be the dropper and who is going to be the catcher first. Remember, you'll be swapping later!

2. Have the catcher hold his or her thumb and pointer finger about 4 cm (1.5 inches) apart to form the letter C. (Use the ruler to measure.) Have the dropper hold the ruler vertically so that 0 cm is at the bottom and between the catcher's fingers.

3. When the dropper lets go of the ruler, the catcher tries to catch it between his or her thumb and pointer finger as quickly as possible. Do three practice runs without recording the results.

4. Now, have the dropper drop the ruler again. This time, the place where the catcher caught the ruler (the marking above his or her finger) is the first measurement. Have the catcher record this measurement in centimeters for Trial #1 in the Ruler Catch Distances data table in Part A on his or her data sheet.

5. Repeat this two more times (trials), having the catcher record the measurements on his or her data sheet.

6. Switch roles and repeat the experiment. The new catcher should record the data on his or her own data sheet.

Ruler Catch Distances (No Distraction)

Trial	My Catch Distance (cm)
Trial #1	
Trial #2	
Trial #3	
My Average Catch Distance*	

*To calculate average catch distance, **add** the three trial distances and **divide** by 3.

Checkpoint A ☐

Part B: Distract and React!

1. Repeat the activity, but this time, have the catcher try to say the alphabet backward while the dropper drops the ruler.

2. Have the catcher record the measurement (where he or she caught the ruler) in the Ruler Catch Distances table in Part B on his or her data sheet. Repeat two more times (trials) and have the catcher record the measurements.

3. Switch roles and repeat the experiment. The new catcher should record the data on his or her own data sheet.

Ruler Catch Distances (Alphabet Distraction)

Trial	My Catch Distance (cm)
Trial #1	
Trial #2	
Trial #3	
My Average Catch Distance*	

*To calculate average catch distance, **add** the three trial distances and **divide** by 3.

Checkpoint B ☐

Part C. Bonus Test!

1. Repeat the activity, but this time, the dropper gets to design a new distraction test! It could be counting backward from 100, having a conversation with the catcher, or something else.

2. Have the catcher record the measurement (where he or she caught the ruler) in the Ruler Catch Distances table in Part C on his or her own data sheet. Repeat two more times (trials) and have the catcher record the measurements. The catcher should write what he or she did for the bonus test.

3. Switch roles and let the new dropper design a test for the new catcher. Do three trials and have the catcher record the data on his or her own sheet.

Ruler Catch Distances (Bonus Test Distraction)

Trial	My Catch Distance (cm)
Trial #1	
Trial #2	
Trial #3	
My Average Catch Distance*	

*To calculate average catch distance, **add** the three trial distances and **divide** by 3.

For my bonus test, I did the following:

_____

_____

_____

_____

_____

_____

Checkpoint C ☐

Now it's time to analyze your results!

1. Does a larger catch distance mean that reaction time was faster or slower? Explain.

   _____

   _____

2. Which of your three tests had the slowest reaction time? Why do you think this was so?

   _____

   _____

   _____

3. How does distraction affect reaction time?

   _____

   _____

   _____

4. What are some examples of when a quick reaction time is important?

   _____

   _____

   _____

Name: _____   Date: _____

Reaction/Distraction

The human **brain** is amazing! Although it weighs only about 3 pounds, it allows you to do all the things that keep you alive. Your brain controls your breathing, movement, feelings, heartbeat, eating, thinking, and learning. Your brain, **spinal cord,** and **nerves** together make up the **nervous system**. The nervous system receives information, or a stimulus, from the environment, processes it, and helps you respond to it. The stimulus can be anything that is noticed by your senses. Things you see, feel, hear, taste, and smell are all stimuli. Your brain gets the information and decides how to respond, then sends the instructions back out to your body. The brain sometimes called is the command center of your body.

How does this work? Imagine that you are about to cross the street. Your eyes scan the street for bicycles and cars, your ears listen for sounds of vehicles, and even your feet receive information about the curb and any bumps in the road. The information is picked up by **sensory receptors** in your **sense organs** (such as your eyes, ears, or nose) and travels along nerves to your brain. Your brain receives all that information, processes it, and sends signals out through your spinal cord to your legs to help you cross safely. If you see a car or bicycle coming toward you, your brain quickly decides whether to walk forward or backward or to stand still! The amount of time this takes is called **reaction time.** Reaction time is involved when a batter decides whether to swing at a ball, a driver brakes when a dog darts into the road, or you decide when to close your hand during the ruler drop test.

Reaction time is important because it can keep us safe. When the brain is **distracted** by too much information, it loses some of its ability to react to different situations. This is a big problem when people try to walk or drive while using cell phones. While you are texting, calling, or taking photos, your brain focuses on that and misses some important information from the environment. It also may respond more slowly than usual. This means that people can walk into cars, fall off curbs, or get hit by bicyclists. Even in your own home, walking while you're using a cell phone can cause accidents such as tripping or falling down steps. These types of accidents have increased a lot since people started using cell phones. It is very important to pay attention when you are on the move so that your brain can focus on keeping you safe!

Questions

1. Why is the brain sometimes called the command center of the body?

   _____
   _____
   _____

2. How does your nervous system keep you safe?

   _____
   _____
   _____

3. Complete the following paragraph using each term in the word bank only once:

The brain, _____, and nerves are all part of the nervous _____. When information is picked up by sensory _____ in our sense organs, the information travels along _____ to the brain. The brain processes the information and sends instructions back out through the spinal cord and nerves to our body so that we can respond and survive.

Word Bank

receptors	system
spinal cord	nerves

Name: _____ Date: _____

Wikki Stix Nervous System Modeling

In the space below, shape and press the Wikki Stix to create a tactile (touchable) model of the nervous system that shows how your body responds to stimuli. Your model must be labeled and include the parts listed in the box.

Your model must include the following:

- Sense organ (eyes, ears, nose, tongue, or skin)
- Nerves
- Brain
- Spinal cord
- Body part that responds (for example, arms, legs, or hands)

Name: _____ Date: _____

Wikki Stix Nervous System Modeling

Shape and press the Wikki Stix to create a tactile (touchable) model of the nervous system that shows how your body responds to stimuli.

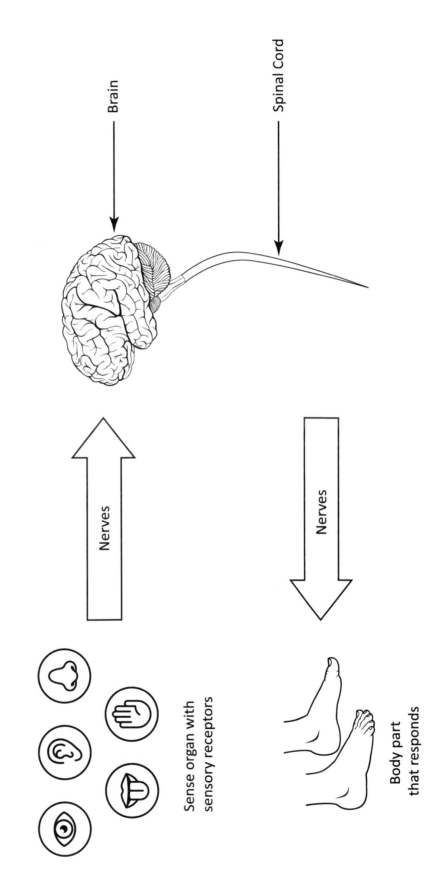

IT'S STILL DEBATABLE! USING SOCIOSCIENTIFIC ISSUES TO DEVELOP SCIENTIFIC LITERACY, K–5

Name: _____ Date: _____

Should Distracted Walking Be Against the Law?

Several cities and towns in the United States have passed laws making distracted walking illegal. In some places, people who are caught texting, talking on cell phones, playing video games, looking at Google Maps, listening to music with headphones in their ears, or using any electronic devices in a crosswalk can be fined. These laws have been introduced because of the rise of pedestrian accidents over the last several years.

Supporters of the laws think that people will be less likely to look at their electronics when they cross the street. They think that the rise in pedestrian accidents is directly linked to more people using their electronics while they walk. They also say, even if *all* the pedestrian accidents aren't from distracted walking, it's clear that walking while looking at your electronic device is dangerous. Opponents of the laws say that the fact that more pedestrians are getting hurt each year doesn't necessarily mean that they are all caused by distracted walking. They also say that the government doesn't have the right to control what people do with their devices. In addition, they think that it's going to be difficult to enforce the law, and that doing so will take police away from enforcing more serious laws.

Where do you stand on this issue? Should distracted walking be against the law?

Name: _____ Date: _____

Should Distracted Walking Be Against the Law? Agree/Disagree T-Chart

Claim: Distracted walking should be against the law.

Agree	Disagree

Name: _____ Date: _____

My Distracted Walking Project Checklist

Your challenge is to develop a project that can both educate people about the issue of distracted walking and communicate your opinion on whether distracted walking should be against the law.

Check ✓ the boxes below when you have completed each step!

> ☐ I have explained what distracted walking is.
>
> ☐ I have explained how distraction affects the brain and reaction time.
>
> ☐ I have included evidence about distraction from my Ready, Set, React! experiment (the ruler test) and evidence from at least one video.
>
> ☐ I have included at least two (2) arguments for and against distracted walking laws.
>
> ☐ I have given my opinion on distracted walking laws and included the reasons for my opinion.

My project is **C**reative, **A**ccurate, and **N**eat (CAN): 5 4 3 2 1

Give yourself 2 pts. for each ✓ and add your CAN score.

My total self-evaluation score is _____ .

Unit VI
Developing Your Own SSI Lessons

So, you've tested some of the lessons provided in this book and feel you're ready to take the plunge into developing your own socioscientific issues (SSI) lessons. Great! You can do it! There is nothing magical about SSI lesson plans; in fact, I would bet that you have already touched on many socioscientific issues in your teaching experiences but may not have realized it at the time. Now that you have a better understanding of what SSI is, you can capitalize on those interesting societal questions without worrying that they're not "real science." The beauty of SSI is that there are many issues that can be connected to any science topic and, conversely, many science topics that can be connected to a given societal issue. This means that SSI can be integrated into your existing curriculum quite easily. I think that the best way to get started is with small steps, working pieces of SSI pedagogy, such as debates, town hall meetings, yes/no argument lines, and the like, into your existing curriculum. This approach will help you become comfortable with the tools and strategies that make SSI unique. Then, you can venture into designing your own original SSI lessons from scratch. And because SSI is so flexible, you can exercise tremendous creativity in designing lessons to suit your curriculum and the needs and interests of your students (Kahn and Hartman 2018; Zeidler and Kahn 2014).

Here are some tips for developing your own SSI lessons:

- **Identify big ideas related to your existing curriculum.** Begin by thinking about the big ideas that you want to address in your curriculum and making a list. If you are using the *Next Generation Science Standards* (*NGSS*)

as your standards, look at the disciplinary core ideas (DCIs) to help guide your learning objectives. Then, research societal issues related to those ideas, looking at news reports, books, television, movies, science education journals, and so on. For example, when I was teaching fourth-grade science, one of my big ideas was to develop a series of SSI lessons on energy and natural resources because this is a DCI ("Energy and fuels that humans use are derived from natural sources, and their use affects the environment in multiple ways") in the *NGSS* that supports fourth-grade performance expectations (4-ESS3-1). Then, I researched related societal issues. Some of the issues and questions I began to think about were "What types of alternative energies are best?" and "Are 'clean energies' really clean?"

- **Don't forget to look in your own backyard.** Some of the best SSI topics are local issues or other issues that are very relevant to your students' lives. For example, if your town or city is grappling with environmental decisions about waste management or water conservation, you can tie these local issues into topics related to ecosystems and the water cycle. If your school is evaluating policies on student cell phone use, you can make connections to engineering design or waves and information transfer. There are local and global controversies related to almost any science topic. At the time I was developing my fourth-grade energy unit, I not only found several interesting media resources, including *60 Minutes* episodes and magazine articles, but I also learned of a local dispute about whether wind turbines could be placed on top of our city's apartment buildings. This local issue really got my students debating! We also brought the issue right into our classroom by examining the question "Are solar or wind energies feasible for our school?"

- **Think across disciplines.** Chances are, as an elementary teacher, you teach other subjects besides science. How can you connect SSI to your other subjects and vice versa? Think about your other standards (e.g., social studies, Engish language arts [ELA], and math) to help you take advantage of the amazingly interdisciplinary nature of SSI. For example, my colleagues and I were elated when we realized that our energy/natural resources unit in science related beautifully to our fourth-grade social studies unit on the westward movement of early settlers in the United States. We were then able to integrate readings from ELA and incorporate problems about energy usage into math. Now that's efficient planning—and teamwork!

- **Think about your students.** This is one of the most obvious yet overlooked steps in lesson planning. What are your students' language and literacy needs? What cooperative skills would be most beneficial for your class? What are your students' assets, and how can you capitalize on them in your planning? What do your students love to do? What organizational strategies seem to work best in your classroom? What barriers need to be removed to ensure full participation for your students? Thinking about these questions in advance of your planning will prevent the need for a lot of retrofitting when you implement the lesson. For example, when I was planning my fourth-grade energy unit, I thought about the fact that I had several students who would have difficulty with activities that involved extensive verbal demands, like debates and role-playing, because of shyness or expressive language challenges. But I also knew that those students were very comfortable with technology. I decided to have an online discussion board to supplement our in-class activities to give all students additional opportunities to express themselves. They were able to log on at other points during the day in school or if they had internet access at home. I was gratified that thinking about this simple Universal Design for Learning (UDL) strategy in advance made our unit that much more accessible and enjoyable for all my students.

- **Start collecting resources as you come across them.** Have you ever heard or read about an issue and thought, "Wow, that would be an interesting topic to discuss with my students, but how or where does it fit?" Save it! Write it down, bookmark it, file it, or do whatever you do to keep ideas and resources accessible. Think to yourself or ask your colleagues, "Is there a socioscientific issue here?" When I read news articles about protestors opposing the use of horse-drawn carriages in our city, I didn't see an immediate connection to our science curriculum, as I already had a wealth of lessons on animals, but I thought it was an interesting issue that could engage my students. I set aside resources, and several months later, I thought, "Aha! Forces and motion!" I began to devise a series of lessons including one that involved pushing and pulling forces using toy horses for my first-grade class. This ultimately became part of a unit on working animals that was included in the original *It's Debatable!* book. So while many of the ideas for your SSI lessons will come from your science standards, some may also begin with societal issues that grab your attention. Save the resources now and think, "Is there a science connection here?"

- **Use a variety of balanced resources.** In any controversy, there are different perspectives. Try to provide students with materials that represent those different perspectives as much as possible. For example, in the lesson Blast

From the Past (pp. 347–390), students read about the pros and cons of a national space program, and in Finders Keepers? (pp. 320–346), students consider the perspectives of four different stakeholders with interests in dinosaur bones. I've found that particularly for my youngest students, I've often had to create my own readings to ensure grade-level appropriateness. Of course, one of the most challenging aspects of collecting resources is recognizing that we each have our own opinions and biases on debatable issues. These biases may lead us to favor materials that support our own opinions, particularly when issues are emotionally charged. To counteract this, I try to have an equal or close to equal number of arguments presented on each side of an issue (remembering that there can be more than two sides!) so that students are influenced more by the quality of the evidence than by the number of reasons they can list, and so I can keep a reasonable check on my own biases.

- **Consider when, where, and how to introduce the issue.** One of the questions that you will no doubt grapple with is when to introduce the controversial issue: at the beginning of the lesson as part of Engage, or perhaps somewhere in the middle of the lesson after students have a bit more schema as part of Elaborate, or even as part of Evaluate. Some practitioners favor starting SSI lessons with the controversial issue or question because it primes students for learning, allows the teacher to assess prior knowledge, and is an authentic way of helping students recognize what information they need to answer the question or investigate the issue. When I taught students at the middle school, high school, and college levels, that was the main way I approached SSI because my students typically had enough background knowledge on issues (albeit occasionally replete with misconceptions) to begin to discuss them using a common vocabulary. On the other hand, in my elementary teaching, I often found that it was beneficial to position some SSI as part of Elaborate in a 5E lesson plan to provide students with background on the science content through different activities and investigations, because my young students had no schema for formulating positions. For example, the first time I did the Leave It to Beavers lesson (pp. 56–74), I showed my first-grade students a video about a town having a dispute about a beaver dam that was causing flooding. I asked my students their thoughts on what should be done about the dam. To my surprise, several of my students didn't understand that the dam was actually built by the beavers (they were envisioning human-made dams *used* by beavers) or what beavers had to do with dams in the first place! While I worked hard to help the students develop questions that could be

Developing Your Own SSI Lessons

investigated so that they could reason through the issue, it was clear that the lack of any schema on this topic was causing distress and confusion. The next time I taught the lesson, I did it the way it is written in this book, with the issue introduced after students have some understanding of beaver life histories and how dams hold back water, thereby making ponds larger and deeper. They were then able to discuss the issue using evidence from their investigations. While this is definitely a compromise in the sense that it isn't as authentic as having students grapple with issues cold, I found that I needed to be flexible and consider the content associated with the issue and the age of the students. That's why the SSI questions and issues are introduced at different points in different lessons. As you develop your own lessons, you can experiment with this and develop your own rhythm. No matter where you introduce the issue in the lesson, be sure to engage students using different media, such as books, magazines, newspapers, advertisements, YouTube videos, photos, or models. Using a broad range of materials ensures that most students will be drawn into the topic and ready to engage in your lesson!

- **Set some ground rules.** Be sure to develop ground rules for respecting different opinions, taking turns during debates and discussions, and emphasizing evidence to support arguments. You can use many of the tools provided in this book for this purpose. For example, to develop a classroom culture based on trust and mutual respect, discuss and model desired language and behaviors, and use scaffolds such as the sentence frames in Figure 3.13 on page 26 for student success. Engaging students in the development of ground rules is one of the most powerful ways of ensuring buy-in and fostering a sense of community.

- **Help students evaluate sources of evidence.** When time is short (as it always is in the elementary classroom!), it is easy to overlook this critical step, as it may not seem as important as the evidence itself. But the only way for students to develop media literacy is to have them use (and reuse) tools like the CARS rubric introduced on page 27. Having students regularly hear (and apply) the CARS concepts of Credibility, Accuracy, Reasonableness, and Support will allow this type of evaluation to become ingrained in their thinking. Keep asking questions like "What do we know about this author/website/video?" "Does the information make sense?" "Do other sources back it up?" "Do you sense any bias?"

- **Assess learning in a variety of ways and at a variety of points.** Remember that students can show their learning through a variety of ways, including debates and town hall meetings, group discussions (in person and online), letter writing, public service announcements, poster boards, written reports, brochures, cartoons and comic strips, performances, traditional tests, and videos. Providing students with choices in the way they can present information gives them a sense of autonomy and is very motivating. Of course, it is critical that you have specific objectives in mind regardless of the assessment type. Following are some key considerations when developing your assessments:

- What is your objective?
- How will you know when it is achieved?
- What are your judgment criteria?

If you are using the NGSS as your standards, thinking about three-dimensional assessment (i.e., practices, core ideas, and crosscutting concepts) is key, as each performance expectation is built around these three dimensions. Fortunately, SSI is an ideal framework for teaching DCIs and crosscutting concepts while modeling science and engineering practices, including evidence-based argumentation and use and development of models. Assessments should allow students to demonstrate learning of not only content (the "what" of science) but also the practices (the "how" of science and engineering). Use the tools presented in Unit 3 (pp. 15–38) and the NGSS performance expectations to get some additional ideas.

- **Take chances and have fun!** SSI is one of the most creative, contemporary, and vibrant frameworks that will ever cross your desk. It can easily be just as enjoyable for you as for your students. Once you have become comfortable with some of the key strategies (e.g., facilitating classroom discussions and debates, evaluating media sources, and promoting evidence-based argumentation), you may just find that SSI is your favorite way to teach science!

References

Kahn, S., and S. L. Hartman. 2018. Debate, dialogue, and democracy through science! *Science and Children* 56 (2): 36–44.

Zeidler, D. L., and S. Kahn. 2014. *It's debatable! Using socioscientific issues to develop scientific literacy, K–12.* Arlington, VA: NSTA Press.

Unit VII

For Teacher Educators: Including *It's Still Debatable!* in Your Preservice and Inservice Elementary Science Courses

First, thank you for your interest in including this book in your teacher education courses. I know that myriad wonderful resources are available to you, and there is never enough time to cover all you want to include in the context of a university course or professional development program. At the risk of sounding melodramatic, though, I can honestly say that implementing socioscientific issues (SSI) at the elementary level as both a teacher and a teacher educator transformed my thinking about scientific literacy and captured the imagination of both my elementary-age students and my teacher candidates. I am hopeful that you and your students will find it to be a compelling framework as well.

SSI is truly a natural fit for elementary teaching, as it is interdisciplinary, has strong emphases in language arts and literacy, and promotes formation of character and conscience (for an extensive review of research related to SSI, see Zeidler [2014]). Given the small amount of time typically allocated for science within elementary classrooms, SSI can prove to be a very efficient option, as teachers can address multiple standards across a variety of disciplines. For these reasons alone, SSI is a worthwhile framework to introduce to elementary teachers, future and current. I believe, however, that there is an equally persuasive reason that I have rarely heard mentioned: SSI provides real-world connections to science within contexts that often appeal to teachers whether they are attracted to science or not. Many elementary teachers and teacher candidates feel trepidation about teaching science. SSI provides a comfortable context for science by connecting to familiar social

studies and English language arts (ELA) concepts and processes quite naturally. For those who are comfortable with or eager to teach science, SSI allows for unlimited expansion, as the topics and pedagogies that are in concert with SSI are limited only by one's imagination. In sum, SSI has something for everyone and may well prove to be your students' (and your) favorite framework!

I thought that the best way to approach this chapter was to share with you the outline and materials for the class on SSI that I include in my early childhood science methods course. I offer this not because I think my approach is particularly inspired or robust, but because it is the product of several iterations of tweaking and testing with considerable feedback and evaluation from preservice and inservice teachers. Following are the goals for teacher candidates and teachers in my three-hour class:

- Become familiar with the SSI framework and supporting research

- Model strategies for engaging students in SSI via argumentation and discourse

- Employ a rubric to assess student performance and evaluate sources of information within the SSI classroom context

- Engage in collaborative planning for implementation of SSI activities

I have provided a lesson plan on pages 492–493, as well as a PowerPoint that is available at *www.nsta.org/stilldebatable*. The lesson plan is essentially a BSCS 5E Instructional Model (5E) lesson embedded within a 5E lesson. In this way, teachers and teacher candidates participate in an SSI experience as their students would, but they also debrief it as teachers to analyze the various lesson components. I address the 5Es in this manner: After I Engage my students in an overview of the SSI framework and supporting research, I have students Explore with an abbreviated 5E elementary SSI lesson. Our Explain segment is a debriefing of the purpose that each lesson component served and how it might be expanded on or modified to better meet their or their students' needs. The Elaborate component is an opportunity for students to collaboratively design and present their own SSI lesson frames (since there isn't enough time for a full lesson plan), and then they collaboratively Evaluate by providing feedback on each other's work. Of course, you can expand this plan to include completion and implementation of a full SSI lesson plan or delivering a lesson in field placements from this book. In sum, the 5E lesson plan for a methods course looks like this:

For Teacher Educators: Including *It's Still Debatable!* in Your Preservice and Inservice Elementary Science Courses

> **In Advance of Class**
>
> - Reading assignment: Some highly accessible introductions to elementary SSI include the prelude and introduction to this book, Zeidler and Kahn (2014), Kahn and Hartman (2018), and Dolan, Nichols, and Zeidler (2009).
>
> **In Class**
>
> (See lesson plan on pp. 492–493; the PowerPoint is available at *www.nsta.org/stilldebatable*.)
>
> - Engage: Introduce the SSI framework and research.
> - Explore: Model an abridged SSI lesson from this book.
> - Explain: Debrief the lesson to analyze, critique, and extend SSI elements.
> - Elaborate: Collaborate to develop and present an SSI lesson frame.
> - Evaluate: Provide class feedback on SSI lesson frame.
>
> **After Class**
>
> (optional; your students may do one or the other)
>
> - Implement a lesson in field placement from this book.
> - Develop and implement an original SSI lesson.

Following are some key points you may wish to emphasize in your class:

- Introducing SSI Theory and Research (see the PowerPoint provided at *www.nsta.org/stilldebatable*.)

  - SSI is a research-based framework that uses complex, societal issues related to science as the context for teaching and learning.

  - Research suggests that SSI supports students' learning, retention, and transfer of science content, as well as argumentation skills, empathy, understanding of the nature of science, moral reasoning, and character development (in particular, understanding the consequences of actions and inactions).

For Teacher Educators: Including *It's Still Debatable!* in Your Preservice and Inservice Elementary Science Courses

- SSI addresses the conceptual shifts identified in the *Next Generation Science Standards* (*NGSS*) by applying science to real-world contexts, integrating science and engineering practices, aligning with *Common Core* subjects, and preparing students for college, careers, and participatory citizenship.

- This book combines the SSI framework, the 5Es (to promote inquiry), and Universal Design for Learning (UDL) (to promote inclusion). It is the sequel to the book *It's Debatable! Using Socioscientific Issues to Develop Scientific Literacy, K–12* (Zeidler and Kahn 2014).

• Modeling SSI in Your Methods Course or Professional Development Program

- Choose a lesson from this book that relates to a science standard that you would like to address.

- Decide which elements of the lesson should be modeled and which can simply be talked through. I would suggest modeling one hands-on science activity and one SSI element (for example, an abbreviated town hall meeting, a congressional hearing, an opinion letter, a brochure or pamphlet, or any of the included templates that require students to develop arguments for a position on a debatable topic). If the lesson includes trade books, include an abridged read-aloud to introduce students to those works.

- Share and review the scoring rubric with students to familiarize them with evaluation of SSI lessons.

- Emphasize nature of science (NOS). Each of the lessons in this book includes NOS connections. Explicit discussion of NOS in your class will help teacher candidates and teachers recognize the connections between SSI and NOS.

- Consider including the Evaluating Media Sources template on page 28 as a valuable resource for their own and their students' research.

- Debrief the lesson using the SSI Lesson Guiding Questions and Checklist on page 495.

• Elaborate and Evaluate: Designing Their Own SSI Lesson Frames

- Have teacher candidates and teachers collaboratively develop an SSI lesson frame using the template on page 494. This is obviously not a

For Teacher Educators: Including *It's Still Debatable!* in Your Preservice and Inservice Elementary Science Courses

full lesson plan, but rather a first step at thinking about how to connect science standards to controversial issues. Dividing students by current teaching grade level is helpful, as they will be more familiar with the grade-level standards. Allow students to peruse this book for ideas.

- Allow students to share their lesson frames with the class, inviting suggestions and ideas.

• Encouraging Teacher Candidates and Teachers to Implement SSI in Their Classrooms

- Encourage your students to use lessons from this book or to design their own by connecting science content to a real-world societal issue. Remind them that it's OK to start small! An issue can be something related to everyday life. Some examples from this book include "What's the best way to clean up spills?" "What makes a great playground?" and "Is football too dangerous for kids?" Or, they can choose to tackle broader issues like "Are bees disappearing?" or "Do we still need a space program?"

- Emphasize argument and discourse by asking, "How are your students going to practice collecting evidence (from science investigations, reading, videos, and so on) and using it to develop and share arguments?" Debates, town hall meetings, opinion letters, brochures/pamphlets, and agree/disagree activities are all great options!

- Use templates from this book to help students organize their thinking and writing.

References

Dolan, T. J., B. H. Nichols, and D. L. Zeidler. 2009. Using socioscientific issues in primary classrooms. *Journal of Elementary Science Teacher Education* 21 (3): 1–12.

Kahn, S., and S. L. Hartman. 2018. Debate, dialogue, and democracy through science! *Science and Children* 56 (2): 36–44.

Zeidler, D. L. 2014. Socioscientific issues as a curriculum emphasis: Theory, research, and practice. In *Handbook of Research on Science Education*, vol. 2, ed. N. G. Lederman and S. K. Abell, 697–726. New York: Routledge.

Zeidler, D. L., and S. Kahn. 2014. *It's debatable! Using socioscientific issues to develop scientific literacy, K–12.* Arlington, VA: NSTA Press.

For Teacher Educators: Including *It's Still Debatable!* in Your Preservice and Inservice Elementary Science Courses

Lesson Plan for Elementary/Early Childhood Science Methods Course or Professional Development Program

Time	5Es	Activities	Rationales	Notes
10 minutes	Engage	"What's the Connection?" Students view a variety of pictures (in PowerPoint) that relate to SSI, such as "Ban GMOs!" "Stop Fracking!" "Close Zoos!" and respond to the question "What's the connection between these pictures?"	Starting with a gamelike activity sparks students' interest and primes them for learning. Makes the connection between science and social issues.	PowerPoint is available at *www.nsta.org/stilldebatable*.
10 minutes		Introduce the theory and research supporting SSI (in PowerPoint).	Provides a context and motivation for students' engagement with SSI.	In PowerPoint.
60 minutes	Explore	Model an abridged SSI lesson from this book. Select a hands-on science element and an associated SSI element (such as town hall or congressional meeting, opinion letter, brochure or pamphlet, or any activity that involves presenting an evidence-based argument to support a position) from a lesson to complete in class. Talk through the rest of the lesson (although if a lesson includes trade books, an abridged read-aloud is also beneficial). Share the assessment rubric (p. 30) with students so that they gain understanding of SSI assessment.	Giving your students the opportunity to engage in a science investigation activity and an associated issues-based argumentation activity increases their understanding of the science-society connection as well as their comfort level about implementing SSI in their own classrooms. If time allows, having an abridged read-aloud of any trade books used in the lesson helps familiarize students with the literature and models read-alouds, guided reading, and organization techniques.	Choose any lesson from this book that relates to a science standard that you would like to address. Include any associated handouts. You may also want to include the full *NGSS* alignment provided in the appendix.

Continued

492 NATIONAL SCIENCE TEACHING ASSOCIATION

For Teacher Educators: Including *It's Still Debatable!* in Your Preservice and Inservice Elementary Science Courses

Lesson Plan for Elementary/Early Childhood Science Methods Course or Professional Development Program *(continued)*

Time	5Es	Activities	Rationales	Notes
10 minutes		Break		
30 minutes	Explain	Debrief and discuss the elements of the lesson using the SSI Lesson Guiding Questions and Checklist. Brainstorm extensions or modifications that students can make to better serve their students.	Having students reflect on their experience with the lesson and analyze the lesson elements aids in their understanding of the lesson structure and awareness of possible challenges that they may experience in teaching this or similar lessons.	Use SSI Lesson Guiding Questions and Checklist.
30 minutes	Elaborate	Students collaboratively review this book to identify a lesson they would like to implement, or they may develop an original preliminary lesson. Have students use the frame and guiding questions checklist for both options to help them focus on the key elements of the lesson plan.	Encouraging students to connect SSI to their own teaching through collaboration, either by developing original lesson frames or by identifying lessons from this book that they'd like to teach, demystifies the framework and allows for different levels of expertise (i.e., inservice versus preservice).	Use SSI Lesson Planning Frame template and SSI Lesson Guiding Questions and Checklist.
30 minutes	Evaluate	Student groups present their lesson frames to class and evaluate elements of each other's frames using the frame template and guiding questions checklist, with understanding that the frames are still quite rudimentary.	Having students present their lesson frames allows for meaningful peer evaluation, reinforces SSI elements, emphasizes areas in need of attention, and helps students commit to implementation.	Use SSI Lesson Planning Frame template and SSI Guiding Questions and Checklist.

For Teacher Educators: Including It's Still Debatable! in Your Preservice and Inservice Elementary Science Courses

SSI Lesson Planning Frame

Use this frame to begin thinking about how you might implement SSI in your classroom. Work with colleagues to brainstorm a preliminary outline.

Lesson Name:
Guiding Question(s):
Grade Level:
Science Standard(s) Addressed (*NGSS* or state standards):
Societal Issue(s):
Time Needed:
5E Lesson Sequence (preliminary ideas):
• Engage
• Explore
• Explain
• Elaborate
• Evaluate (if using *NGSS*, look at the three dimensions of the performance expectation)
Resources Needed (people, materials, time, space):
How will I ensure this lesson is accessible to all my students?

SSI Lesson Guiding Questions and Checklist

- Is a grade-appropriate science standard addressed?

- If using *Next Generation Science Standards* (*NGSS*), are science and engineering practices, crosscutting concepts, and disciplinary core ideas reflected in the lesson and assessment?

- Is a societal issue being addressed? (The issue can be ethical, political, economic, religious, or social in nature.)

- Are students engaging in inquiry? Are you using the BSCS 5E Instructional Model?

- Are students engaging in argument (making a claim) using evidence? (Evidence can be data from their own or someone else's investigation or experiment, information from books or videos, observations, or experiences.)

- Are students evaluating the sources of their evidence? If so, how?

- Are you connecting to *Common Core State Standards*? If so, which ones?

- Is the nature of science (NOS) being addressed? If so, how?

- Are there opportunities for students to communicate their arguments and collaborate with others (through writing, drawing, speaking, investigating, researching)?

- Are there formative and summative assessments? Do the assessments match the lesson objectives?

- Have you used Universal Design for Learning (UDL) or another approach to ensure equitable opportunities for all students? If so, how?

Unit VIII
Finale: Embracing the Controversy in Your Classroom

If you've made it this far, congratulations! You have taken the very brave step of expanding your already-robust repertoire of teaching tools to include socioscientific issues (SSI). Or, perhaps you're already an SSI veteran and are using this book to supplement your SSI curriculum. No matter what your level of experience is with SSI, taking your teaching to the next level can seem daunting. You might find, as I often did, that you have some trepidation. I often worried: "What if my students just argue?" "What might their parents say about these issues?" "What should I do if my students ask me about *my* stance on different issues?" "What if things get out of control?" "What if my principal walks by and sees students engaging in activities that don't look like science to her?" "What if my colleagues just think this is weird?"

While every school, classroom, and teacher is different, I would like to offer a few final tips that helped me navigate some of these choppy waters:

- **Embrace the controversy.** You know from the research set out in the early chapters of this book that evidence-based argumentation is a key practice across all your core subjects. If you keep your students focused on making thoughtful claims, providing evidence and reasoning, and evaluating their sources, you will be making a tremendous contribution to the success of all their future learning. Also, remember that SSI provides an outstanding context for practicing social skills; use the cooperative learning strategies suggested in the book and watch your students (mostly!) argue productively and courteously. There is some debate in the SSI research community about whether teachers should share their opinions on various issues with their students. My general rule of thumb was that I did not share my opinions

with my elementary-age students. If they asked, I would tell them that I preferred not to share my opinion because I didn't want it to influence theirs. I strongly believed (and still believe) that the role of the teacher is to facilitate *how* students think, not *what* they think. Given that young children often see their teachers as experts on so many things, I didn't want to unwittingly sway them or make them feel that their ideas were right or wrong based on their alignment with mine.

However, I can think of a few instances when students asked my opinion, and I weighed the seriousness of the issue and the age of the students against whether there could be a benefit to modeling my thinking for them. For example, when fourth-grade students asked me whether I thought sponges and cloths were superior to paper towels, I shared my opinion after they shared theirs because it allowed me to model my thinking for them without, in my estimation, swaying theirs. I said something like this: "Based on our research findings, I think there are a lot of jobs for which sponges and cloths are better than paper towels, such as [I gave some examples], but there are some jobs, especially when I'm on the go, where paper towels might be more convenient or sanitary. Can you offer any arguments to change my mind?" (They offered many!) But, when first-grade students asked me whether I thought we needed zoos, I told them that I had an opinion on it but preferred not to share it because I didn't want my opinion to influence them. I made this decision in part because of the age of the students and the fact that this was a more emotionally charged issue than the paper towel question that I investigated with the older students. This was my more typical approach in elementary teaching. I think it is worth reflecting on this question as you embark on SSI. But don't be afraid of it—embrace it!

- **Communicate with parents.** If SSI is new to you, be sure to let parents know that you are trying something new. In my experience, students were often excited about SSI topics and frequently discussed them at home. At the beginning of the year, send home a letter or discuss it at back-to-school or open-house nights and parent-teacher conferences. Or better yet, include parents by having students interview them to find out their experiences and perspectives on socioscientific issues. Moreover, because the range of SSI is so sweeping, you can probably find parents with some expertise on most of the issues. Invite them to speak to your class and share their insights on local, historical, or global scientifically related societal issues.

- **Communicate with administrators.** One of the most powerful aspects of SSI is that it aligns so well with nature of science (NOS) tenets. Many SSI

Finale: Embracing the Controversy in Your Classroom

topic provide the context for easy connections to the empirical, tentative, and human features of scientific endeavors. In addition, SSI lends itself quite readily to the three-dimensional learning framework outlined in the *NGSS*, as SSI incorporates and promotes the knowledge, practices, and habits of mind of scientists. That said, not all school administrators are aware of the contemporary shifts that have taken place in science education. For those more familiar with traditional science teaching approaches, the SSI classroom may seem foreign, as seeing such things as T-charts and yes/no argument lines may not look like science to them at all. Taking the time to share the very solid research base on SSI with your school administrators can make a big difference in gaining their support and reducing any anxiety that you might have. Consider offering the first few chapters of this book to them, along with some of your ideas on implementation, and getting their feedback. You may find that a little extra effort up front yields big dividends in the form of strong administrative support as you move along your SSI trajectory.

- **Find collaborators.** I can't say it any more clearly: SSI screams for collaboration! Every single lesson in this book could easily benefit from collaborations within grade levels, among grade levels, and with subject specialists. And don't forget librarians, educational technology specialists, and intervention specialists at your school. All these colleagues can contribute meaningfully to your work in SSI. Of course, collaborations aren't limited by school walls. Colleagues from middle and high schools, parents, civic leaders, local college and university faculty, and local residents representing a range of different backgrounds, vocations, and perspectives can help bring SSI to life for your students—and for you!

- **Don't just talk, take action!** SSI curriculum can be implemented quite successfully by simply having students discuss and debate the various issues and leaving it at that. However, I believe that such an approach would miss precious opportunities to have your students take initiative, demonstrate leadership, and become more civically engaged. Almost every lesson in this book makes some connection to action, either in the main lesson or in the extensions. Action can take many forms, such as making presentations to other classes; educating the school community through written communication, oral or video presentations, or other means; writing letters; volunteering; protesting; sharing research findings with members of the community; investigating local issues of importance; fund-raising; or meeting with civic leaders. Children's time in elementary school is precious and fleeting, so make it count!

- **Listen to your students.** The most gratifying thing about SSI, in my opinion, is watching your students find their voices, both literally and figuratively. I have consistently found that it is often the shyest, quietest students or those least excited about science who become most animated and empowered through SSI instruction. Providing options for the products of their learning, such as letters, debates, comic strips, or commercials, allows every student to shine, while using rubrics for project requirements maintains rigor and consistency. And be sure to ask students what issues matter to them. Children may not see the science connections that can be made to different issues, but you can and will. Let your students help guide your SSI planning … and then watch them take flight!

Thank you again for your interest in this book and for all you do to shape the future of our world. I wish you all the best and look forward to hearing from you!

—Sami Kahn

Appendix: *NGSS* Lesson Plan Alignment Matrices

Appendix: *NGSS* Lesson Plan Alignment Matrices

TABLE A.1.
Connecting to the *NGSS* for Lesson 1: Leave It to Beavers

Performance Expectations

K-ESS2-2. Earth's Systems
Construct an argument supported by evidence for how plants and animals (including humans) can change the environment to meet their needs.

K-2-ETS1-2. Engineering Design
Develop a simple sketch, drawing, or physical model to illustrate how the shape of an object helps it function as needed to solve a given problem.

Science and Engineering Practices	Disciplinary Core Ideas	Crosscutting Concepts
Engaging in Argument From Evidence • Construct an argument with evidence to support a claim. (K-ESS2-2) Developing and Using Models • Develop a simple model based on evidence to represent a proposed object or tool. (K-2-ETS1-2)	ESS2.E: Biogeology • Plants and animals can change their environment. (K-ESS2-2) ETS1.B: Developing Possible Solutions • Designs can be conveyed through sketches, drawings, or physical models. These representations are useful in communicating ideas for a problem's solutions to other people. (K-2-ETS1-2)	Systems and System Models • Systems in the natural and designed world have parts that work together. (K-ESS2-2) Structure and Function • The shape and stability of structures of natural and designed objects are related to their function(s). (K-2-ETS1-2)

Source: NGSS Lead States. 2013. *Next Generation Science Standards: For states, by states.* Washington, DC: National Academies Press. *www.nextgenscience.org/next-generation-science-standards.*

Appendix: *NGSS* Lesson Plan Alignment Matrices

TABLE A.2.
Connecting to the *NGSS* for Lesson 2: Swingy Thingy

Performance Expectations

K-PS2-1. Motion and Stability: Forces and Interactions
Plan and conduct an investigation to compare the effects of different strengths or different directions of pushes and pulls on the motion of an object.

K-PS2-2. Motion and Stability: Forces and Interactions
Analyze data to determine if a design solution works as intended to change the speed or direction of an object with a push or a pull.

K-2-ETS1-1. Engineering Design.
Ask questions, make observations, and gather information about a situation people want to change to define a simple problem that can be solved through the development of a new or improved object or tool.

Science and Engineering Practices	Disciplinary Core Ideas	Crosscutting Concepts
Planning and Carrying Out Investigations • With guidance, plan and conduct an investigation in collaboration with peers. (K-PS2-1) Analyzing and Interpreting Data • Analyze data from tests of an object or tool to determine if it works as intended. (K-PS2-2) Asking Questions and Defining Problems • Ask questions based on observations to find more information about the natural and/or designed world(s). (K-2-ETS1-1)	PS2.A: Forces and Motion • Pushes and pulls can have different strengths and directions. Pushing or pulling on an object can change the speed or direction of its motion and can start or stop it. (K-PS2-1-2) PS2.B: Types of Interactions • When objects touch or collide, they push on one another and can change motion. (K-PS2-1) ETS1.A: Defining and Delimiting Engineering Problems • A situation that people want to change or create can be approached as a problem to be solved through engineering. Such problems may have many acceptable solutions. (K-2-ETS1-1)	Cause and Effect • Simple tests can be designed to gather evidence to support or refute student ideas about causes. (K-PS2-1-2) Structure and Function • The shape and stability of structures of natural and designed objects are related to their function(s). (K-2-ETS1-2)

Source: NGSS Lead States. 2013. *Next Generation Science Standards: For states, by states.* Washington, DC: National Academies Press. *www.nextgenscience.org/next-generation-science-standards.*

Appendix: *NGSS* Lesson Plan Alignment Matrices

TABLE A.3.

Connecting to the *NGSS* for Lesson 3: Take a (Farm) Stand

Performance Expectations

K-LS1-1. From Molecules to Organisms: Structures and Processes
Use observations to describe patterns of what plants and animals (including humans) need to survive.

1-LS3-1. Heredity: Inheritance and Variation of Traits
Make observations to construct an evidence-based account that young plants and animals are like, but not exactly like, their parents.

2-LS2-1. Ecosystems: Interactions, Energy, and Dynamics
Plan and conduct an investigation to determine if plants need sunlight and water to grow

Science and Engineering Practices	Disciplinary Core Ideas	Crosscutting Concepts
Analyzing and Interpreting Data • Use observations (firsthand or from media) to describe patterns in the natural world in order to answer scientific questions. (K-LS1-1) Constructing Explanations and Designing Solutions • Make observations (firsthand or from media) to construct an evidence-based account for natural phenomena. (1-LS3-1) Planning and Carrying Out Investigations • Plan and conduct an investigation collaboratively to produce data to serve as the basis for evidence to answer a question. (2-LS2-1)	LS1.C: Organization for Matter and Energy Flow in Organisms • All animals need food in order to live and grow. They obtain their food from plants or from other animals. Plants need water and light to live and grow. (K-LS1-1) LS2.A: Interdependent Relationships in Ecosystems • Plants depend on water and light to grow. (2-LS2-1) LS3.B: Variation of Traits • Individuals of the same kind of plant or animal are recognizable as similar but can also vary in many ways. (1-LS3-1)	Patterns • Patterns in the natural and human designed world can be observed and used as evidence. (K-LS1-1) (1-LS3-1) Cause and Effect • Events have causes that generate observable patterns. (2-LS2-1)

Source: NGSS Lead States. 2013. *Next Generation Science Standards: For states, by states.* Washington, DC: National Academies Press. www.nextgenscience.org/next-generation-science-standards.

TABLE A.4.

Connecting to the *NGSS* for Lesson 4: Monkey Business

Performance Expectations

1-LS1-2. From Molecules to Organisms: Structures and Processes
Read texts and use media to determine patterns in behavior of parents and offspring that help offspring survive.

1-LS3-1. Heredity: Inheritance and Variation of Traits
Make observations to construct an evidence-based account that young plants and animals are like, but not exactly like, their parents.

Science and Engineering Practices	Disciplinary Core Ideas	Crosscutting Concepts
Constructing Explanations and Designing Solutions • Make observations (firsthand or from media) to construct an evidence-based account for natural phenomena. (1-LS1-2) Obtaining, Evaluating, and Communicating Information • Read grade-appropriate texts and use media to obtain scientific information to determine patterns in the natural world. (1-LS3-1)	LS1.B: Growth and Development of Organisms • Adult plants and animals can have young. In many kinds of animals, parents and the offspring themselves engage in behaviors that help the offspring to survive. (1-LS1-2) LS3.A: Inheritance of Traits • Young animals are very much, but not exactly, like their parents. Plants also are very much, but not exactly, like their parents. (1-LS3-1)	Patterns • Patterns in the natural and human designed world can be observed, used to describe phenomena, and used as evidence. (1-LS1-2) (1-LS3-1)

Source: NGSS Lead States. 2013. *Next Generation Science Standards: For states, by states.* Washington, DC: National Academies Press. *www.nextgenscience.org/next-generation-science-standards.*

TABLE A.5.

Connecting to the *NGSS* for Lesson 5: Soaky Doaky

Performance Expectations

2-PS1-2. Matter and Its Interactions

Analyze data obtained from testing different materials to determine which materials have the properties that are best suited for an intended purpose.

Science and Engineering Practices	Disciplinary Core Ideas	Crosscutting Concepts
Analyzing and Interpreting Data • Analyze data from tests of an object or tool to determine if it works as intended. (2-PS1-2)	PS1.A: Structure and Properties of Matter • Different properties are suited to different purposes. (2-PS1-2)	Cause and Effect • Simple tests can be designed to gather evidence to support or refute student ideas about causes. (2-PS1-2)

Source: NGSS Lead States. 2013. *Next Generation Science Standards: For states, by states.* Washington, DC: National Academies Press. *www.nextgenscience.org/next-generation-science-standards.*

Appendix: *NGSS* Lesson Plan Alignment Matrices

TABLE A.6.

Connecting to the *NGSS* for Lesson 6: Bee-ing There for Bees

Performance Expectations

2-LS2-2. Ecosystems: Interactions, Energy, and Dynamics
Develop a simple model that mimics the function of an animal in dispersing seeds or pollinating plants.

K-2-ETS1-2. Engineering Design
Develop a simple sketch, drawing, or physical model to illustrate how the shape of an object helps it function as needed to solve a given problem.

Science and Engineering Practices	Disciplinary Core Ideas	Crosscutting Concepts
Developing and Using Models • Develop a simple model based on evidence to represent a proposed object or tool. (2-LS2-2) (K-2-ETS1-2)	LS2.A: Interdependent Relationships in Ecosystems • Plants depend on animals for pollination or to move their seeds around. (2-LS2-2) ETS1.B: Developing Possible Solutions • Designs can be conveyed through sketches, drawings, or physical models. These representations are useful in communicating ideas for a problem's solutions to other people. (K-2-ETS1-2)	Structure and Function • The shape and stability of structures of natural and designed objects are related to their function(s). (K-2-ETS1-2)

Source: NGSS Lead States. 2013. *Next Generation Science Standards: For states, by states.* Washington, DC: National Academies Press. *www.nextgenscience.org/next-generation-science-standards.*

Appendix: *NGSS* Lesson Plan Alignment Matrices

TABLE A.7.

Connecting to the *NGSS* for Lesson 7: Weather or Not

Performance Expectations

K-ESS3-2. Earth and Human Activity
Ask questions to obtain information about the purpose of weather forecasting to prepare for, and respond to, severe weather.

3-ESS3-1. Earth and Human Activity
Make a claim about the merit of a design solution that reduces the impacts of a weather-related hazard.

K-2-ETS1-1. Engineering Design
Ask questions, make observations, and gather information about a situation people want to change to define a simple problem that can be solved through the development of a new or improved object or tool.

3-5-ETS1-1. Engineering Design
Define a simple design problem reflecting a need or a want that includes specified criteria for success and constraints on materials, time, or cost.

Science and Engineering Practices	Disciplinary Core Ideas	Crosscutting Concepts
Asking Questions and Defining Problems • Ask questions based on observations to find more information about the designed world. (K-ESS3-2) (K-ETS1-1) • Define a simple design problem that can be solved through the development of a new or improved object or tool. (K-2-ETS1-1) • Define a simple design problem that can be solved through the development of an object, tool, process, or system and includes several criteria for success and constraints on materials, time, or cost. (3-5-ETS1-1)	ESS3.B: Natural Hazards • Some kinds of severe weather are more likely than others in a given region. Weather scientists forecast severe weather so that the communities can prepare for and respond to these events. (K-ESS3-2) ESS3.B: Natural Hazards • A variety of natural hazards result from natural processes. Humans cannot eliminate natural hazards but can take steps to reduce their impacts. (3-ESS3-1)	Cause and Effect • Cause and effect relationships are routinely identified, tested, and used to explain change. (3-ESS3-1) • Events have causes that generate observable patterns. (K-ESS3-2) Influence of Science, Engineering, and Technology on Society and the Natural World • People's needs and wants change over time, as do their demands for new and improved technologies. (3-5-ETS1-1)

Continued

Appendix: *NGSS* Lesson Plan Alignment Matrices

Table A.7. Connecting to the *NGSS* for Lesson 7: Weather or Not (*continued*)

Science and Engineering Practices	Disciplinary Core Ideas	Crosscutting Concepts
Engaging in Argument from Evidence • Make a claim about the merit of a solution to a problem by citing relevant evidence about how it meets the criteria and constraints of the problem. (3-ESS3-1)	ETS1.A: Defining and Delimiting Engineering Problems • A situation that people want to change or create can be approached as a problem to be solved through engineering. (K-2-ETS1-1) • Asking questions, making observations, and gathering information are helpful in thinking about problems. (K-2-ETS1-1) • Before beginning to design a solution, it is important to clearly understand the problem. (K-2-ETS1-1) ETS1.A: Defining and Delimiting Engineering Problems • Possible solutions to a problem are limited by available materials and resources (constraints). The success of a designed solution is determined by considering the desired features of a solution (criteria). Different proposals for solutions can be compared on the basis of how well each one meets the specified criteria for success or how well each takes the constraints into account. (3-5-ETS1-1)	

Source: NGSS Lead States. 2013. *Next Generation Science Standards: For states, by states.* Washington, DC: National Academies Press. *www.nextgenscience.org/next-generation-science-standards.*

Appendix: *NGSS* Lesson Plan Alignment Matrices

TABLE A.8.

Connecting to the *NGSS* for Lesson 8: *Eggstreme Sports*

Performance Expectations

3-PS2-1. Motion and Stability: Forces and Interactions
Plan and conduct an investigation to provide evidence of the effects of balanced and unbalanced forces on the motion of an object.

4-LS1-1. From Molecules to Organisms: Structures and Processes
Construct an argument that plants and animals have internal and external structures that function to support survival, growth, behavior, and reproduction.

3-5-ETS1-3. Engineering Design
Plan and carry out fair tests in which variables are controlled and failure points are considered to identify aspects of a model or prototype that can be improved.

Science and Engineering Practices	Disciplinary Core Ideas	Crosscutting Concepts
Planning and Carrying Out Investigations • Plan and conduct an investigation collaboratively to produce data to serve as the basis for evidence, using fair tests in which variables are controlled and the number of trials considered. (3-PS2-1) (3-5-ETS1-3) Engaging in Argument From Evidence • Construct an argument with evidence, data, and/or a model. (4-LS1-1)	PS2.A: Forces and Motion • Each force acts on one particular object and has both strength and a direction. An object at rest typically has multiple forces acting on it, but they add to give zero net force on the object. Forces that do not sum to zero can cause changes in the object's speed or direction of motion. (3-PS2-1) LS1.A: Structure and Function • Plants and animals have both internal and external structures that serve various functions in growth, survival, behavior, and reproduction. (4-LS1-1) ETS1.B: Developing Possible Solutions • Tests are often designed to identify failure points or difficulties, which suggest the elements of the design that need to be improved. (3-5-ETS1-3) ETS1.C: Optimizing the Design Solution • Different solutions need to be tested in order to determine which of them best solves the problem, given the criteria and the constraints. (3-5-ETS1-3)	Cause and Effect • Cause and effect relationships are routinely identified. (3-PS2-1) Systems and System Models • A system can be described in terms of its components and their interactions. (4-LS1-1)

Source: NGSS Lead States. 2013. *Next Generation Science Standards: For states, by states.* Washington, DC: National Academies Press. www.nextgenscience.org/next-generation-science-standards.

Appendix: *NGSS* Lesson Plan Alignment Matrices

TABLE A.9.

Connecting to the *NGSS* for Lesson 9: Marsh Madness

Performance Expectations
3-LS4-4. Biological Evolution: Unity and Diversity
Make a claim about the merit of a solution to a problem caused when the environment changes and the types of plants and animals that live there may change.

Science and Engineering Practices	Disciplinary Core Ideas	Crosscutting Concepts
Engaging in Argument From Evidence • Make a claim about the merit of a solution to a problem by citing relevant evidence about how it meets the criteria and constraints of the problem. (3-LS4-4)	LS2.C: Ecosystem Dynamics, Functioning, and Resilience • When the environment changes in ways that affect a place's physical characteristics, temperature, or availability of resources, some organisms survive and reproduce, others move to new locations, yet others move into the transformed environment, and some die. (secondary to 3-LS4-4) LS4.D: Biodiversity and Humans • Populations live in a variety of habitats, and change in those habitats affects the organisms living there. (3-LS4-4)	Systems and System Models • A system can be described in terms of its components and their interactions. (3-LS4-4)

Source: NGSS Lead States. 2013. Next Generation Science Standards: For states, by states. Washington, DC: National Academies Press. www.nextgenscience.org/next-generation-science-standards.

TABLE A.10.

Connecting to the *NGSS* for Lesson 10: Finders Keepers?

Performance Expectations

3-LS4-1. Biological Evolution: Unity and Diversity

Analyze and interpret data from fossils to provide evidence of the organisms and the environments in which they lived long ago.

Science and Engineering Practices	Disciplinary Core Ideas	Crosscutting Concepts
Analyzing and Interpreting Data • Analyze and interpret data to make sense of phenomena using logical reasoning. (3-LS4-1)	LS4.A: Evidence of Common Ancestry and Diversity • Some kinds of plants and animals that once lived on Earth are no longer found anywhere. *(Note: Moved from K–2.)* (3-LS4-1) • Fossils provide evidence about the types of organisms that lived long ago and also about the nature of their environments. (3-LS4-1)	Scale, Proportion, and Quantity • Observable phenomena exist from very short to very long time periods. (3-LS4-1)

Source: NGSS Lead States. 2013. *Next Generation Science Standards: For states, by states.* Washington, DC: National Academies Press. *www.nextgenscience.org/next-generation-science-standards.*

TABLE A.11.

Connecting to the *NGSS* for Lesson 11: Blast From the Past

Performance Expectations

5-PS1-2. Matter and Its Interactions
Measure and graph quantities to provide evidence that regardless of the type of change that occurs when heating, cooling, or mixing substances, the total weight of matter is conserved.

3-PS2-1. Motion and Stability: Forces and Interactions
Plan and conduct an investigation to provide evidence of the effects of balanced and unbalanced forces on the motion of an object.

5-PS2-1. Motion and Stability: Forces and Interactions
Support an argument that the gravitational force exerted by Earth on objects is directed down.

3-5-ETS1-3. Engineering Design
Plan and carry out fair tests in which variables are controlled and failure points are considered to identify aspects of a model or prototype that can be improved.

Science and Engineering Practices	Disciplinary Core Ideas	Crosscutting Concepts
Using Mathematics and Computational Thinking • Measure and graph quantities such as weight to address scientific and engineering questions and problems. (5-PS1-2) Planning and Carrying Out Investigations • Plan and conduct an investigation collaboratively to produce data to serve as the basis for evidence, using fair tests in which variables are controlled and the number of trials considered. (3-PS2-1) (3-5-ETS1-3)	PS1.A: Structure and Properties of Matter • The amount (weight) of matter is conserved when it changes form, even in transitions in which it seems to vanish. (5-PS1-2) PS1.B: Chemical Reactions • No matter what reaction or change in properties occurs, the total weight of the substances does not change. (5-PS1-2)	Scale, Proportion, and Quantity • Standard units are used to measure and describe physical quantities such as weight, time, temperature, and volume. (5-PS1-2) Cause and Effect • Cause and effect relationships are routinely identified and used to explain change. (3-PS2-1) (5-PS2-1)

Continued

Appendix: *NGSS* Lesson Plan Alignment Matrices

Table A.11. Connecting to the *NGSS* for Lesson 11: Blast From the Past *(continued)*

Science and Engineering Practices	Disciplinary Core Ideas	Crosscutting Concepts
Engaging in Argument From Evidence • Support an argument with evidence, data, or a model. (5-PS2-1)	PS2.A: Forces and Motion • Each force acts on one particular object and has both strength and a direction. An object at rest typically has multiple forces acting on it, but they add to give zero net force on the object. Forces that do not sum to zero can cause changes in the object's speed or direction of motion. (3-PS2-1) PS2.B: Types of Interactions • Objects in contact exert forces on each other. (3-PS2-1) • The gravitational force of Earth acting on an object near Earth's surface pulls that object toward the planet's center. (5-PS2-1) ETS1.B: Developing Possible Solutions • Tests are often designed to identify failure points or difficulties, which suggest the elements of the design that need to be improved. (3-5-ETS1-3) ETS1.C: Optimizing the Design Solution • Different solutions need to be tested in order to determine which of them best solves the problem, given the criteria and the constraints. (3-5-ETS1-3)	

Source: NGSS Lead States. 2013. *Next Generation Science Standards: For states, by states.* Washington, DC: National Academies Press. *www.nextgenscience.org/next-generation-science-standards.*

Appendix: *NGSS* Lesson Plan Alignment Matrices

TABLE A.12.

Connecting to the *NGSS* for Lesson 12: "Mined" Your Own Business

Performance Expectations

5-ESS3-1. Earth and Human Activity
Obtain and combine information about ways individual communities use science ideas to protect the Earth's resources and environment.

5-PS1-3. Matter and Its Interactions
Make observations and measurements to identify materials based on their properties.

Science and Engineering Practices	Disciplinary Core Ideas	Crosscutting Concepts
Obtaining, Evaluating, and Communicating Information • Obtain and combine information from books and/or other reliable media to explain phenomena or solutions to a design problem. (5-ESS3-1) Planning and Carrying Out Investigations • Make observations and measurements to produce data to serve as the basis for evidence for an explanation of a phenomenon. (5-PS1-3)	ESS3.C: *Human Impacts on Earth Systems* • Human activities in agriculture, industry, and everyday life have had major effects on the land, vegetation, streams, ocean, air, and even outer space. But individuals and communities are doing things to help protect Earth's resources and environments. (5-ESS3-1) PS1.A: *Structure and Properties of Matter* • Measurements of a variety of properties can be used to identify materials. (5-PS1-3)	Systems and System Models • A system can be described in terms of its components and their interactions. (5-ESS3-1) Scale, Proportion, and Quantity • Standard units are used to measure and describe physical quantities such as weight, time, temperature, and volume. (5-PS1-3)

Source: NGSS Lead States. 2013. *Next Generation Science Standards: For states, by states.* Washington, DC: National Academies Press. *www.nextgenscience.org/next-generation-science-standards.*

TABLE A.13.

Connecting to the NGSS for Lesson 13: Fueling Around

Performance Expectations

3-5-ETS1-1. Define a Simple Design Problem
Define a simple design problem that can be solved through the development of an object, tool, process, or system and includes several criteria for success and constraints on materials, time, or cost.

4-PS3-1. Energy
Use evidence to construct an explanation relating the speed of an object to the energy of that object.

4-ESS3-1. Earth and Human Activity
Obtain and combine information to describe that energy and fuels are derived from natural resources and their uses affect the environment.

5-ESS3-1. Earth and Human Activity
Obtain and combine information about ways individual communities use science ideas to protect the Earth's resources and environment.

Science and Engineering Practices	Disciplinary Core Ideas	Crosscutting Concepts
Constructing Explanations and Designing Solutions • Use evidence (e.g., measurements, observations, patterns) to construct an explanation. (4-PS3-1) Obtaining, Evaluating, and Communicating Information • Obtain and combine information from books and other reliable media to explain phenomena. (4-ESS3-1) (5-ESS3-1)	PS3.A: Definitions of Energy • The faster a given object is moving, the more energy it possesses. (4-PS3-1) ESS3.A: Natural Resources • Energy and fuels that humans use are derived from natural sources, and their use affects the environment in multiple ways. Some resources are renewable over time, and others are not. (4-ESS3-1) ESS3.C: Human Impacts on Earth Systems • Human activities in agriculture, industry, and everyday life have had major effects on the land, vegetation, streams, ocean, air, and even outer space. But individuals and communities are doing things to help protect Earth's resources and environments. (5-ESS3-1)	Energy and Matter • Energy can be transferred in various ways and between objects. (4-PS3-1) Cause and Effect • Cause and effect relationships are routinely identified and used to explain change. (4-ESS3-1) Systems and System Models • A system can be described in terms of its components and their interactions. (5-ESS3-1)

Continued

Appendix: *NGSS* Lesson Plan Alignment Matrices

Table A.13. Connecting to the *NGSS* for Lesson 13: Fueling Around *(continued)*

Science and Engineering Practices	Disciplinary Core Ideas	Crosscutting Concepts
Asking Questions and Defining Problems • Define a simple design problem that can be solved through the development of an object, tool, process, or system and includes several criteria for success and constraints on materials, time, or cost. (3-5-ETS1-1)	ETS1.A: Defining and Delimiting Engineering Problems • Possible solutions to a problem are limited by available materials and resources (constraints). The success of a designed solution is determined by considering the desired features of a solution (criteria). Different proposals for solutions can be compared on the basis of how well each one meets the specified criteria for success or how well each takes the constraints into account. (3-5-ETS1-1)	Influence of Science, Engineering, and Technology on Society and the Natural World • People's needs and wants change over time, as do their demands for new and improved technologies. (3-5-ETS1-1)

Source: NGSS Lead States. 2013. *Next Generation Science Standards: For states, by states.* Washington, DC: National Academies Press. *www.nextgenscience.org/next-generation-science-standards.*

TABLE A.14.

Connecting to the *NGSS* for Lesson 14: Watch Your Step

Performance Expectations

4-LS1-1. Construct an argument that plants and animals have internal and external structures that function to support survival, growth, behavior, and reproduction.

4-LS1-2. Use a model to describe that animals receive different types of information through their senses, process the information in their brain, and respond to the information in different ways.

Science and Engineering Practices	Disciplinary Core Ideas	Crosscutting Concepts
Developing and Using Models • Use a model to test interactions concerning the functioning of a natural system. (4-LS1-2) Engaging in Argument From Evidence • Construct an argument with evidence, data, and/or a model. (4-LS1-1)	LS1.A: Structure and Function • Plants and animals have both internal and external structures that serve various functions in growth, survival, behavior, and reproduction. (4-LS1-1) LS1.D: Information Processing • Different sense receptors are specialized for particular kinds of information, which may be then processed by the animal's brain. Animals are able to use their perceptions and memories to guide their actions. (4-LS1-2)	Systems and System Models • A system can be described in terms of its components and their interactions. (4-LS1-1) (4-LS1-2)

Source: NGSS Lead States. 2013. *Next Generation Science Standards: For states, by states.* Washington, DC: National Academies Press. *www.nextgenscience.org/next-generation-science-standards.*

Index

Page references in **boldface** *indicate information contained in figures and tables.*

A

About Habitats: Wetlands (Sill), 287, 288, 294
Absorbancy, 163–164. *See also* Soaky Doaky lesson plan
Accessibility. *See* Swingy Thingy lesson plan
Accomodations for students with disabilities, 10, 47
Action and expression, 10, **11**
Administration communication, 498–499
Advantages/disadvantages graphic organizer, **168**
Agree/Disagree argument lines, 34
Agree/Disagree T-charts, **23**, 23–24, **469, 479**
Allocation of resources. *See* Blast From the Past lesson plan
Altruism. *See* Monkey Business lesson plan
Anemometers, **231**
Animal rights. *See* Leave It to Beavers lesson plan; Monkey Business lesson plan
Assessment, 51–54

B

Beans, 115. *See also* Take a (Farm) Stand lesson plan
Beavers, 60–62. *See also* Leave It to Beavers lesson plan
The Bee Book (Milner), 191, 198–199, 203–205
Bee-ing There for Bees lesson plan, 186–214
    additional resources, 194–195
    background for teachers, 192–194
    CCSS connections, 187–188
    C3 Framework, 188
    driving questions, 186
    5E lesson plan, 195–208, **197**
        elaborate, 203–207
        elaborate and evaluate, 200–203
        engage, 195–197
        evaluate, 207–208
        explain, 198–200
        explore, **197**, 197–198
    going deeper, 208
    handouts, 209–214
    materials, 189–191
    media, 191–192
    misconception alert, 198
    nature of science, 187
    NCSS connection, 188
    NGSS connection, 186–187, **507**
    overview, 186
    rubric for "Bee" an Engineer!, 214
    safety notes, 191
    societal issues, 187
    suggested grade levels, 186
    suggested schedule and sequence, 189
    UDL Toolkit, 189
Bees, 192–194. *See also* Bee-ing There for Bees lesson plan
Biological Sciences Curriculum Study. *See* BSCS 5E Instructional Model
Biomass energy, 450
Blast From the Past lesson plan, 347–390
    additional resources, 356
    background for teachers, 354–356, **355**
    CCSS connections, 349
    C3 Framework, 350
    driving questions, 347
    elaborate, 372–375, **373**
    5E lesson plan, 356–376, **357, 369, 371, 373**
        engage, 356–358, **357**
        evaluate, 361–362, 369–371, **371**, 375–376

Index

explore and explain, 358–361, 362–369
going deeper, 376
handouts, 353, 377–390
materials, 351–353
media, 354
misconception alert, 365
nature of science, 348
NCSS connections, 349
NGSS connections, 347–348, **513–514**
overview, 347
safety notes, 353
societal issues, 348
suggested grade levels, 347
suggested schedule and sequence, 351
UDL Toolkit, 350
Born in the Wild: Baby Mammals and Their Parents (Judge), 137–138
Brain function, 464. *See also* Watch Your Step lesson plan
BSCS 5E Instructional Model, **9,** 9–10, 50, **50**. *See also specific lesson plans*
Butternut Hollow Pond (Heinz), **294**

C

California Gold Rush, 395–396. *See also* "Mined" Your Own Business lesson plan
The California Gold Rush (Friedman), 395, 397, 400–402
CARS (Credibility, Accuracy, Reasonableness, Support) rubric, 27, **28,** 485
CCSS connections. *See* Common Core State Standards (CCSS)
C3 Framework. *See* College, Career, and Civic Life (C3) Framework for Social Studies State Standards
Choice boards, **31,** 31–32
Claim, Evidence, Reasoning (CER) framework, 21–22, **22**
Claim, Evidence, Source (CES) framework, 22, **22**
College, Career, and Civic Life (C3) Framework for Social Studies State Standards, **44,** 44–45, **45, 46, 47**
  Bee-ing There for Bees lesson plan, 188
  Blast From the Past lesson plan, 350
  *Egg*streme Sports lesson plan, 251
  Finders Keepers? lesson plan, 322
  Fueling Around lesson plan, 415–416
  Leave It to Beavers lesson plan, 58
  Marsh Madness lesson plan, 284
  "Mined" Your Own Business lesson plan, 393
  Monkey Business lesson plan, 128
  Soaky Doaky lesson plan, 156
  Swingy Thingy lesson plan, 77
  Take a (Farm) Stand lesson plan, 97
  Watch Your Step lesson plan, 458
  Weather or Not lesson plan, 217–218
Common Core State Standards (CCSS)
  about, 42–43, **43**
  Bee-ing There for Bees lesson plan, 187–188
  Blast From the Past lesson plan, 349
  *Egg*streme Sports lesson plan, 250–251
  Finders Keepers? lesson plan, 321
  Fueling Around lesson plan, 415
  Leave It to Beavers lesson plan, 57
  Marsh Madness lesson plan, 283
  "Mined" Your Own Business lesson plan, 392
  Monkey Business lesson plan, 127
  Soaky Doaky lesson plan, 155
  Swingy Thingy lesson plan, 76–77
  Take a (Farm) Stand lesson plan, 96–97
  Watch Your Step lesson plan, 457–458
  Weather or Not lesson plan, 217
Concept maps, **19,** 19–20, **134**
Concussions, 255–256. *See also Eggs*treme Sports lesson plan
Congressional debate, rubric for, 29–31, **30,** 455
Constructivist learning, 9
Consumer product advertising. *See* Soaky Doaky lesson plan
Cooperative learning strategies, 32–37, **33, 35, 36**

D

Decision making, 1–2

Index

Digging Up Dinosaurs (Aliki), 325, 332–333
Dino Dig simulation game, 333–335, **334**
Disciplinary core ideas (DCIs), 481–482
Discrimination. *See* "Mined" Your Own Business lesson plan
Dissolution, 365
Distracted walking, 461–463. *See also* Watch Your Step lesson plan
Distribution of resources. *See* Take a (Farm) Stand lesson plan

E

Economic costs of natural hazards. *See* Weather or Not lesson plan
Economics of energy production and consumption. *See* Fueling Around
EF Scale, 230, **230**
*Egg*streme Sports lesson plan, 249–281
    additional resources, 257
    background for teachers, 255–256
    CCSS connections, 250–251
    C3 Framework, 251
    driving questions, 249
    5E lesson plan, 257–266, **259, 260, 261, 263**
        elaborate, 263–266
        engage, 257–259, **259**
        evaluate, 266
        explain, 261–263, **263**
        explore, 259–261, **260, 261**
    going deeper, 266–267
    handouts, 254, 268–281
    materials, 253–254
    media, 254–255
    nature of science, 250
    NCSS connections, 251
    NGSS connections, 249–250, **510**
    overview, 249
    safety notes, 254
    societal issues, 250
    suggested grade levels, 249
    suggested schedule and sequence, 252
    UDL Toolkit, 252
Elaborate, **9,** 9–10, 50, **50**
Energy, 419–420. *See also* Fueling Around lesson plan
Energy Island (Drummond), 419, 421–422
Engage, **9,** 9–10, 50, **50**
Engagement, 10, **11**
Environmental concerns. *See* "Mined" Your Own Business lesson plan
Environmental costs/benefits. *See* Fueling Around
Environmental impact of humans. *See* Bee-ing There for Bees lesson plan
Environmental justice. *See* Fueling Around
Environmental stewardship. *See* Leave It to Beavers lesson plan
Environmental/sustainability concerns. *See* Marsh Madness lesson plan; Soaky Doaky lesson plan
Evaluate, **9,** 9–10, 50, **50**
Evaluating Media Sources template, 27, **28**
Explain, **9,** 9–10, 50, **50**
Explore, **9,** 9–10, 50, **50**

F

Fair negotiations. *See* Swingy Thingy lesson plan
Farming methods, contemporary versus traditional. *See* Bee-ing There for Bees lesson plan
Finders Keepers? lesson plan, 320–346
    additional resources, 327
    background for teachers, 326–327
    CCSS connections, 321
    C3 Framework, 322
    driving question, 320
    5E lesson plan, 327–338
        elaborate, 333–337, **334**
        engage, 327–328, 330
        evaluate, 337–338
        explain, 331–333
        explore, **330**, 330–331, **331**
        explore and explain, 328–329
    going deeper, 338
    handouts, 325, 339–346
    materials, 323–325
    media, 325–326
    nature of science, 321
    NCSS connections, 322
    NGSS connections, 320, **512**
    overview, 320
    safety notes, 325
    societal issues, 321
    suggested grade levels, 320

Index

suggested schedule and sequence, 323
UDL Toolkit, 322
5E lesson plans, **9,** 9–10, 50, **50**. *See also specific lesson plans*
Fossils, 326–327. *See also* Finders Keepers? lesson plan
Fossils Tell of Long Ago (Aliki), 325, 327–328
Four corners, **33,** 33–34
Four Square Writing Method, **20,** 20–21, **21**
A Framework for K–12 Science Education, 2
Freedom. *See* Monkey Business lesson plan
Fueling Around lesson plan, 413–455
    additional resources, 420–421
    background for teachers, 419–420
    CCSS connections, 415
    C3 Framework, 415–416
    driving questions, 413
    5E lesson plan, 421–434, **425, 426, 429, 431**
        elaborate, 430–433, **431**
        elaborate and evaluate, 426–430, **429**
        engage, 421–422
        evaluate, 434
        explore and explain, 422–426, **425, 426**
    handouts, 418
    materials, 417–418
    media, 419
    nature of science, 414
    NCSS connections, 415
    NGSS connections, 413–414, **516–517**
    overview, 413
    rubric for congressional debate, **30,** 455
    safety notes, 418
    societal issues, 414
    suggested grade levels, 413
    suggested schedule and sequence, 416–417
    UDL Toolkit, 416

G
Gas properties, 365
Geothermal energy, 449

Gold rush, 395–396. *See also* "Mined" Your Own Business lesson plan
Government control. *See* Watch Your Step lesson plan
Government regulation. *See Egg*streme Sports lesson plan
Graphic organizers, **165, 168**
Green City (Drummond), 221, 238

H
Here Is the Wetland (Dunphy), 287, **294,** 295–297
High-stakes testing, 51–54
The Honeybee (Hall & Arsenault), 191, 196
Honeybees, 192–194. *See also* Bee-ing There for Bees lesson plan
How Animal Babies Stay Safe (Fraser), 136–137
How Do Parachutes Work? (Boothroyd), 354, 367–368
Human-animal conflicts. *See* Leave It to Beavers lesson plan
Hydropower, 451

I
IBiome-Wetland, 291, **292,** 292–293
Individual freedom versus public good. *See* Watch Your Step lesson plan
Inservice elementary science instruction
    about, 487–491
    lesson guiding questions and checklist, **494**
    lesson plan for, **492–493**
    lesson planning frame, **494**
Interdisciplinary standards-based learning, 12, **12**

J
Jigsaw, 36, **36,** 426–430

K
KLEW charts, 17, **17,** 226
Know, Want to Know, Learned (KWL) graphic organizers, 15–16, 196, 210, 357

L
Land use. *See* Marsh Madness lesson plan; Swingy Thingy lesson plan
Leave It to Beavers lesson plan, 56–74

Index

additional resources, 62
background for teachers, 60–62
beaver OWL chart, 70
CCSS connections, 57
C3 Framework, 58
driving questions, 56
5E lesson plan, 62–69, **65, 66, 67**
    elaborate, 67–68
    engage, 62–63
    evaluate, 68–69
    explain, **66,** 66–67, **67**
    explore, 63–65, **65**
going deeper, 69
materials, 59–60
media, 60
nature of science, 57
NCSS connections, 57–58
NGSS connections, 56–57, **502**
overview, 56
rubric for My Beaver Proposal, 74
safety notes, 60
societal issues, 57
suggested grade levels, 56
suggested schedule and sequence, 58–59
UDL Toolkit, 58
Lesson plans guide
    additional resources, 49
    assessment, 51–54
    background for teachers, 49
    CCSS connections, 42
    C3 Framework, **44,** 44–45, **45, 46, 47**
    connecting to the *NGSS,* 40–41
    driving questions, 40
    5E lesson plans, 50, **50**
    going deeper, 51
    lesson overview, 40
    lesson title, 39
    materials, 48
    media, 49
    misconception alert, 50
    National Curriculum Standards for Social Studies (NCSS), 43–44
    nature of science, 42
    safety notes, 48–49
    societal issues, 41
    suggested grade levels, 40
    suggested schedule and sequence, 47–48
    UDL Toolkit, 47

M
Maddi's Fridge (Brandt), 101
Manifest destiny. *See* Blast From the Past lesson plan
Marsh Madness lesson plan, 282–319
    additional resources, 288
    background for teachers, 287–288
    CCSS connections, 283
    C3 Framework, 284
    driving questions, 282
    5E lesson plan, 288–301, **289, 290, 292, 294, 298, 299**
        elaborate, 297–299, **298, 299**
        engage, 288–289
        evaluate, 299–301
        explain, 295–297
        explore, 291–295, **292, 294**
        explore and explain, 289–291, **290, 291**
    going deeper, 301–302
    handouts, 286, 303–319
    materials, 285–286
    media, 287
    nature of science, 283
    NCSS connections, 283
    NGSS connections, 282–283, **511**
    overview, 282
    picture books about wetlands, **294**
    rubric for Town Hall Meeting, **30,** 318
    rubric for Wetlands flip-book, 319
    safety notes, 286–287
    societal issues, 283
    suggested grade levels, 282
    suggested schedule and sequence, 285
    UDL Toolkit, 284
Meadowlands: A Wetlands Survival Story (Yezerski), **294**
"Mined" Your Own Business lesson plan, 391–412
    additional resources, 397
    background for teachers, 395–396
    CCSS connections, 392
    C3 Framework, 393
    driving questions, 391
    5E lesson plan, 397–402, **398, 399, 400**
        elaborate, 401–402
        engage, 397, **398**
        evaluate, 402
        explain, **400,** 400–401

Index

explore, **398**, 398–399, **399**
going deeper, 402–403
handouts, 404–412
materials, 394
media, 395
NCSS connections, 392–393
NGSS connections, 391–392, **515**
overview, 391
safety notes, 395
societal issues, 392
suggested grade levels, 391
suggested schedule and sequence, 393
UDL Toolkit, 393
Mining, 395–396. *See also* "Mined" Your Own Business lesson plan
Models, 201
Mohs Hardness Scale, **399**
Monkey Business lesson plan, 126–153
additional resources, 131
background for teachers, 130–131
CCSS connections, 127
C3 Framework, 128
driving questions, 126
5E lesson plan, 131–142, **134, 136, 140**
elaborate, 138, 139–142
engage, 131–134, **134**
evaluate, 139
explain, 136–138
explore, 134–135, **136**
going deeper, 142–143
handouts, 129, 144–153
materials, 129
media, 130
misconception alert, 138
nature of science, 127
NCSS connections, 127–128
NGSS connection, 126–127, **505**
overview, 126
societal issues, 127
suggested grade levels, 126
suggested schedule and sequence, 129
UDL Toolkit, 128
Moonshot (Floca), 354, 356–358, **357**, 361
My Visit to the Zoo (Aliki), 131–133

N

National Curriculum Standards for Social Studies (NCSS), 43–44
Bee-ing There for Bees lesson plan, 188
Blast From the Past lesson plan, 349
*Egg*streme Sports lesson plan, 251
Finders Keepers? lesson plan, 322
Fueling Around lesson plan, 415
Leave It to Beavers lesson plan, 57–58
Marsh Madness lesson plan, 283
"Mined" Your Own Business lesson plan, 392–393
Monkey Business lesson plan, 127–128
Soaky Doaky lesson plan, 155–156
Swingy Thingy lesson plan, 77
Take a (Farm) Stand lesson plan, 97
Watch Your Step lesson plan, 458
Weather or Not lesson plan, 217
NCSS connections. *See National Curriculum Standards for Social Studies (NCSS)*
Near One Cattail (Fredericks), **294**
Next Generation Science Standards (NGSS), 2
Bee-ing There for Bees lesson plan, 186–187, **507**
Blast From the Past lesson plan, 347–348, **513–514**
*Egg*streme Sports lesson plan, 249–250, **510**
Finders Keepers? lesson plan, 320, **512**
Fueling Around lesson plan, 413–414, **516–517**
Leave It to Beavers lesson plan, 56–57, **502**
Marsh Madness lesson plan, 282–283, **511**
"Mined" Your Own Business lesson plan, 391–392, **515**
Monkey Business lesson plan, 126–127, **505**
Soaky Doaky lesson plan, 154, **506**
and SSI implementation, 3–5, **4**, 481–482
Swingy Thingy lesson plan, 75–76, **503**

Index

Take a (Farm) Stand lesson plan, 95–96, **504**
Watch Your Step lesson plan, 456–457, **518**
Weather or Not lesson plan, 215–216, **508–509**
NGSS connections. *See* Next Generation Science Standards (NGSS)
Nuclear energy, 447

O
Observe, Wonder, Learn (OWL) charts, 15–16, **16,** 70
Opinion letter, 28, **29**
Otis and the Tornado (Long), 221, 225–226
OWL charts. *See* Observe, Wonder, Learn (OWL) charts

P
Paper towel testing, 159–160. *See also* Soaky Doaky lesson plan
Parachutes, 367–368. *See also* Blast From the Past lesson plan
Parent communication, 498
Personal autonomy. *See Egg*streme Sports lesson plan; Watch Your Step lesson plan
Pioneering. *See* Blast From the Past lesson plan
Planetary protection. *See* Blast From the Past lesson plan
Plants, 100–102. *See also* Take a (Farm) Stand lesson plan
Playground design, 80–81. *See also* Swingy Thingy lesson plan
Pollination, 192–194. *See also* Bee-ing There for Bees lesson plan
Poverty, addressing. *See* Take a (Farm) Stand lesson plan
Preservice elementary science instruction
    about, 487–491
    lesson guiding questions and checklist, **494**
    lesson plan for, **492–493**
    lesson planning frame, **494**
Professional development, 487–491
Property rights. *See* Finders Keepers? lesson plan; Marsh Madness lesson plan; "Mined" Your Own Business lesson plan
Public versus private goods. *See* Finders Keepers? lesson plan

R
Reading and Analyzing Nonfiction (RAN) charts, 18, **18**
Renewable energy. *See* Fueling Around
Representation, 10, **11**
Risk assessment. *See* Weather or Not lesson plan
Rocket launches, 354–356, **355**. *See also* Blast From the Past lesson plan

S
School regulation. *See Egg*streme Sports lesson plan
Scientific discoveries, ownership of and access to. *See* Finders Keepers? lesson plan
Scientific literacy, 2
The Secret Pool (Ridley), **294**
Sentence Frames for Arguments, 25–26, **26**
Soaky Doaky lesson plan, 154–185
    additional resources, 160
    background for teachers, 159–160
    CCSS connections, 155
    C3 Framework, 156
    driving questions, 154
    elaborate, **167,** 167–168, 167–170, **168**
    5E lesson plan, 160–171, **162, 165, 167, 168**
        engage, 160–161
        evaluate, 170–171
        explain, 164–167, **165**
        explore, 163–164
        explore and explain, 161–163, **162**
    going deeper, 171
    handouts, 158, 172–185
    materials, 157–158
    media, 159
    nature of science, 155
    NCSS connection, 155–156
    NGSS connection, 154, **506**
    overview, 154

Index

safety notes, 158
societal issues, 155
suggested grade levels, 154
suggested schedule and sequence, 157
UDL Toolkit, 156
Socioscientific issues (SSI) curriculum
    about, 2–3, 7–9, **8**
    administration communication, 498–499
    benefits of, 5
    cross-disciplinary topics, 482
    cross-discipline collaborators, 499
    developing own SSI lessons, 481–486
    embracing controversies, 497–498
    features of, **8**
    ground rules, 485
    key elements of, **5**
    lesson guiding questions and checklist, **494**
    lesson plan for methods course on, **492–493**
    lesson planning frame, **494**
    and local issues, 482
    and *NGSS*, 3–5, **4,** 481–482
    parent communication, 498
    in preservice and inservice instruction, 487–491
    resources, 483–484
    student voices in, 500
    taking action, 499–500
    timing of introduction of issues, 484–485
    and UDL strategies, 483
Solar energy, 446
Song Of the Water Boatman (Sidman), **294**
Source evaluation, 27, **27, 28,** 485
Sources of Evidence template, 26, **27**
Student-centered inquiry, 9
Swingy Thingy lesson plan, 75–94
    additional resources, 81
    background for teachers, 80–81
    CCSS connections, 76–77
    C3 Framework, 77
    driving question, 75
    5E lesson plan, 82–91, **88**
        elaborate, 86–91, **88**
        engage, 82
        evaluate, 83, 85, 91
        explain, 83–84
        explore, 82–83
        explore and explain, 84–85
    going deeper, 91
    materials, 78–79
    media, 80
    nature of science, 76
    NCSS connections, 77
    NGSS connections, 75–76, **503**
    overview, 75
    safety notes, 79
    societal issues, 76
    suggested grade levels, 75
    suggested schedule and sequence, 78
    UDL Toolkit, 77

T

Take a (Farm) Stand lesson plan, 95–125
    additional resources, 102
    background for teachers, 100–102
    CCSS connections, 96–97
    C3 Framework, 97
    driving questions, 95
    5E lesson plan, 103–114, **106, 107, 108**
        elaborate, 105, 112–113
        elaborate and evaluate, 113–114
        engage, 103
        evaluate, 105–109, **106, 107, 108,** 111
        explain, 104–105
        explain and elaborate, 111
        explore, 103–104, 109–111
    going deeper, 114
    handouts, 115–124
    materials, 98–99
    media, 100
    misconception alert, 104
    nature of science, 96
    NCSS connections, 97
    NGSS connections, 95–96, **504**
    overview, 95
    rubric for plant journal, 125
    safety notes, 99–100
    societal issues, 96
    suggested grade level, 95
    suggested schedule and sequence, 98
    UDL Toolkit, 97

Index

Talking chips or sticks, 36, **37**
T-charts, **23,** 23–24, 228, **229,** 262, **263,** 469
Teacher instruction, 487–491
Texting and walking. *See* Watch Your Step lesson plan
Think-pair-share strategy, 35, **35**
Tornadoes, 221–224, **222**. *See also* Weather or Not lesson plan
Tornadoes (Gibson), 221, 228, 234
Town hall meeting, rubric for, 29–31, **30,** 318

U
The Ugly Vegetables (Lin), 103
Universal Design for Learning (UDL), 10–11, **11**. *See also* UDL Toolkit in *specific lessons*

V
Venn diagrams, 18–19, **19, 66,** 400

W
Walking, distracted. *See* Watch Your Step lesson plan
Watch Your Step lesson plan, 456–480
   additional resources, 463
   background for teachers, 461–463
   CCSS connections, 457–458
   C3 Framework, 458
   driving questions, 456
   5E lesson plan, 463–470, **469**
     elaborate, 468–470, **469**
     engage, 463–464
     evaluate, 470
     explain, 465–468
     explore, 464–465
   going deeper, 470
   handouts, 460–461, 471–480
   materials, 460–461
   media, 461
   misconception alert, 464
   nature of science, 457
   NCSS connections, 458
   NGSS connections, 456–457, **518**
   overview, 456
   safety notes, 461
   societal issues, 457
   suggested grade levels, 456
   suggested schedule and sequence, 459

Weather or Not lesson plan, 215–248
   additional resources, 224–225
   background for teachers, 221–224, **222**
   CCSS connections, 217
   C3 Framework, 217–218
   driving questions, 215
   5E lesson plan, 225–238, **227, 229, 230, 231**
     elaborate, 234–235
     elaborate and evaluate, 232–234, 236–238
     engage, 225–226
     evaluate, 235–236
     explain, 228–230, **229, 230**
     explore, 226–228, **227**
     explore and explain, **231,** 231–232
   going deeper, 238–239
   handouts, 220, 240–248
   materials, 219–220
   media, 221
   misconception alert, 235
   nature of science, 216
   NCSS connections, 217
   NGSS connections, 215–216, **508–509**
   overview, 215
   safety notes, 220–221
   societal issues, 216
   suggested grade levels, 215
   suggested schedule and sequence, 218
   UDL Toolkit, 218
Weighing the Evidence template, 25, **25**
Wetland Food Chains (Kalman & Burns), 287, 293, **294**
Wetlands, 287–288. *See also* Marsh Madness lesson plan
Wind energy, 448

Y
Yes/No argument lines, 34, **35**
Yes/No T-charts, **23,** 23–24, **140, 373**

Z
Zoos, need for. *See* Monkey Business lesson plan